# SKY WALKING

# SKY WALKING

## An Astronaut's Memoir

## TOM JONES

To ANA —
THANKS + KEEP FLYING
HIGH!
BEST WISHES —
Tom Jones
STS-59-68-80-98

Smithsonian Books

Collins

An Imprint of HarperCollinsPublishers

HarperCollins books may be purchased for educational, business, or sales promotional use. For information please write: Special Markets Department, HarperCollins Publishers, 10 East 53rd Street, New York, NY 10022.

Designed by Iva Hacker-Delaney

Library of Congress Cataloging In Publication Data:

Jones, Thomas D.
    Sky walking : an astronaut's memoir / Thomas D. Jones
        p.  cm.
    Includes bibliographical references and index.
    ISBN-10: 0-06-085152-X
    ISBN-13: 978-0-06-085152-1
    1. Astronauts—United States—Biography.   2. Space flights.
    3. Columbia (Spacecraft)   4. Extravehicular acitivity
(Manned space flight).   5. Space stations.   I. Title.

TL789.85.J66.A3    2006
629.45'0092—dc22.
    [B]

06 07 08 09 10 WBC/RRD 10 9 8 7 6 5 4 3

To the NASA family,
who brought me safely home to mine.

# CONTENTS

# FOREWORD

*John W. Young*

The human exploration of space is one of the most important endeavors of the twenty-first century. But why do humans explore? We do so to preserve our species over the long haul. We want to expand our knowledge of our environment, to know what is out there, and to find opportunity over the horizon.

On this trip into orbit with space scientist and former astronaut Tom Jones, we learn how today's astronauts prepare for work on the frontiers of space. From the aftermath of the 1986 *Challenger* disaster to the loss of *Columbia* in 2003, Tom takes us through the astronauts' training, through success and disappointment in space, and finally to the creation of one of the most promising cooperative undertakings in history, the International Space Station.

The Space Station may very well be the stepping-stone to new knowledge and new worlds. Right now it provides the best laboratory for performing experiments in zero gravity to ensure that humans can be fit and productive after long-duration trips in deep space. Should research prove that zero gravity's effects are too worrisome for extended journeys, we can use the station to test, for example, a promising concept for generating artificial gravity, called a human rotation bicycle. This astronaut-powered centrifuge could easily be fitted into one of the inflatable habitats now being developed commercially. As the rider pedals, he or she would power the bike in a tight circle, effectively creating a small-diameter centrifuge and applying a "gravitational acceleration" that increases with speed. Humans, who adapt readily and well to zero gravity, will likely have no problems whirling themselves around on a bike, just as pilots adapt to high rates of roll in fighter or attack aircraft. In this and other experiments, the Space Station that Tom Jones helped build can be a problem-solving laboratory that gets us on the way to the planets.

The Space Station's unique zero-gravity environment makes possible scientific research that offers great promise for us Earth people. Promising results from *Columbia*'s last mission suggest that a space bioreactor may enable us to study growing cancer tumors or cultivate human body tissues such as heart muscle for transplant use. Small improvements gained through space research in the combustion efficiency of automobile and jet engines will save enormous amounts of fossil fuels back on Earth. Investigations into small, weightless flames on the shuttle have already given us new insight into combustion physics that should lead to the development of cleaner, more fuel-efficient engines. Learning how to grow plants efficiently in space will be vital for getting us to the Moon and Mars. Using hydroponic farming, elevated carbon dioxide levels, and intensified light, we have grown wheat in laboratories on Earth yielding more than 1,000 bushels an acre. Perfecting those techniques in space on the station should reduce the amount of food we'll need to launch from Earth and improve our agricultural practices on Earth, too.

Earth observation from the Space Station is just one way that NASA is improving the understanding of our own planet. Tom's first two missions with the Space Radar Laboratory furnish a good example. In 2000 the same shuttle-based radars produced a digital three-dimensional map of 80 percent of Earth's terrain. Already in use by the US military, this global data set will eventually be used in aircraft cockpits to help eliminate one of the main causes of aviation accidents: controlled flight into terrain. Future space-based radars will use the techniques explored on Tom's missions to monitor earthquake zones and active volcanoes for the telltale motions that signal an impending tremor or eruption. If we can predict these disasters worldwide, we can get folks evacuated in time to save thousands of lives.

I've known since before flying on *Gemini 3* that working out there in space is risky business, but at stake is humanity's survival down here on planet Earth. Looking back at Earth from the Moon, I saw how our world is unique in this vast universe. The space technologies we must develop to explore the Moon, Mars, and the rest of the solar system are the same tools that will protect our planet when the unexpected inevitably happens; catastrophes such as asteroid or comet impacts and the eruption of supervolcanoes can threaten our very civilization. In this book, Tom discusses how visits to asteroids on the way to the Moon and Mars will give us the means to deal with one of these threats. The more technology we have in hand to prevent or cope with such hazards, the better our chances for survival as a species.

In describing his experiences as an astronaut on Earth and in space, Tom Jones writes in beautiful detail about how he and his crewmates got to orbit and home again on four different shuttle missions. Tom has an incredible knack for accurately recounting the highlights of these adventures. Through his account of a decade-long journey toward the International Space Station, we gain a rare insight into how astronauts and their families deal with the risks of spaceflight.

Tom's story is a wonderful journey for the reader, who will learn about much of the good work accomplished in space and the challenges to be overcome if we are to explore the solar system with humans and our machines. On those voyages that lie ahead, we will learn much more about our universe and about the courage that propels humanity's drive to explore it.

# ACKNOWLEDGMENTS

I would never have had the privilege of representing the United States of America in space without the active support of thousands of people across this country of explorers. The people of NASA and their colleagues in industry and academia who reach into space are national treasures, all determined to extend our understanding and curiosity into the solar system.

This book was a long time in the making, and many have furnished support and encouragement along the way. Among those who provided advice were Michael Benson, William Burrows, Michael Cassutt, Andrew Chaikin, Michael Collins, June A. English, Jerry Linenger, Clay Morgan, and the late Edwards Park. My agent, Jake Elwell, encouraged the concept of a book about the shuttle, the International Space Station, and our future in space, and I'm grateful to both him and Smithsonian Books editor Jeff Hardwick for bringing the idea to commercial reality.

The notes, diaries, and memories I used for the book were supplemented by the contributions of dozens of colleagues. My crewmates— Jay Apt, Kevin Chilton, Rich Clifford, Linda Godwin, Sid Gutierrez, Mike Baker, Dan Bursch, Steve Smith, Terry Wilcutt, Jeff Wisoff, Ken Cockrell, Tammy Jernigan, Kent Rominger, Story Musgrave, Bob Curbeam, Marsha Ivins, Mark Lee, and Mark Polansky—refreshed my memory through interviews and patient answers to my countless questions.

My NASA coworkers, instructors, and flight controllers answered many questions and volunteered their own experiences on the ground during my four flights. Thanks go to Andy Algate, Sally Davis, Dom Del Rosso, Kerri Knotts, Glenda Laws, and Daryl Schuck.

Eileen Hawley at the Johnson Space Center Public Affairs Office and Lucy Lytwynsky in the Astronaut Office also provided valuable assistance. Duane Ross at the Astronaut Selection Office tracked down some original material from my astronaut interview, which gave us both a good laugh. Bernadette Hajek in the Astronaut Office helped me verify technical information about the space shuttle and astronaut training. The staff at the NASA Flight Medicine Clinic furnished details about the physiological effects of spaceflight.

At the Lyndon B. Johnson Space Center, Susan Erskin, Mike Gentry, and Kathy Strawn located high-quality originals of the photographs from Earth and sky. Thanks also go to the staff of the John F. Kennedy Space Center's Multimedia Gallery for their help in locating countdown and launch imagery.

At Smithsonian Books, my editor, John Ross, helped me transform a sprawling story into the book that exists today. His suggestions and advice on improving the content and flow of the book were indispensable. Joanne S. Ainsworth and Mary Dorian contributed their expert copy editing skills. Shubhani Sarkar and Iva Hacker-Delaney created the book's pleasing design and T.J. Kelleher, senior editor, carried *Sky Walking* through production with decisiveness and focus. I am grateful to all.

Thanks go to my family, who launched me on these adventures. My mother, Rosemarie Jones, found the strength to attend four launches at the Cape and "let me go" each time. My siblings and their families came time after time to Florida to cheer me on. All offered their love and support to Liz, Annie, and Bryce.

Each time I left Earth, I hated to say good-bye to Annie and Bryce. Each time I returned, I realized how much I had missed them, and Liz, my unshakable partner in life. She provided me with the strong foundation I needed to learn, to train, and to fly. Without her, both spaceflight and writing this book would have been impossible.

# PART ONE
# EARTH

# 1

# No Way Out

*Obstacles cannot crush me. Every obstacle yields to stern re-*
*solve. He who is fixed to a star does not change his mind.*
LEONARDO DAVINCI (1452–1519), NOTEBOOKS

November 28, 1996—Thanksgiving evening. The space-suit fan whirred quietly behind my head, pumping the weightless oxygen from backpack to helmet as I drifted in the airlock. The shuttle *Columbia* had completed 144 orbits of Earth. By this tenth day of our mission, we had recaptured the Wake Shield Facility, our space-based computer chip factory, and berthed it safely into the shuttle's payload bay, its mission complete. The other satellite we had deployed on this science mission, an ultraviolet astronomical telescope called ORFEUS-SPAS, trailed us in orbit by about thirty miles, busily surveying the heavens. In the four days before we chased down the telescope, we had an exciting new task to perform. It was time to prove that the National Aeronautics and Space Administration's (NASA's) plans for building a space station were practical and workable in the weightless environment of orbit. For our five-person crew on the eightieth mission of the Space Transportation System, or STS-80, this traditional American holiday was no day off. I had trained hard for six years for this moment.

Sealed inside the cramped canister-shaped compartment at the rear of *Columbia*'s crew cabin, Tammy Jernigan and I were still connected to the orbiter's power and oxygen supplies by a pair of life-support umbilicals. Tammy, our crew's lead space walker, pushed aside a tangle of drifting tools and grasped the chamber's depressurization valve with a weightless glove. Cleared by Mission Control, she rotated the black knob to its open position: the life-giving air surrounding us roared through the valve directly into the vacuum of space. The only thing between us and space was the thin metal of the airlock's rear hatch. It was hard to imagine that there were no more practice sessions in NASA's Weightless Environment Training Facility (WETF) standing between us and our two planned space walks. But here we were, suited up and ready, not submerged in a big swimming pool but immersed in the real thing at last. Hovering above Tammy's backpack, I glanced down at the digital readout on my own suit and saw something never seen in our space-walk training under water: the airlock pressure was creeping toward zero.

The countdown to this first extravehicular activity (EVA) had gone perfectly so far. Story Musgrave, our veteran crewmate and in-cabin partner, had marched us through the preparation checklist, and as usual, his practiced eye had missed nothing. He had double-checked every detail of suit-up, and about thirty minutes earlier he had closed the hatch leading from Columbia's middeck, isolating Tammy and me from the crew cabin. Now, as my fingertips tingled with anticipation, we were executing the "Airlock Depress" checklist, with hatch opening just ahead.

Racing through my mind were the details of the six hours of work awaiting us outside. In our 130 hours of underwater drills and endless tabletop rehearsals, Tammy and I had burned into our memories every key task of our upcoming weightless ballet. Only by sticking with our tightly scripted timeline could we make it through the space walk's long list of Space Station assembly tests. All the WETF (pronounced Wet-F) rehearsals, the hours spent working out the orbital choreography, the space-suited vacuum chamber tests, and our final checklist review in the cabin last night lay behind us now. Outside, the new Space Station cargo crane waited for us in its payload bay cradle. Opposite the crane on the orbiter sidewall was a station solar array battery swathed in white thermal insulation, and the new station tools were nestled in their storage lockers down on the cargo bay floor. There was no time to methodically go over the whole plan again. Instead, I focused on the first few steps that would get me out the door; once there, I would rely on slipping into the comfortable groove that repetition had worn into our memories.

When the airlock pressure dropped down from the sea-level value of 14.7 pounds per square inch (psi) to 5, Tammy halted the depressurization for a planned space-suit leak check. Her suit pressure was running slightly high, but her breathing would soon consume enough oxygen to bring the reading within operational limits. My gauge reading was perfect at 4.3 psi, the pressure at which the suit's pure oxygen atmosphere would fully charge our blood with the life-giving gas. Satisfied, Tammy twisted the depressurization valve wide open, and the remaining air molecules fled into the void outside. Under the dim fluorescent lights of the airlock, we drifted in the hard vacuum conditions of space, where the pressure was less than one ten-billionth of that at sea level.

Still connected to the orbiter by our suit umbilicals, we got the critical "Go" from Story to open the airlock hatch. In a moment we would be out the door, "on stage" at last. The butterflies in my stomach were in full zero-g flight; feeling something akin to stage fright, I was more fearful of making a mistake in front of my colleagues than of the EVA hazards of micrometeoroids, searing temperatures, radiation, and hard vacuum.

With her gloved left hand gripping the yellow handrail rimming the outer hatch, Tammy reached out, grabbed the hatch handle with her other hand, and spun the crank clockwise. About 30 degrees through its single revolution the handle stopped unexpectedly. When she tried again, this time with more muscle, the force of her effort swung her weightless body back up toward me. I could hear her determined breathing over the intercom, but the handle resisted her efforts to advance it further. After straining through a half-dozen attempts with no success, she called for reinforcements. "Tom, it won't budge. Swap places and you have a go at it."

I squeezed by Tammy, pushed some floating tools out of the way, and stationed myself in front of the hatch. As our helmets passed a few inches apart, she gave me a wry smile and raised eyebrow, positioning herself over my shoulder to stabilize my suit. I grabbed the handrail for leverage, then shoved the forged steel handle clockwise. Thunk! It smacked solidly into some unseen obstacle at the same 30-degree mark. After several more unsuccessful tries, I knew we had a problem: we were nowhere near getting the handle through the full revolution needed to retract the metal rollers that sealed the hatch against the bulkhead. We could see nothing obstructing the handle, yet it felt as if it were jamming against a hard metal stop. In training we had heard of sticky gaskets or lubricants chilled by the cold, but this kind of hardware problem was new to both of us.

Pivoting to face Tammy, I caught her eye through her helmet faceplate. She shook her head in frustration, and both of us mouthed silent oaths of disgust. In words pitched with exasperation, she described our predicament to Story. Our situation had a bit of the surreal about it: just an eighth of an inch of aluminum separated us from the experience of a lifetime, and we couldn't open the damned door!

As Story relayed word down to the Mission Control Center (MCC) in Houston, Texas, I could just imagine the stunned reaction. Our flight director, Al Pennington, would be quizzing Glenda Laws, our instructor, seated behind him at the EVA console: "Can't your team even get the hatch open?" On the floor of the flight control room, I knew that controllers would shift quickly into a problem-solving stance. Engineers would hurry to join the huddle around the EVA console, unfolding blueprints and technical manuals. I later learned that veteran EVA astronauts Jerry Ross and Leroy Chiao gathered with others at the capsule communicator (capcom) console, lending additional experience to the effort. In the Maintenance, Mechanical, Arm, and Crew Systems back room at MCC, other flight controllers began zeroing in on possible ways that the hatch mechanism could fail.

As Houston quizzed us about the stubborn handle, Tammy and I each tried again to exert maximum leverage on the crank. "No joy," she told Story and the ground. I couldn't really blame Capcom Bill McArthur for his next transmission: "Tom, uh, forgive us for asking the obvious, but could you please confirm you're turning the handle *clockwise?*" His apologetic tone was clear, but it didn't make the sting any easier to take. I'm sure he read the agitation in my own voice as I responded with a clipped "Confirmed!"

Within the tight confines of the airlock, the two of us scrambled to find a body position that would deliver more muscle power to the handle. Angling for leverage, I got my suit heads-up, braced my gloved hands against the ceiling, lifted a boot onto the end of the handle, and began straightening my legs. The handle didn't budge, and hearing me describe that technique, an alarmed Houston quickly called me off, fearing that so much force could damage the gears or linkages in the hatch mechanism. Controllers instead had us remove the crank handle from its spindle-like shaft for inspection, but we found nothing out of order.

Temporarily stymied, we took a breather to think things over. Tammy and I were in no danger. We were still plugged into *Columbia's* oxygen, water, and electricity supplies, and we had hours of $CO_2$ scrubber, or carbon dioxide removal, capacity left in our backpacks. But we had spent months sweating underwater in the WETF getting ready for this space walk, and we refused to entertain the possibility that the balky hatch might not open. I was sure MCC would come up with a technical fix to the problem. Although a few space walkers had encountered brief difficulty getting an unlocked hatch free from its sticky rubber seals, no shuttle hatch had *ever* locked a crew out of a space walk.

We had trained under water to free a jammed linkage on the outer, cargo bay surface of the hatch, but Tammy and I never dreamed we would have such a struggle before we got outside. Everything visible within our closet-sized universe was maddeningly in order, so Mission Control asked our copilot Kent "Rommel" Rominger to check for problems outside: he swung *Columbia's* fifty-foot robot arm with its probing, tip-mounted television camera into action. While Tammy and I waited for the exterior inspection, Story repressurized the airlock to 5 psi to flex the airlock's metal walls and perhaps free a warped hatch, but the handle still stubbornly refused to budge. As Tammy prepared to vent the airlock again, I joked about the novel procedures we were following: "I sort of like these contingency [unplanned] EVAs, you know?" Her laughter was good to hear: "Yeah, but I didn't think we'd be doing one inside the airlock!"

To assist Rommel's inspection, we vented a jet of air out through a valve in the jammed hatch, blowing open the protective insulation blanket covering the outside of the door. This gave Rommel, who was maneuvering the end of the arm within a few feet of the hatch exterior, a clear view of the suspect linkage. To our dismay, his video survey of the hatch exterior showed nothing amiss. Our suspicions began to focus on the handle mechanism itself. But sealed as it was behind an airtight metal cover, we couldn't see into or dismantle the handle's inner workings. Our $2-billion space shuttle's doorknob was broken. And we were stumped.

After two hours of close inspection and dozens of futile attempts to get the low-tech door handle to rotate further, even MCC was out of ideas. With their turkey dinners slowly cooling to room temperature in picnic hampers throughout the control center, Houston decided to regroup and attack the problem overnight. Bill McArthur radioed up the bad news: we would hang up our space suits for the day.

Tammy checked that the hatch was still sealed, clearing Story to repressurize the airlock and reopen the hatch into the crew cabin. When he finally popped our gloves and helmets off and we could talk face-to-face, we found he was as mystified and frustrated as we were. As Tammy and I shucked off our suits, Story, Rommel, and Commander Ken "Taco" Cockrell crowded into the airlock to lay hands on the infamous crank themselves. It was little consolation to us that they could find nothing wrong with the handle, either.

Story tried to buck up our spirits a bit, swearing he could think of nothing else we could have done to get the hatch open. He had never heard of a failure like this one. These assurances from our Hubble space walker were good enough for the moment. Tammy and I believed that MCC would figure this out overnight, and success, though delayed, would certainly come tomorrow.

But doubts lingered. That night's Thanksgiving dinner aboard *Columbia* was a gloomy affair, all of us struggling to find reasons for optimism amid a rising sense of frustration. The holiday menu of rehydrated shrimp cocktail, shelf-stabilized turkey dinners, freeze-dried asparagus, and ten-day-old tortillas couldn't mask the sour taste of failure. As we brainstormed possible fixes, I was struck by how this seemingly simple problem resisted all attempts at a solution. A technical answer was nowhere in sight. Like me, Tammy was deeply disappointed by our mission's first significant setback. As our crew's EVA leader, she keenly felt not only her own lost opportunity but also the impact on our colleagues back on Earth: dozens had worked for months on our training and the station test hardware.

I tried to stay optimistic about our prospects. On this and two previous missions, I had known only success in space. This was just a delay; we would certainly find and solve the problem tomorrow. But I had been ready *today*. Tammy and I had doggedly gone into the water again and again, honing our skills for these space walks. Our suits, our tools, were in perfect shape. *Columbia* was, except for the hatch, performing superbly. Our pair of satellites had already returned a double dose of valuable science. Despite the hard work by our crew and thousands on the ground team, we had been turned back. Why?

We had run into the unexpected, and the encounter had taken me by surprise. It shouldn't have. Exploration always proceeds in fits and starts, I knew. Project Apollo had begun in tragedy and ended in triumph. The shuttle had amazed the world for five years; then came the *Challenger* disaster. As a capcom, I had seen missions nearly thwarted by minor technical problems or unforeseen complications. An astronaut crew owed their ultimate success in the crucible of spaceflight not only to how well they executed an oft-rehearsed flight plan but to how effectively they reacted to the unexpected. In space, it was the unanticipated failure that led to defeat.

But I hadn't come this far from Earth to be turned back. For more than thirty years I had dreamed of working off the planet and made it my single goal in life. I had spent thousands of hours in the cockpit, years in graduate school, and half a decade living the demanding life of an astronaut to get out that airlock door. I wanted the payoff of this space walk, and I would use every last bit of my experience to get it.

This EVA was a stepping-stone. Working outside would prepare me to grapple with fresh challenges in space: there was a space station to build, and on the road to that outpost I would encounter obstacles and face barriers that I knew would make this hatch problem seem trivial. If I couldn't beat this, how could I hope to succeed in the tests that surely lay ahead?

Tucked inside my tiny sleep compartment on *Columbia*'s middeck, I worried. The unexpected had stopped us on the verge of success, and it hurt. Sliding the door shut, I turned out the lights on one of the most frustrating days of my life.

# 2

# ASTRONAUTS WANTED: TRAVEL REQUIRED

*Ambition leads me not only farther than any other man has been before me, but as far as I think it possible for man to go.*

CAPTAIN JAMES COOK,
JOURNAL, JANUARY 30, 1774

knew that spaceflight carried risk. But nothing in my flying experience prepared me for the shock of January 28, 1986. A graduate student in planetary sciences at the University of Arizona, I had just returned from NASA's Jet Propulsion Laboratory (JPL) in Pasadena, California, where over the weekend I had shared with colleagues the mind-tingling experience of *Voyager 2*'s first-ever spacecraft encounter with the planet Uranus. Back in Tucson that Tuesday morning, I stepped next door from my campus office to the Flandrau Planetarium to watch the launch of the space shuttle *Challenger*.

With me in the darkened planetarium were a few dozen elementary schoolchildren on a field trip to watch the first teacher rocket into orbit. It wasn't a big crowd; shuttle launches were no longer a novelty. Nevertheless, the planetarium staff and I shared the excitement visible on the young faces as the shuttle soared into a pure blue sky on the blazing exhaust of its twin boosters. We tracked *Challenger*'s upward progress as the camera zoomed in, filling the giant screen with the shuttle's image and bringing looks of wonder and awe to the faces of the youthful crowd. Just seventy seconds after liftoff, our exhilaration turned to stomach-turning shock. Nine miles above Cape Canaveral, the space shuttle disintegrated before our eyes. A failed rubber O-ring seal allowed the escape of burning propellant gases from the side of the right booster; the scorching flame triggered a structural failure that tore apart the boosters, fuel tank, and orbiter. That churning fireball, I knew instantly, sealed the fate of all seven crew members, but with the children I strained for a glimpse of a miracle—the orbiter emerging safe from the flaming debris. Still hoping, a few of the students clapped as a lone parachute carrying a fragment of one of the boosters drifted toward the sea. But we were wishing for the impossible: *Challenger*'s crew wore no parachutes. As the camera focused on wreckage splashing into the empty ocean, I could see my own horror reflected in the shock and disbelief on the young faces around me. No one knew what to say; the sudden tragedy had silenced the room. I offered my only thought to the children and their teachers: "Now would be a good time to say a prayer."

Dazed, I left the planetarium as news of the disaster stunned the country. The loss seemed to affect everyone. The space scientists and graduate students I worked with, no strangers to the harsh physics of space travel, were grief-stricken at losing seven colleagues from the space community. Several of us had met *Challenger*'s commander, Francis R. "Dick" Scobee, when he had visited the campus the previous year; he had encouraged me to apply for a job as an astronaut.

Although the destruction of *Challenger* and her crew was a personal blow, I knew that risk was part of flying, no less in space than it was in the jets I had piloted. I remembered well the 1967 *Apollo 1* fire that had killed three astronauts; just twelve then, I had reacted with the same disbelief and sadness displayed by the schoolchildren at the planetarium. But that tragedy twenty years earlier had not dimmed spaceflight's appeal for me. I thought the best way to honor these fallen astronauts was to follow in their footsteps.

The *Challenger* accident threw the space program into chaos and grounded the shuttle fleet for two and a half years, but America was determined to get back into space. The shuttle roared back into orbit on the resoundingly successful STS-26 mission, launched September 29,

1988. *Discovery*'s return to space buoyed my hopes of competing for an opening in NASA's astronaut corps. I had flown B-52 bombers for the Air Force's Strategic Air Command (SAC) and had just finished a PhD in planetary sciences in Arizona; now working as an engineer for the Central Intelligence Agency (CIA), I thought it was a good time to update my astronaut application. I had first sought the job in 1986 while still working on my doctorate, and I had kept my file updated even as the agency struggled to recover from the *Challenger* disaster. When NASA hired a new astronaut class in 1987, I failed to even make it to the interview stage; with my doctoral studies still incomplete, I simply wasn't competitive. In 1988 NASA announced that beginning with the class of 1990, it would conduct astronaut selections regularly every other year. Thirty-four years old and anxious to try again, I sent in my updated application, hoping that this time NASA might give me a closer look.

By September 1989 I knew NASA was asking some of my professors and old supervisors for letters of reference, kicking off the selection process for 1990. Remembering the agency's perfunctory rejection of my previous application, I tried to keep my expectations low. About 5 percent of all qualified applicants would be interviewed; fewer than 1 percent would be hired. Still, when Duane Ross, chief of the Selection Office at the Lyndon B. Johnson Space Center (JSC), left a message at the CIA asking me to call him, I couldn't suppress my excitement.

Ross wanted to know whether I could be in Houston in ten days for an astronaut interview. I almost shouted my "Yes!" back over the receiver, but a few hours later my calendar forced me to reconsider. I had planned to be away on business that week, and I couldn't cancel my commitments with the CIA on such short notice. I apologized to Ross, and he graciously offered to let me join an interview group in October, about six weeks later. I breathed a silent prayer of thanks. The later date would also give me more time to get ready.

The invitation for an astronaut interview set my mind whirling. How should I prepare? After all, this opportunity might be my one shot at spaceflight: the combination of flying, science, and physical thrill I'd been dreaming about for a lifetime. First things first—that night I started to run.

It was not as if I had been inactive: I had been exercising regularly since leaving the Air Force and through five years in graduate school, and I frequently jogged on the old rail trail behind our CIA offices. But working for the Agency had severely tested my willpower. Our contractors often plied their government visitors with a large tray of doughnuts, bagels, or cookies, and I now needed to lose five pounds. As the countdown to my interview proceeded, my stepped-up conditioning and a stricter, low-fat diet happily began to pay off. At least I'd be ready physically for Houston.

Appearances were, of course, important for any job interview, so I did a little shopping: new casual shirts, a few pairs of slacks, even some new underwear. The day before departure, my wife of eight years, Liz, and I threw everything I needed for the week in Houston on our bed. We looked it over, then put half of it back in the closet: more clothes weren't going to cure my nervousness.

Liz was a little concerned, too. This NASA business was getting a bit more serious than she had bargained for. She had always assumed these applications of mine were harmless exercises in futility. But when I unpacked my suitcase in the slightly tattered Ramada Kings Inn just outside the front gate of the Johnson Space Center, she showed me again I'd been right to marry her. Among my polo shirts and khaki trousers (the unofficial astronaut uniform) she had hidden a space shuttle ruler, an orbiter pencil sharpener, an astronaut pen, shuttle playing cards, and so on. Each was wrapped and tagged with a day of the week and a short note of encouragement. Now I was ready for my close encounter with the space agency.

IT WAS A HEADY FEELING, meeting my interview group on Sunday afternoon, October 22, 1989. Among the candidates were test pilots, scientists, physicians, and a variety of engineers. I already knew one of my colleagues: Allison Sandlin was a space scientist I had met two years earlier while finishing my PhD. The group also included future astronauts Jim Halsell, Rick Searfoss, and Scott Horowitz, although none of us had any idea we would wind up as colleagues. While confident, none of us was sure we were good enough to make the grade. We twenty-two were just one of five groups of prospective astronauts NASA would interview that fall.

## Day One—Sunday

At a NASA contractor office just off-site, we got our schedules and a quick introduction from Duane Ross on the "big picture" for the week. A wiry athlete with a shock of gray hair and a face tanned by the Texas sun, Ross described an astronaut's duties very simply: "Your job is to become an expert in manned spaceflight." Astronauts faced long training hours, extensive travel, lots of time spent in "dull, boring meetings," an occasional detached assignment away from Texas, and frequent public appearances. Duane noted that we wouldn't get rich as government employees, but NASA could at least offer job security. Steve Hawley, an astronomer and one of the original shuttle astronauts chosen in 1978, mentioned that there were currently eighty-five crew members, sixty-eight assigned to upcoming missions. New candidates should expect a wait of about three years from selection to assignment to a space shuttle flight. He emphasized that while "every effort was being made to improve the safety of operations," there was a certain inescapable risk involved. The shuttle was a high-performance vehicle, operating in the hostile environment of space and under tremendous aerodynamic stresses on launch and landing. Although the orbiter *Atlantis* was in orbit as he spoke, having just successfully launched the *Galileo* probe to Jupiter, the grievous wounds of *Challenger*'s loss less than four years earlier were obviously still raw.

After brief introductions, our group of hopefuls buckled down to three hours of psychological screening. These written tests seemed to focus quite a lot on childhood and relationships with siblings, peers, and especially parents. Finally putting down my number-2 pencil, I thought of the gift Liz had packed for me that day. From the National Air and Space Museum, it was a small package labeled "Astronaut Fruits and Nuts."

## Day Two—Monday

The day-long battery of medical tests started at 7:15 a.m. with a blood draw at the NASA Flight Medicine Clinic. I watched with interest as the phlebotomist slid a Texas-sized needle into my arm. When I regained consciousness, I lifted my head to look up at what I correctly guessed were my own knees. Slumped fully forward in the chair, I thought that fainting probably wasn't a good way to inaugurate my medical exam, but the staff seemed unperturbed, and I proceeded along to the next event on NASA's characteristically detailed schedule.

NASA wasn't footing the bill for any rental cars, so we got around the center by walking or taking the on-site shuttle bus. My next stop was in the adjacent Life Sciences building, formerly the Lunar Receiving Laboratory used to quarantine the first three Apollo crews returning from the lunar surface. The research team I met there told me I was to evaluate a small "personnel rescue sphere." Designed to transport astronauts safely across a vacuum, the sphere would be floated along by a space-suited colleague from a crippled space shuttle into the safety of a rescue vehicle's airlock. I personally scoffed at the whimsical impracticality of the idea, but it really had been proposed for shuttle-to-shuttle rescues. Rigged with a heart monitor, I was zipped into what looked like a large beach ball inflated with fresh air and featuring just one tiny window. Although ostensibly an evaluation, it was easy to guess the purpose of the sphere exercise: it was a claustrophobia test. No matter. I found the interior of the womb-like sphere relaxing; it was certainly preferable to those cramped isolation boxes I'd endured years ago during POW training at the US Air Force Academy. After ten minutes comfortably snuggled inside, I nearly dozed off. Pass.

I trooped next to the old JSC centrifuge building, where Gemini and Apollo astronauts had tested their mettle against the daunting physical forces generated by launch and reentry. Now Building 29 housed the WETF, the twenty-five-foot-deep swimming pool used for neutral buoyancy space-walk training. In a back room off the pool deck, I met a team of musculoskeletal researchers whose job was, quite literally, to size me up. Taking measurements with all the precision of a Savile Row tailor, the team's focus was on confirming that I met the size and flexibility requirements for work in the cramped shuttle cabin.

The flight surgeons corralled me next. Back in the Building 8 clinic, they gave me a thorough going-over: eyes, ears, nose, and throat; lungs and heart; reflexes; and more. The medical procedures chewed up most of the afternoon as I moved through hearing tests, an extensive X-ray workup, and finally an ophthalmology exam. NASA's battery of eye tests did everything but turn my eyeballs inside out for inspection. I joined my fellow examinees in wearing disposable dark glasses to shield my dilated pupils from the bright Gulf Coast sun. With our shades and conservative dress, we looked like escapees from a Secret Service convention.

The last event of Day Two was pure aviation fun. At Ellington Field's Hangar 276, a remnant of the former Air Force base five miles north of the space center, we got our introduction to NASA's fleet of astronaut training aircraft. A NASA instructor pilot gave us a glimpse of the Grumman Gulfstream shuttle training aircraft, the high-altitude WB-57 research plane, and last and most interesting to me, the Northrop T-38 Talon. I'd flown the two-seat jet trainer in both pilot training and in SAC, but I'd never seen one decked out like this, gleaming in white paint and blue trim. Sitting in the cockpit of the supersonic jet, I ran my hands over familiar controls. I thought for a moment of my recent flying in single-engine Cessnas (top speed, 130 knots). I wanted this. I wanted it bad.

That night back at the hotel, I tackled the homework assignment Duane Ross had given us on Sunday. He had asked us to compose a short essay—no more than 1,500 words—on the ever-popular topic of "why I would like to be an astronaut." The work and editing took up most of my evening. Yes, I worried over it, but I had been living these words for a long time, and they flowed from my heart. Tomorrow morning I would turn it in at my Selection Board interview.

## Day Three—Tuesday

The staff of the Flight Medicine Clinic greeted us at seven in the morning and handed each of us a "24-hour collection bag," a blue insulated picnic cooler that I would later become very familiar with as a crew member. The docs wanted to analyze a full day of urine output, so for the rest of the day I toted a couple of half-gallon brown plastic jugs discreetly hidden along with an ice pack inside the cooler. A technician next wired me up with a Holter cardiac monitor with a set of seven small electrodes, each taped to a spot on my chest she had vigorously sanded to ensure good conductivity. The leads ended at a bulky Walkman-style cassette recorder that hung from my belt. Cardiologists would examine a day's worth of my heart's electrical activity for signs of arrhythmia and other disqualifying conditions.

I cut a dashing figure lugging those jugs and heart monitor into Building 9, a cavernous hangar-like facility next door to Flight Medicine. There I joined my new friends for a quick tour of the full-sized space shuttle mock-ups, accurate reproductions of the shuttle cockpit and payload bay. We took full advantage of our host, astronaut Bob Parker, by peppering him with questions about his background and experiences. He described the shuttle flight experience as a "campout," showing us the sleeping bags, storage lockers, and galley. I was surprised to learn that it took twenty minutes to clean up after using the toilet and that duct tape was worth its weight in gold in the orbiter cabin. Parker then demonstrated the orbiter's new escape and bail-out system designed after the *Challenger* accident to give astronauts at least some chance to parachute from a gliding orbiter. Then it was on to the mock-up of space station *Freedom* and a look at its laboratory module, habitat, docking node, cupola, rescue vehicle, and robot arm.

That morning in Building 4, home of the Astronaut Office, I found time to talk briefly with veterans Dick Richards, Steve Nagel, Hoot Gibson, Dave Leestma, and Mark Brown. The chance to meet active astronauts was a highlight of the week for me. Their professional and personal insights gave me my best understanding of what the job title "astronaut" really meant.

Just before 11:00 a.m., I presented myself at Building 259 for my Selection Board interview. I was a little self-conscious about my fashion accessories, specifically the thick cable harness emerging from my dress shirt just above the belt line. Pretending not to notice, Duane Ross's assistant, Teresa Gomez, welcomed me, took my essay, and delivered it to the panel. Was that laughter I heard coming from the conference room? I'd soon find out. The door opened and Ross beckoned me in.

Under my breath I mumbled my all-purpose prayer: "Jesus, Mary, and Joseph, help me!" Across the center of a long conference table, I faced panel chair Don Puddy, chief of the Flight Crew Operations Directorate (FCOD), who supervised the Astronaut Office and JSC's aircraft operations. Also on the board were Duane Ross; Dr. Joe Atkinson, a NASA personnel officer; Dr. Carolyn Huntoon, head of the Life Sciences Directorate; and astronauts Hoot Gibson, Jim Buchli, Jay Apt, Kathy Sullivan, and John Young (Good God! He'd walked on the Moon!). With all the space experience in that room, I was more than a little nervous. How tough was this going to be?

If I had a strategy going in, it was just to relax and be myself. I had the basic qualifications; now wasn't the time to try to invent a new,

NASA-oriented personality. The board would meet me—warts and all. I was more than a bit worried about that, but at least I knew I could play myself in this little stage production.

Things started off well. Puddy broke the ice by asking if, as an Air Force Academy graduate, I had noted the results of the recent Air Force–Navy football game. I responded that I had enjoyed attending the game a few weeks earlier in Annapolis, which Air Force had won with a lopsided 35–7 score. This prompted a laugh from Puddy and a groan from Navy grad Buchli; I quickly added that the game had been well played and much closer than the score would indicate. So far, so good. Puddy next asked me to summarize my background and education. I wanted to be concise and interesting. I certainly didn't want to leave the impression that I was a boring, windy know-it-all.

We then discussed my Air Force flying, my NASA-sponsored research on asteroids, and what little I could say about my CIA job. The board members were pleasant, respectful, and seemed genuinely interested in what I had to say. We discussed my opinions on US space policy; how to justify space spending in the face of the nation's domestic needs; my hardest challenge as a crew commander in the Air Force; and concerns that Liz might have about my becoming an astronaut. I don't recall any tough questions, but they did ask my opinion of women serving at the academies, in the military, and in combat. I answered that I had supervised women cadets at the Academy, worked closely with female officers in SAC, and believed that women could perform effectively in combat positions, particularly in aviation. Puddy invited me to add any other comments about my qualifications that the board might find useful. We had already gone over my work experience and interests, so I paused for a moment, wondering how to close.

What had I left out? I decided to be straightforward. Space, I told them, was my life's overriding passion. I had been reading about and studying spaceflight since I was ten years old. I had wanted to be an astronaut ever since John Young, sitting across the table to my left, had been flying the Titan rockets built in my hometown. I felt comfortable taking on the technical end of the job. The final paragraph from my essay captured my feelings well:

> To sum up . . . I want to join the astronaut corps now because it
> is, to my mind, the best job to take full advantage of my profes-

*sional skills and experience, and because it is the job where I could contribute the most to NASA in its execution of the country's space policy. I would find it a great honor to join a group of professionals who have raised teamwork to a new level, made excellence a given, and challenged conventional thinking with courage and imagination. I'm ready, and I want to be a part of that team.*

That was it. There was no immediate feedback; the board members displayed nothing except cool politeness. Puddy thanked me. I stood up, heaved a mental sigh of relief, and tried to retain my dignity as I hitched up the sagging heart monitor cable, grabbed my blue cooler, and strode out of the room.

The board members followed me out, headed for their lunch break. John Young caught up with me at the door to the parking lot. "Hey, Tom, I sure enjoyed your interview." *Columbia's* first skipper was being polite. "Can I give you a lift somewhere?" Who's going to turn down a ride from a Moon-walker? It lasted only five minutes, but the chance to talk with *Apollo 16's* commander made all the anxiety and medical indignities of the week worthwhile.

Anything following this experience was bound to be a letdown. A pair of psychiatrists spent the better part of three hours with me in a roundabout conversation whose purposes were unclear. Call it NASA's attempt to categorize my personality. My notes record that we discussed my life history, my likes, dislikes, personality, character, fantasies, my self-image compared with others, guilt—the whole gamut. I remember how relieved I was to hear the shrinks tell me that the psychology team didn't have veto power over a candidate.

## Day Four—Wednesday

Wednesday proved to literally be the most gut-wrenching day of the week. I dumped the Holter monitor and the "little brown jug" at Flight Medicine. Relieved at not having to spend another sleepless night in bed worrying about being strangled in a tangle of electrodes, I hustled over to the Shuttle Mission Simulator in Building 5.

My first visit to Building 5 had occurred eleven years earlier, when as a young pilot I had paid a weekend visit to Houston. Building 5 then

housed an Apollo command and service module, a relic of the previous decade of spaceflight training. The early versions of the shuttle's cockpit simulators had just been installed. But now Building 5 was completely revamped for space shuttle crew training—this was where astronauts found out if they had the right stuff. Outside the security checkpoint we met Sid Gutierrez, a tall and trim shuttle pilot dressed in khakis and a polo shirt, his friendly face topped by thick dark hair just beginning to show some gray. He had joined NASA with the astronaut class of 1984, and his first spaceflight would be STS-40, a Spacelab microgravity research mission scheduled to lift off in the fall of 1990.

We would be unable to take a tour through Building 5, he explained, because the simulators were currently in use rehearsing an upcoming classified shuttle mission. But he did run us through the small JSC museum and the Mission Control Center, cheerfully enduring our constant stream of questions.

In Building 30's famous control room, now a national historic landmark, we sat at the consoles where legendary flight directors Chris Kraft, Gene Kranz, Jerry Griffin, and a dozen other spaceflight pioneers had directed the missions paving the way for the first lunar landing. This was the place that had captured my imagination as a boy growing up in the Baltimore suburb of Essex. Neil Armstrong's words "The Eagle has landed" were first heard in this room; this was where the NASA team pulled *Apollo 13*'s crew back from the brink of a lonely death between Earth and the Moon. Scanning the many mission patches lining the walls of the flight control room (FCR, pronounced "ficker"), I tried to imagine what it must be like to be at the other end of the invisible bond between Mission Control and spacecraft, connected to the home planet by only the tenuous links of radio or television.

As Sid described the flight control center and how orbiting astronauts worked with the controllers, I never imagined that I might one day join him in space. I began to realize for the first time the enormous complexities involved in manned spaceflight. I had wanted to do this for as long as I could remember, but could I handle the challenge?

During some free time, I poked my head into a few offices in Building 4. Among the astronauts who shared a few moments with me were Kathy Thornton, Sonny Carter, Tammy Jernigan, and Vance Brand (who had been in Houston since 1966). All provided a surprisingly candid glimpse into astronaut life, including the downsides of the job: fre-

quent travel, long workdays, family separations, and the intense pressure accompanying long months of training. Their openness reinforced the impact of their frank opinions. These were people with whom I would enjoy working.

Unfortunately, the heady experience of talking to astronauts gave way to close encounters of an entirely different sort. Flight Medicine filled my afternoon with a series of comprehensive tests. Breezing through neurological and dental exams, I came to the centerpiece of the fitness evaluation—the exercise tolerance test.

In a sterile-looking lab with bare walls softened only by a few shuttle launch photos, I changed into shorts and running shoes. Dangling a new crop of EKG electrodes, I faced the industrial-strength treadmill occupying most of one wall. As I stepped onto the running deck, I noticed that the back of the lab door was decorated with several dozen Polaroids of current and former astronauts, all dressed just like me, their chests trailing enough wires to jump-start a used car lot. The nurse applied a blood pressure cuff and data harness to my left arm, clamped my nostrils shut, and crowned me with a skull cap lifted straight from a Frankenstein movie. Breathing through a flexible hose dangling from the head band, I would walk, then run, at an ever-faster pace until I had reached my target heart rate of 160 beats per minute, giving the EKG machine a clear picture of my heart under stress. The breathing tube gripped in my teeth would measure my lungs' oxygen intake as the workload ramped up. I gave a thumbs-up, and the belt began to move.

We began with an easy walking pace, but every two minutes the belt speed ramped up and the incline steepened. After about ten minutes, with the treadmill at a steep 30 degrees, I was running nearly flat out and approaching that magic threshold of 85 percent of maximum heart rate. Huffing and puffing into the breathing hose, I resembled an oversized lab rat on a running wheel. With my heart slamming against my chest wall, I was glad that I had spent those extra six weeks getting in better shape. Above the din from the motor, the cardiologist called out that I had reached the target rate and asked if I wanted to continue. Trying to ignore my heaving lungs and screaming calf muscles, I decided to run for one more minute but no more: Why give the docs more time to look for something wrong with my ticker? My eyes hungrily tracked that second hand around to the 12 on the wall clock a few feet in front of me; I jerked a fist thumbs-down at the cardiologist and called it quits. While I recovered, the doctor scanned his EKG traces and told me that he saw no obvious problems. I would play lab rat to the treadmill again and again in the years ahead, but none of those encounters would be as novel as the first. I still have the Polaroid they took of me.

There were no souvenir photos of what came next: the proctosigmoidoscopy exam, administered by the incredibly named Dr. Hind.

## Day Five—Thursday

Thursday's schedule included an eyesight exam, an electroencephalogram session in which scalp electrodes recorded my waking and resting brain waves (I aced the sleep portion), and an echocardiogram. Watching ultrasound video images of my pumping heart, which revealed a slight leak in my right ventricle's pulmonary valve, I realized that I was getting a world-class physical exam, all courtesy of the taxpayers. Even if NASA didn't hire me, I was certainly getting one first-rate annual checkup. The battery of tests made my Air Force flight physical look like a quick trip to Jiffy Lube.

Our group finally got a chance to unwind a bit on Thursday evening at Pe-Te's Cajun BBQ, a zydeco dance joint and eatery that had been an Ellington Field fixture for a decade. The dance hall was lined with Formica booths, the walls covered with a huge collection of signed astronaut photos and an eclectic assortment of space mementos. Over a cold beer and a spicy buffet of beef barbecue, beans, and rice, our poked-and-prodded group of hopefuls met casually with a wider group of astronauts and most of the Selection Board. With a last opportunity to schmooze with them and Don Puddy, our prospective boss, we hoped good food and small talk would erase any doubts lingering from our formal interviews.

We had learned that some previous interview groups had performed skits or song parodies at the Pe-Te's evening, and a few of my colleagues suggested we try something similar. This struck me as a bad idea. We knew almost nothing of the realities facing the astronaut corps, and attempting to spoof its members could backfire unpredictably. I wasn't willing to risk my own chances for this job by starting a new career as a stand-up comedian. The dinner passed amiably enough, and all left with good feelings about our visit to JSC. I was simply grateful for an evening of good conversation with people who were, in fact, my heroes. As for my prospective classmates, some of the friendships that sprouted over the course of that week have lasted more than a decade, growing only stronger with time.

## Day Six—Friday

Friday, and almost done. Going over the lengthy tests I had endured five days earlier, the psychologist assured me that I did indeed possess a personality, then turned me over to the flight surgeons. No problems had surfaced during my extensive tests, although Flight Medicine had informed a few other candidates of problematic or disqualifying medical conditions.

And so my week of astronaut "fantasy camp" was over. I had escaped the interview with no glaring faux pas, and the flight surgeons had given me a clean bill of health. I felt I had done well, but I knew that my individual odds weren't promising. More than 2,000 applicants had sent NASA their resumés, civil service forms, and letters of reference. From that initial pool, about 500 had survived the first cut. From this smaller group, the board had chosen to interview just over a hundred. Out of the interview groups, only twenty or so would actually be offered a position as an astronaut candidate. Duane and Teresa sent us off with a promise that we would hear something shortly after the new year. The wait for news over the coming weeks would try my patience, but having waited years for this opportunity, how could I complain about an extra couple of months?

Some of the interviewees I had met were so personally invested in the outcome that a rejection would be a terrible blow to their egos. I didn't see it that way. Although the astronaut job was my ultimate goal, I was building a satisfying career track in the sciences, and I knew I had given the process my personal best. Back home in Virginia, I continued my travel and work for the CIA while trying to keep abreast of science developments in the planetary community. I made contact with a colleague in NASA headquarters' Solar System Exploration Division, the outfit that ran the country's planetary exploration program. Wangling an invitation to visit, I began to attend the regular lunch seminars there as a way of keeping up with goings-on in space science. That soon led to an unexpected opportunity.

Don Davis, a fellow asteroid scientist I had worked with in Tucson, introduced me to Harvey Feingold, a senior scientist and Don's manager at the Chicago offices of Science Applications International Corporation (SAIC), which provides contract technical services to the government. Harvey and I had lunch on his next visit to Washington, and I was soon on a plane to Chicago to discuss a possible position working for Harvey on SAIC's contract with NASA headquarters.

Harvey thought I could help strengthen SAIC's support of NASA's solar system exploration planning, and after our Chicago interview, he invited me to join the firm's Washington office as a senior scientist doing advanced program planning for the Solar System Exploration Division. It was a timely opportunity. Not that I was unhappy with the CIA—far from it. But the SAIC job would put me squarely back on the planetary science career path, involve me in strategic exploration planning at NASA headquarters, and possibly lead to more responsible positions at SAIC or NASA. I would add to my remote sensing and laboratory experience some insight into how NASA developed new planetary spacecraft missions, and I would have the opportunity to revive my own research efforts into the nature and composition of asteroids. I missed the hands-on telescope and spectroscopy work that I had been forced to give up inside the CIA's "black" universe.

There was one little detail, however, that I thought my prospective boss deserved to know. "Harvey," I began, "I just wrapped up an astronaut interview in Houston. The Selection Board is finishing its remaining interviews and will shortly choose a new astronaut candidate class. If NASA offers me the job, I would very much like to accept that opportunity, but it would mean leaving SAIC just a few months after joining you." Hearing this frank admission that I would drop SAIC like a hot potato for the job in Houston, who could blame Harvey if he decided to postpone or even withdraw his offer?

He paused and digested this information. "I think I'll take the risk," Harvey said. "What are the odds now, one in five or six? If you get an astronaut slot, SAIC will just have to take it as a compliment."

I could see few downsides to Harvey's offer. I would get a raise, travel less, and be able to talk about my work again. The only negative was the long commute to downtown DC each day. I decided to go with SAIC. Although it would be hard to say good-bye to "Virgil S.," "Gary E.," and my other CIA colleagues, I tendered my resignation to my boss. My last day with the Agency would be January 12, 1990. After a week's vacation, I would start with SAIC on Monday, January 22.

On January 16, Liz and I took the kids on a day-long outing in the Shenandoah Valley and into West Virginia. When we arrived home late in the afternoon, I got a message that Duane Ross was trying to reach me. I was disappointed: word was that Duane called the losers, while the new astronaut candidates heard from Don Puddy. While Liz got the kids started on supper, I went up to our bedroom to phone the Selection Office. The butterflies took off in my gut as I punched the numbers into the phone.

"Tom, thanks for calling us back." Duane sounded casual, cheery. "We were hoping to reach you today. Mr. Puddy would like to speak with you, if you can hold on a moment."

My heart began to race, and I imagined this was what the onset of weightlessness felt like. Don Puddy came on the line.

"Tom, it's good to talk to you again. Hope you've been well since our meeting at JSC." After a few more pleasantries, he got to the point. "If you're still interested in the job, how would you feel about moving here next summer to join our new astronaut candidate class?"

I nearly dropped the phone! In words that are just a blur to me now, I managed to stammer out an acceptance. Puddy gave me a few dates to keep in mind and suggested I follow up with Duane Ross about the paperwork details. The chief of Flight Crew Operations (my new organization) congratulated me and again welcomed me to NASA. The conversation took less than five minutes.

Having put food in front of the children and unable to bear the suspense, Liz had joined me in the bedroom, listening to my side of the conversation. We were both more than a bit stunned. Although Ross had emphasized that the news should remain confidential until the press release was issued the following day, Liz and I figured that it wouldn't hurt to notify both sets of parents. All were both pleased and amazed. Sharing the news with my mom and dad, who had known of their crazy kid's dream for decades, was particularly satisfying, although my mother saw it as a mixed blessing. She was happy at my achievement, but she was sure that my flying with NASA would only give her more cause to worry.

My classmates and their spouses have ever since referred to that fateful January 16 communication as "the call." Puddy's summons would turn our lives upside down and lead me inexorably to four incredible adventures, replete with exhilaration, anxiety, satisfaction, disappointment, amazement, and danger.

THE NEXT DAY, AS previously planned, Liz and I dropped the kids at my parents' home in Baltimore, then continued on for a winter visit to the beach in Ocean City, Maryland. Walking along the boardwalk and the Atlantic shore in a chilly breeze, we wondered what the new job in Houston meant for us and our family. There would be the usual disruptions that come with any relocation. Eventually, we would have to contend with the excitement—and the dangers—of a spaceflight.

I knew this dream of mine would inevitably involve Liz in the astronaut business. With few of the personal rewards, she would take on all the risks and worries of a spaceflight career. We met in 1978 when Liz was working in Baltimore as a certified public accountant for Ernst & Young, a public accounting firm, and a mutual friend had arranged a blind date while I was back home in Maryland on leave from the Air Force. We dated for two years, with me in Fort Worth and her in Baltimore, before we were married in September 1981. Liz transferred with her firm to Fort Worth, then to Tucson two years later when I left the service to begin graduate school.

Just a year before my NASA selection, we had returned from the West and thought we were back home for good; now we would be pulling up stakes and leaving the East Coast, moving far from family. Once again, as in Texas and Arizona, she would be making a move so that I could pursue my own career goals. This time we would be moving with two small children: four-year-old Annie and one-year-old Bryce. I tried to downplay the impact, not to deceive her but to somehow find an aspect of the move that could ease the sacrifices she would be making. "Here's how I see it," I told her after some quick mental arithmetic. "A year of training, then a couple of flights in the next three years. I'll be satisfied with that. Four years, two flights, and we're out. Promise."

My proposal was sincere. It seemed a fair compromise to ease Liz's worries and put an upper limit on our stay in Houston. One spaceflight would be a dream come true; two, a fantastic privilege.

There was just one more phone call to make—to Harvey Feingold.

"Harvey, this is Tom Jones in DC. I have some good news and some bad news." (Unoriginal, but it was the best I could do.) "The good news is I'm really looking forward to coming to work for you on Monday. The bad news is I can only stay for six months. You remember our fall conversation about the Astronaut Office? Well, NASA called Tuesday and offered me the job, and I'm afraid I can't pass up the opportunity."

For someone who had just been told his company had, in effect, finished in second place, Harvey was gracious to a fault. He said that although he was sorry to lose me, he was happy that my dream was moving closer to reality.

SAIC proved to be excellent preparation for my new job in Houston. I enjoyed the NASA headquarters insights into the inner workings of the space program. The Solar System Exploration Division managed robotic spaceflight, a bit of ground-based astronomy, and ongoing data analysis. My experiences there would complement my future education in the much larger shuttle and station programs in Houston. As an SAIC scientist, I helped push forward ideas for asteroid and comet exploration and proposals for missions to Saturn and Pluto. I was particularly interested in a novel concept for a network of Mars surface stations, known as the Mars Environmental Survey mission, to be emplaced on the rocky Martian surface using a giant cushion of inflated airbags. That proposal later evolved into the very successful *Pathfinder/Sojourner* lander and rover, which bounced safely down on Mars on July 4, 1997.

My brief taste of NASA headquarters made walking away from SAIC at the end of June a tough proposition. As our move to Houston approached, it was difficult to say good-bye to my parents and siblings in Baltimore. My kids would be without their grandparents, aunts, and uncles. Our once-busy house now empty, we turned our two cars out of the driveway for the long journey south.

In May we had visited Houston just long enough to put a contract on a new house about ten minutes from JSC's back gate. But it was still under construction and wouldn't be ready until late August, so we settled in for what we hoped would be a short stay in some corporate apartments just outside the space center.

July 16, 1990, was a bright and typically stifling Houston Monday morning. I put on a suit and drove to the Selection Office for my first day at work as an astronaut candidate. Stepping from my car, I greeted the first of my new classmates, a freckled redhead in Air Force blue, Captain Susan Helms. She was already a pioneer, marching onto the terrazzo at the Air Force Academy in 1976 with the first women cadets. Although we had overlapped a year at the Academy, we had never met. Helms and I walked in together to meet the rest of Astronaut Group XIII and begin our long journey toward the launch pad—and the stars beyond.

# 3

# WHO'S GOT
# THE RIGHT STUFF?

*Training is everything. The peach was once a bitter almond;*
*cauliflower is nothing but cabbage with a college education.*
MARK TWAIN, *PUDD'NHEAD WILSON*, 1894

*A* scans (rhymes with "glass cans"). We would hear that word a lot. It was the Astronaut Office's designation for us "astronaut candidates." Don Puddy welcomed us to JSC that muggy July morning, and if he didn't use the term right off the bat, one of the astronauts in our welcoming committee certainly did. Whatever our previous qualifications, that morning we were back at the bottom of the organization chart.

Our new boss was Dan Brandenstein, chief of the Astronaut Office. With him were two people we were to get to know very well in the next year: Jim Voss, a member of the 1987 astronaut class, who would be our immediate supervisor for the year of training, and Merri Sanchez, Brandenstein's technical assistant and our day-to-day training coordinator.

The official welcome for Group XIII took place at the astronaut crew quarters in a quiet back corner of the space center. There were twenty-three new hires in the astronaut class of 1990: sixteen of us were mission specialists and seven were pilot astronauts. The twenty-two new colleagues I met that morning became my professional family for the next year. I would have the privilege of flying in space with several, and many have become lifelong friends. The seven pilot astronaut candidates were experienced military test pilots headed for the shuttle front seats. The rest of us were mission specialists—the scientists, physicians, and engineers essential to any shuttle mission. Five were women, one of whom would become the first female shuttle commander. My class ran the gamut of astronaut personalities: the jocks, the freewheeling bachelors, the fighter pilots, the intense scientists, the veteran NASA engineers, and even three regular guys from Cleveland.

Back in January I had seen NASA's press release listing our class members (see table). Rick Searfoss and Jim Halsell had been in my interview group in October; we had at least become acquaintances during our week of being poked and prodded. Another familiar face was Charlie Precourt, my classmate from the Air Force Academy, which we affectionately called "the Zoo." I remembered Charlie from a course or two during our four years there. He had gone on to become a test pilot at Edwards Air Force Base, California, and though we hadn't seen each other since 1977, that common bond was reassuring. Seven of our group, in fact, turned out to be "Zoomies."

Whatever our backgrounds, though, none of us was completely prepared for our coming year of astronaut training. That first morning at JSC was an introduction to the new experiences we might expect. First was our formal introduction to the press.

In front of a full-scale orbiter mock-up in Building 9, we introduced ourselves one by one. It immediately became clear that the reporters weren't interested in our individual names or backgrounds. The space agency was again under fire following a recent spate of bad news. Six weeks before our arrival, on May 30, hydrogen leaks in the engine compartment had scrubbed the astronomy mission of the shuttle *Columbia*. Engineers were forced to haul the orbiter back to the Vehicle Assembly Building (VAB) to troubleshoot the leak. A fix was proving elusive. The leaks were followed by public reprimands to two veteran astronauts who had been grounded for recent flying safety violations. Worst of all, NASA's triumphant launch in April of the long-awaited Hubble Space Telescope had turned into a public relations nightmare: a manufacturing and testing failure had produced a seriously flawed 2.4-meter main mirror. What NASA had hoped would be the most capable astronomy instrument in history was now portrayed by the press as a blurry-eyed $2 billion

## Astronaut Group XIII

*July 1990*

| | | |
|---|---|---|
| LCdr. Daniel W. Bursch | US Navy | Mission Specialist |
| Leroy Chiao, PhD | Civilian | Mission Specialist |
| Maj. Michael R. U. Clifford | US Army | Mission Specialist |
| Kenneth D. Cockrell | Civilian | Pilot |
| Maj. Eileen M. Collins | US Air Force | Pilot |
| Capt. William G. Gregory | US Air Force | Pilot |
| Maj. James D. Halsell Jr. | US Air Force | Pilot |
| Bernard A. Harris Jr., MD | Civilian | Mission Specialist |
| Capt. Susan J. Helms | US Air Force | Mission Specialist |
| Thomas D. Jones, PhD | Civilian | Mission Specialist |
| Maj. William S. McArthur Jr. | US Army | Mission Specialist |
| James H. Newman, PhD | Civilian | Mission Specialist |
| Ellen Ochoa, PhD | Civilian | Mission Specialist |
| Maj. Charles J. Precourt | US Air Force | Pilot |
| Capt. Richart A. Searfoss | US Air Force | Pilot |
| Ronald M. Sega, PhD | Civilian | Mission Specialist |
| Capt. Nancy J. Sherlock | US Army | Mission Specialist |
| Donald A. Thomas, PhD | Civilian | Mission Specialist |
| Janice E. Voss, PhD | Civilian | Mission Specialist |
| Capt. Carl E. Walz | US Air Force | Mission Specialist |
| Maj. Terrence W. Wilcutt | US Marine Corps | Pilot |
| Peter J. K. Wisoff, PhD | Civilian | Mission Specialist |
| David A. Wolf, MD | Civilian | Mission Specialist |

piece of space junk. The Hubble problems had surfaced just a week before our arrival, and the press was anxious to hear comments on the fiasco from the agency's newest spokespersons, who happened to be us.

The reporters directed most of their questions to the women in our group. Major Eileen Collins was NASA's first prospective female shuttle pilot, and physicist Ellen Ochoa was heralded as the nation's first Hispanic woman in space. The five women were inundated with requests for sound bites; I got away with "name, rank, and serial number."

It was a relief when the press event ended and we headed up to the third, and highest, floor of Building 4, home to our new offices. The space heroes of my youth had shared these offices during the heyday of the Gemini and Apollo programs, and now I was walking the same halls. I felt a profound humility along with an undeniable and genuine pleasure at being there.

Still grinning like a kid in a candy store, I found my office across the hall from the elevators. It was a small room jammed with government-issue steel desks, bookcases, and filing cabinets lit by a wall of windows opposite the door. But I was dazzled by my new office mates: Kathy Thornton, a scientist from the 1984 astronaut class; Kevin Chilton, another Zoomie and test pilot who had been an astronaut for three years; and Bud Ream, a stocky research pilot in his sixties who had been a fixture out at Ellington for decades. Bud had been assigned with Merri Sanchez to be our class's unofficial chaperones for the next year.

Jim Voss and Merri Sanchez didn't waste time easing us into the demanding Ascan schedule. Classroom sessions on the space shuttle and the basics of spaceflight began right away and occupied about six hours a day at first, plus additional hours spent reading the background material from the training division. Early mornings and afternoons also featured sessions with an instructor in the Single Systems Trainer (SST), where we could put our newly acquired knowledge into practice in a simple functional mock-up of the shuttle cockpit. To ready us for a shuttle crew assignment, our teachers would take us from classroom to SST and eventually to the high-fidelity Shuttle Mission Simulator (SMS).

Erlinda Stevenson, our lead scheduler, ruled our Ascan lives, deciding what time we got up, what we would learn each day, and when we might go home at night. Each week Erlinda, whom we soon came to treat with the respect due our own mothers, produced a computer printout with our training and classroom calendar. That schedule soon began to fill up with field trips. Just a week after starting at JSC, we headed to Vance Air Force Base for ejection seat and parachute training, a prerequisite to flying the T-38 jet trainers at Ellington.

At Vance, where I had been a student pilot in 1977, the Air Force instructors taught us ejection seat operation, demonstrated the right way to roll out of a parachute landing fall, and soon had us practicing our landings with short leaps into beds of pea gravel. The graduation exercise was a full-up ride in a parachute canopy towed behind an Air Force pickup. From 300 feet up, the truck was a toy streaming a trail of dust across the pasture as the red and yellow nylon rippled softly overhead. Soon I was descending smoothly earthward, an instructor shouting up his critique with his bullhorn. The ground rose up and *Smack!* Twelve years after leaving Vance, I was qualified to use an ejection seat again.

While I was relishing every aspect of the busy training environment, Houston was proving less enjoyable for Liz and the kids. Forced inside by the torrid weather and restricted by the schedules of two small children, Liz found the apartment claustrophobic. Our unit looked across a steaming parking lot to a stagnant drainage ditch winding its way to Clear Lake. Most afternoons the oppressive heat gave way to a brief thundershower, followed by even more humidity.

Liz knew virtually no one in town. The cramped two-bedroom apartment was no place for two active young children, but the summer heat precluded going out much. And the outdoors had its own hazards. In the first few days in the complex, both Bryce and Annie stumbled into fire ant mounds, and their legs and ankles were soon spotted with tender red welts.

Worse, Bryce took a tumble off our bed and knocked out two of his four lower teeth. A pediatric dentist attempted to save them, wiring them back in place. Bryce soon learned a new word: pudding. Two days later, I departed for Spokane, Washington, and land survival training. I didn't like leaving after Bryce's accident, but I didn't have the option of missing this trip. At Fairchild Air Force Base we headed into the woods for a survival refresher. That first night, Rich Clifford sliced deeply into his hand with a sharp knife while constructing a sleeping shelter, earning the nickname, "The Blade."

When I came out of the forest several days later and called Liz, the news wasn't good. Bryce's teeth hadn't made it. After a couple of days, the dentist had advised removal, and Liz had reluctantly agreed. Bryce apparently didn't miss the teeth, but their loss was traumatic for Liz. The accident just added to her general unhappiness with our first weeks in Houston. My travel and the construction delays on our new house meant there were few bright spots on the horizon. I could only tell her, "Hold on. Things will be better when the house is ready, this travel is over, Bryce heals up," and so on. I hoped I was right.

I joined the other Ascans in trying to figure out the mysterious workings of the Astronaut Office. Those few of our class who had worked at JSC tried to fill us in. After a Navy career, Ken Cockrell had flown as a research pilot out at Ellington. Rich Clifford was an Army test pilot who had worked with JSC's vehicle integration and test team (VITT). Don Thomas had been a JSC mechanical engineer and shuttle microgravity experimenter. And Jim Newman had taught astronauts as a space shuttle systems instructor. With their insights, we started to piece together that elusive big picture.

As chief of FCOD, Don Puddy, a former shuttle flight director, was responsible for providing qualified crew members to meet shuttle and space station mission requirements. His most famous predecessor had been Deke Slayton. Puddy appointed Dan Brandenstein as the chief of the Astronaut Office, with mission specialist Jim Buchli as deputy. Those astronauts not training for a flight were detailed by the Office to dozens of technical assignments, providing flight safety and operations experience to shuttle managers, engineers, and flight controllers.

Brandenstein had told us our job as Ascans was to absorb the fundamentals of human spaceflight and learn the intricacies of the space shuttle inside out. Assuming we could meet that challenge over the coming months, he promised we would have our own technical jobs at the end of our first year. Next would come what all of us wanted: assignment to a flight crew. However, we were at a loss to explain how that happened, and the chief hadn't yet bothered to give us the details. The flight assignment process and the many possible factors that affected it were a constant focus of our conversations, and we continually buttonholed our more senior office mates for insights.

When not on the road that first summer, our typical training day began at 8:00 a.m. with a technical lecture in our big third-floor conference room. These classes on spaceflight fundamentals or shuttle systems were taught by JSC engineers and scientists, supplemented with practical information from shuttle training instructors and astronaut experts. For example, a JSC thermal systems engineer would discuss the principles of spacecraft thermal control. An hour later, a shuttle systems instructor would introduce us to the shuttle's thermal control systems and normal operations; after that, an astronaut would present the best techniques for recognizing system problems and handling emergency situations in the simulator.

After a lunch break, Erlinda would fill our afternoons with hands-on shuttle training, usually in the SST. This classroom trainer was a low-fidelity mock-up of the shuttle cockpit, with plywood switch panels and basic computer displays. While we sat in office chairs in front of the instrument panels, an instructor would review our morning lessons and apply them to normal and emergency operation of a single orbiter system, going step-by-step through the shuttle checklists. Reading these documents, or Flight Data File (FDF), was a daunting challenge. Although similar in format to aircraft checklists, the language was entirely alien, written in a NASA and Mission Control shorthand that is pure gibberish to the uninitiated. For example, the steps for responding to an orbiter cabin leak are shown in the table. Long before I would be eligible to fly, I would have to learn to read that specialized language and perform those steps like a veteran. The only acceptable standard in execution was perfection.

**CAB PRESS LEAK**
1.    CAB RELIEF A – CL (pause)
                 B – CL
2.    √Tabs/Visors – CL/LES O2 – ON
3.    If RTLS/TAL: O2/N2 CNTLR VLV SYS 2 – OP
4.    O2 TK1,TK2 HTRS A,B (four) – AUTO
⇒  5.       TK3 HTRS A,B (two) – AUTO
If possible:
      6.    Remove Inner A/L Hatch Caps (two)
      7.    Equal vlv (two) – EMER
8.    If V > 15K: MLS (three) – OFF;
                 TACAN (three) – OFF
    At V = 10K: TACAN 2,3 (two) – GPC
    At M = 2.7:
          If reqd: Use two MLS, GNC I/O RESET
When CAB PRESS < 10 psia, if dP/dT EQ < 0.67:
    9.    √O2/N2 CNTLR VLV SYS 2 – OP >>
When CAB PRESS < 6.5 psia:
    10.  √O2/N2 CNTLR VLV SYS 2 – OP
    11.  CAB FAN (two) – ON

The summer weeks raced by in a series of class trips designed to equip us with the basics needed to function as T-38 crew members, novice astronauts, and public representatives of the space agency. After parachute and survival training, we hit a string of NASA installations in quick succession. At Pensacola, Florida, the Navy provided a water survival refresher. I became familiar again with parasailing, life rafts, and open-water pickups by helicopter. Each hour our class spent together gave us valuable time to learn about each other and planted seeds of trust and friendship that would later become important in training and in space.

What should Group XIII call itself? This was no small question, for if we didn't choose our own group name, our Office colleagues would be only too happy to invent one of their own, with long-lasting and embarrassing results. For example, the 1978 class, the first shuttle-era group, called themselves "Thirty-Five New Guys," and the 1984 group christened themselves "the Maggots," after former Navy SEAL Bill Shepherd's constant references to that boot camp term during their land survival training. Their group patch, featuring a brood of space-suited larvae emerging from a shuttle orbiter, was a stroke of genius.

Brandenstein was already calling us the "Excedrin Group," saying we would solve all his manpower headaches, so we had to move fast. We wanted to capitalize on our "lucky number," to somehow embrace and thus disarm the superstition surrounding 13. This line of reasoning led us down the road to broken mirrors, walking under ladders, and finally black cats. Somehow the cat idea spun off in the direction of "funny things associated with cats." Bingo! The intellectual purity and mature appeal of the term *hairballs* swept away all opposition—and it eliminated the very real danger of being called "the Pussies." We adopted "the Hairballs" as our name, with a chest-pounding, hacking cough as our class motto. To satisfy NASA's public affairs apparatus, Rick Searfoss created an official-looking class emblem, but our black cat design soon showed up on shirts, baseball caps, and coffee mugs.

The Hairballs embarked on a series of autumn familiarization trips to NASA field centers. Our visit to the John C. Stennis Space Center in southern Mississippi coincided with the test firing of a space shuttle main engine (SSME). Three of these hydrogen/oxygen–burning engines power the space shuttle off the pad, and before an SSME propels a shuttle to orbit, it is qualified by a full-thrust engine run-up on a Stennis test stand.

Engineers scheduled the test for just after sunset. From atop the blockhouse a thousand feet away, we glimpsed amber and red warning lights pulsing atop the massive reinforced concrete test stand surrounded by miles of undeveloped pine forest. The distant roar of a flare stack burning off excess hydrogen gas finally gave way to the rush of thousands of gallons of water flooding into the flame bucket, ready to both cool the structure and muffle the damaging noise of the engine run. Ten seconds before ignition, a fountain of golden sparks played beneath the engine bell to burn away any stray hydrogen fuel, followed by the rising whine of the two turbopumps forcing tons of liquid fuel and oxidizer into the combustion chamber. Ignition! Orange-red flames rattled the flaring nozzle, instantly giving way to the brilliant blue-white blowtorch of a hydrogen-fueled engine building to 100 percent thrust. A second or two passed as a billowing white steam cloud, eerily lit from within, burst silently from the flame bucket toward the dark forest. Then the sound hit us.

It was as much a physical as an aural assault. A simultaneous shriek and earsplitting roar buffeted us, wrapping us in a wave of sound, and cutting me off from conversation with friends standing just a foot away. My body reverberated from the shock of 394,000 pounds of sea-level thrust, a power output that put the B-52s I used to fly to shame; this single engine matched the power output of four of my old bombers at full thrust.

Scarcely a minute into its simulated shuttle flight profile, the SSME ramped up to a full 104 percent of its rated thrust. Visible in the nearly transparent blue flame streaking from the nozzle was a train of glowing shock diamonds, a necklace of visible shock waves illuminating the sheer power disappearing into the flame bucket. After nine minutes, the controller upped the thrust output to 109 percent. My only thought as I squeezed my hands over battered eardrums was that if my dream of flying ever came, I'd be riding *three* of these demons, not a thousand feet away, but just a hundred feet beneath my seat. Was strapping into an orbiter a good idea?

Astronaut Sonny Carter escorted us the next day through the Michoud Assembly Facility near New Orleans, Louisiana, where we watched workers assembling the huge shuttle external fuel tanks (ETs). On the bus back to New Orleans from the ET factory, we blitzed him with nonstop questions about his training and spaceflight experiences.

Sonny told us about the lunch his STS-33 crew had with Vice President and Mrs. Quayle. Their meeting took place in early 1990, just after the *Columbia* STS-32 crew's spectacular retrieval of the Long-Duration Exposure Facility, a hefty satellite the size of a school bus. Unfortunately, according to Sonny, the vice president had been poorly briefed that day. Mr. Quayle started the conversation by congratulating his guests: "You gentlemen did a heckuva job bringing back that big satellite." The crew's silence was just getting awkward when Sonny's commander, Fred Gregory, answered, "Uh, sir, that wasn't us. That was the *Columbia* crew, the flight following ours. Our mission involved a Department of Defense payload."

The vice president recovered quickly: "Oh, yes, right, and it was a great job you all did, too." The excellent lunch quickly erased the awkward moment, and too soon the vice president rose to thank his guests. As all stood to part company, Quayle announced he had a few souvenirs to remind the crew of their visit to the White House. Taking a handful of lapel pins bearing the vice-presidential seal from his pocket, he began tossing them at the guests around the table, landing one of the mementos squarely in a brimming cup of coffee.

Sonny also told us how he had lost his wristwatch in orbit. He had briefly removed it in zero-g and the watch had drifted away. Despite searching *Discovery*'s cabin high and low for the rest of the five-day mission, it never turned up, and Cape Canaveral technicians failed to locate the watch after landing. Sonny's timepiece was apparently gone for good.

But spaceflight is full of surprises. On the anniversary of his flight in November 1990, the crew of STS-31, who had next flown *Discovery* and delivered the Hubble Space Telescope to orbit, called Sonny to the front of the Monday morning meeting. Steve Hawley told the conference room of astronauts and Ascans how during a routine filter cleaning he had removed an access plate on the flight deck's port side and found a hairbrush, a packet of salt tablets, and—a watch floating gently in midair. Hawley then formally presented Sonny with his well-traveled timepiece, which had racked up twice as much space time as its owner.

A few weeks later the Hairballs visited the shuttle launch site, the

John F. Kennedy Space Center (KSC), Florida. From Cape Canaveral, just to the south, Alan Shepard and John Glenn had ridden their tiny Mercury capsules into space. Just north of the Cape, Gus Grissom, Ed White, and Roger Chaffee had perished in a 1967 Apollo spacecraft fire. From the twin pads that now hosted the shuttle, Neil Armstrong, Buzz Aldrin, and Mike Collins had left Earth for the Moon. The ghost of *Challenger* still haunted the place. I was eager to see it as an insider.

We Ascans got the VIP tour, starting with the orbiter processing facilities. The three identical hangars each housed a space shuttle orbiter wrapped in scaffolding and tended by a skilled corps of technicians and engineers. In 1990 there were three ready orbiters: *Columbia, Discovery,* and *Atlantis;* a fourth, *Endeavour, Challenger's* replacement, was under construction in California. For the first time I stood close to a real spaceship, the black-tiled bottom of the orbiter *Discovery* stretching above me like the low ceiling of a coal mine. Massive wings swept back from a blunt, gray-tipped nose, and three main engine nozzles jutted powerfully from beneath the towering vertical tail. This ship had traveled to space and back many times, and I marveled at the idea I might one day ride this orbiter or one of her sisters.

Adjacent was the cathedral-like interior of the VAB, where Wernher von Braun's Saturn Vs had been assembled more than twenty years before. Here, workers mated the orbiter with its external tank and solid rocket boosters before the entire massive assembly was moved to the launch pad. From the VAB's lofty upper levels, we looked down on a soaring tank-and-booster "stack" and later inspected at close range the booster components that had failed so disastrously four years earlier. Technicians explained the design changes made to prevent a recurrence of the *Challenger* accident.

The KSC Astronaut Crew Quarters were five miles south, serving not only as the astronauts' home while working at the center but also as their quarantine facility for the last few days prior to launch. Tucked away on the top floor of the Apollo-era Operations and Checkout (O&C) Building, the quarters housed a conference room, offices, a dozen or so tiny bedrooms, shared bathrooms down the hall, a kitchen and dining room, and a small gym. During a brief lunch stop, I explored the facility with the rest of the Hairballs. Surrounded by the spotless 1960s-era furniture and decor, it was easy to imagine Neil Armstrong or John Young padding down the hall to suit up for their Moon shots.

Finally, to the launch pad. Launch Complex 39 is a monument to human exploration. Atop its twin concrete and fire-brick pads, Apollo and shuttle crews boarded ships bound for the stars. Exploring the towering gantry at Pad 39A, we strode across the narrow swing arm bridging the gap between the gantry and the orbiter. Inside the White Room, the vestibule just outside the hatch, I got my first look inside a space shuttle. Peering through the orbiter hatch into the cramped confines of its high-tech cabin, I wondered how it would feel to see that hatch sealed behind me. Each of us, I knew, wanted to be first in line to find out.

THE LATTER HALF OF 1990 was an uncertain time in the Astronaut Office. With Hubble's troubles still a staple of Johnny Carson's *Tonight Show* monologue and *Columbia* still grounded by SSME hydrogen leaks, NASA's reputation continued to suffer. The ongoing delays made for increasing uncertainty over the flight schedule as summer turned to fall.

In today's era of e-mail, voicemail, and the Internet, our 1990 intraoffice communications now seem laughably old-fashioned. My phone messages appeared on little yellow slips left on my desk by Kiki Chaput, my secretary. My office mates shared a single computer used mainly for viewing and printing out the most basic of e-mail messages; we would check those a couple of times a week. Our crucial information-sharing tool was Monday morning's office meeting. Dan Brandenstein would relay the latest from JSC and FCOD management. Astronauts would brief us on the latest news from the Cape, report on shuttle troubleshooting or safety developments, and we would even hear occasionally about a nascent program called the space station. Every astronaut felt free to comment or ask questions. The Hairballs knew enough to keep our mouths shut and listen while we figured out how things worked.

Brandenstein was regarded as one of the best of the shuttle commanders. Before his three shuttle flights, he had flown Grumman A-6 attack jets in Vietnam. His living-room coffee table sported the bullet-fractured front windscreen of one Intruder he had landed safely aboard his carrier after a particularly hairy mission. The chief combined professional skill with an easygoing office manner, and he took pains to keep us Ascans informed about our long-range job prospects.

He called us together a few months into the year to explain his policy on the dominant topic of all Hairball conversations—flight assignments. Brandenstein kept it simple. He periodically compared notes with Mission Operations on the shuttle mission manifest, getting a "start training" date for each projected shuttle crew. A few months before the start date, he would privately pencil in possible crew members, matching experience, technical background, and job performance to the required crew positions of commander, pilot, and up to five mission specialists. Once he had drafted his crew nominations, he would forward them on to Don Puddy for final approval. Puddy would then notify the new crews, and NASA would issue the corresponding press release.

The numbers game worked out like this, Brandenstein said: NASA was flying about six shuttle flights a year, with an average crew size of six. Even if he put a couple of us rookies on each flight, it would still take two years to get all twenty-three Hairballs off the ground. As he put it, "Someone has to be first, and someone has to be last." Those going late needn't worry about their career progression, he assured us. (We noted, for example, that Curt Brown, an '87 astronaut, was still waiting for his first flight assignment more than three years after his arrival. But Brown later flew six missions in just over seven years. Even Neil Armstrong was last in his class to fly.) Brandenstein assured us the wait would be worth it. Still, all of us wanted to fly early. How did we get to the front of the line? There was no magic formula, the chief told the Hairballs. Do your best in Ascan training, tackle your technical jobs energetically, and time will take care of the rest.

So all of us redoubled our efforts to hit the books and earn brilliant reputations in academic and skills training. I plowed through an endless series of softbound workbooks written by the training division to introduce the shuttle systems knowledge we would use in the SST and, later, in the simulator. By the end of 1990 we had progressed far enough in the SSTs to begin our first sessions in the Shuttle Mission Simulator. Its flight deck an exact copy of the real orbiter cockpit, the SMS accurately replicated all the shuttle systems operations and displayed a realistic depiction of orbital scenes outside its television "windows." With a qualified astronaut along for advice, three Ascans would take turns in the flight deck seats as we explored the most elementary problems in shuttle operations. From their consoles outside the simulator, which replicated even the motion and vibrations of flight, our SMS instructors could monitor our performance and introduce the failures we had been briefed to expect. Normal, or "nominal," launches and landings would progress to simple failures, giving us a chance to practice the checklist steps for, say, a main engine failure.

As a mission specialist I would never launch or land at the controls of the shuttle, but in one of these early SMS lessons I got one of my few chances to land the (simulated!) shuttle orbiter. The orbiter had the familiar heft of my old B-52 as I banked into the steeply descending spiral toward the KSC runway. The video picture out the front windscreen showed the Cape landscape in startling and rapidly growing detail. The breakneck rate of descent, nearly 12,000 feet per minute, put me in mind not of the B-52 but one of its Mark 82 500-pound bombs—miss the preflare down low and before you know it you're a smoking hole in the alligator-ridden swamp surrounding the runway. Maybe it was due to my early experience as a glider pilot, but I did get the orbiter safely on the ground in each of my landing attempts. The orbiter's flying qualities gave me an enduring respect for the shuttle pilots, who through hundreds of practice approaches manage to bring the orbiter down precisely despite the curveballs thrown at them by devilishly imaginative instructors.

We rookies usually pulled the SMS graveyard shift, the 8:00 p.m.-to-midnight session disdained by assigned shuttle crews. One night we found that John Young would serve as our instructor in the commander's seat—quite an honor for us new guys. At age sixty, Young still kept his hand in with regular simulator sessions and frequent sorties in the T-38 and shuttle training aircraft. In the SMS that night, a Hairball was in the right (pilot's) seat. I sat behind in the Mission Specialist 1 (MS-1) seat, and classmate Jim Newman served as flight engineer, or MS-2, sitting between and aft of the pilots behind the center console. We had already survived a couple of ascent runs as we counted down for another launch. The cab of the motion base simulator had rotated us to the "extended pitch" or launch position, the hydraulic jacks rocking us on our backs as the orbiter's nose pointed skyward. As the final seconds of the count unwound, the main engines rumbled to life, the boosters jolted us free of the pad, and we were off.

John and our pilot handled a few minor malfunctions with ease, and the first minute of the flight went well. Newman, strapped in on my left, gave me a nudge and whispered, "Watch this!" What was he up to? Jim grabbed the slender aluminum "swizzle stick," a wand that enabled the backseat mission specialists to reach switches and push buttons otherwise inaccessible while strapped in. He inched the stick out over the center console toward the three SSME shutdown push buttons. I shot him an alarmed "What the hell are you doing?" expression, but Jim just grinned and nodded toward the console. We were passing Mach 5 when he mashed the center engine shutdown button, then pulled the stick quickly out of view. The loud warble of the master alarm burst through the cockpit; error messages flashed on three computer screens. "Center engine down!" announced the pilot, running the engine failure check-list. John Young calmly selected the proper abort mode and hit the push button to send the orbiter toward North Africa. "What the hell happened?" he asked the cockpit at large. Newman was successful at keep-ing a straight face, but I was appalled. Why would you want to kill an engine on the most senior astronaut, the first man to command the shut-tle? One word from John Young to Don Puddy and we would never get closer to space than this simulator.

Descending into Ben Guerir, Morocco, John asked the instructors what had caused the engine failure; they ruefully admitted they didn't know. Young put the orbiter down smoothly on the desert runway, con-tinuing to puzzle over what had taken out his center engine. Finally, Newman lost it; he burst out laughing as he confessed that he was the real culprit. Newman, the former SMS instructor, had just been keeping his hand in, feeding Young one last malfunction. What surprised me was that John loved the practical joke. Far from being upset at the gag, John wryly gave Newman credit for pulling it off.

I eagerly looked forward to another chance to fly, this time in a real airplane. My reintroduction to the elegant, needle-nosed T-38 was a highlight of my Ascan year. NASA's Ellington Field ground crews main-tained the thirty or so Talons in superb shape, parking them wingtip to wingtip in front of Hangar 276. In the Air Force I had racked up more than 2,000 hours of flying time, and handling the controls of the nimble jet after more than seven years away was pure joy. Although as a mission specialist I flew only in the Talon's backseat, it was a small sacrifice in exchange for flying privileges. I always relished the prospect of flying with a NASA instructor, a veteran shuttle pilot, or, best of all, one of the Hairballs. Booming out of Ellington on twin afterburners, I found my hour or two of weekly flying time the perfect antidote to a stack of workbooks or another late session flipping switches in the SST.

On one such flight, when we were forty miles out over the Gulf of

Mexico, Bob Cabana, a Marine pilot just back from his STS-41 mission, offered to demonstrate what a real shuttle launch felt like. Bob put our T-38 into a near-supersonic dive, bottomed out just above 10,000 feet, and began a Public Affairs Office (PAO) narration of our launch: "and we have liftoff, liftoff of the shuttle *Discovery*!" At over 500 knots, Cabana hauled the stick back and zoomed into a vertical climb, the g's crushing us down into our seats. "Houston, roll program!" he intoned seriously, delicately pivoting the jet on its tail to mimic the shuttle stack's slow roll to the proper ascent heading. Blue ocean and white clouds whirled by as the jet rolled obediently under Bob's right hand. Arrowing up through 20,000 feet at full throttle, Bob laughed, called "$P_C$ less than 50!" (booster chamber pressure less than 50 psi, the indication of impending booster burnout and jettison), and unloaded just enough to ease us out of our seats, simulating the waning thrust from our solid rocket boosters. Still pointing skyward, we zoomed through 25,000 feet, rapidly running out of airspeed. I had just managed to drag my stomach back into the cockpit when "Cabana Bob" popped the stick forward and pushed the nose over to the horizon—we floated weightless against our straps. "Houston, *Discovery*: MECO! Main engine cutoff!" he shouted, his grin coming right through the headphones into my helmet. I was grinning a mile wide myself. "Still want to go to space, Tom?" He knew the answer before I could say, "Sign me up!"

The Office had lined up a series of distinguished speakers to introduce the Hairballs to the history of the human spaceflight program. We heard from Chris Kraft on the early days of the Mercury and Gemini projects, Tom Stafford about Gemini rendezvous and Apollo, Gene Kranz on the first lunar landing, Joe Kerwin about Skylab 2, and John Young about the first shuttle flight. But the briefing that impressed me most was Sonny Carter's detailed chronicling of the *Challenger* accident investigation. After reviewing the causes of the accident—O-ring failure and subsequent burn-through of the solid rocket booster (SRB) casing—he traced the path of *Challenger* from breakup to the ocean floor. Sonny put on his old flight surgeon hat and ticked off the forensic evidence recovered from the Atlantic, weaving together a story of human physiology, captured Luftwaffe data on high-altitude survival, and orbiter structural engineering that indicated the crew survived *Challenger*'s initial breakup. Recovered cabin debris showed the crew had activated three of their emergency airpacks, perhaps reacting to an explosive cabin decompression. But the airpacks were designed to prevent smoke inhalation, not for high-altitude survival, and without pressure suits the crew would have lost consciousness within seconds. Investigators could find no further evidence that the crew took any other switch or control actions, leading the accident board to conclude that the astronauts probably never regained consciousness during the nearly three-minute fall to the Atlantic. The ocean impact was not survivable.

Unlike the *Challenger* astronauts, we Hairballs would have pressure suits, parachutes, and a bail-out capability on our flights aboard the shuttle. But Sonny's professional yet grim accounting forced us to consider the slim margin between success and disaster when operating in space. I came away realizing that my survival depended on the level of precision I could bring to the job—and something else. The complexity of space flight meant that I had no choice but to rely on the dedication and skill of others, from my crewmates to the most junior technician turning a bolt at the Cape. Trust and risk are inseparable elements of space exploration.

The risk was ever-present. On April 5, 1991, three of us had just stepped out of the motion base simulator in Building 5. We were still asking questions of John Young, our instructor pilot, when Curt Brown intercepted us in the parking lot. As was his habit, Curt ignored us Ascans and spoke directly to Young. "John, we just got the call . . . we lost Sonny Carter." Sonny had been traveling to a NASA public relations appearance on a small commuter plane, which had rolled over and crashed into a field near New Brunswick, Georgia. There were no survivors. A chill swept over me despite the warmth of the Texas sun. Just like that, Sonny Carter—flight surgeon, fighter pilot, test pilot, and astronaut—was gone, leaving behind his wife, Dana, and two daughters.

His memorial service packed the pews of University Baptist Church, obvious testament to Sonny's professional and personal reputation. As we mourned this young husband and father, Liz recalled how Dana had once sat down to chat with a group of Hairball wives at an astronaut spouses' function. "Are you all close to each other?" Dana asked the women gathered around the table. "Not especially," was the collective answer. "You might want to develop those relationships," said Dana. "You might find you'll need them someday."

The news of Sonny's death didn't slow our relentless training pace.

Next up was space walking, or extravehicular activity, beginning with a lecture from Story Musgrave, one of Sonny's STS-33 crewmates. His lecture to the Hairballs was to focus on EVA, but his experience stretched across decades of spaceflight history. With an athletic build that belied his fifty-five years, Story was the office's Renaissance man.

Chosen as a scientist-astronaut at the height of Apollo, Story had waited sixteen years for his first spaceflight. That fact alone put me in awe of him. An enlisted Marine in Korea and later a medical doctor, he had worked on the Apollo and Skylab programs, and helped develop the design for the shuttle space suit, known as the Extravehicular Mobility Unit (EMU). Aboard *Challenger* on STS-6 in 1983, Story had led the first shuttle space walk. His trademark shaven head, crow's-feet, and wizened grin gave him an undeniable resemblance to Yul Brynner, star of the old movie musical *The King and I*. His talk was as wide-ranging as his twenty-three years with NASA.

Story had four spaceflights under his belt, but he admitted that he was frightened before each and every one. Spaceflight, he told us, was unavoidably risky. There was nothing to do but confront the danger and deal with it. Story broke problems down, then solved them piece by piece. That approach showed in his EVA philosophy. He had found that most of the fatigue and physical workload in the space suit came from fighting against the suit itself or from a body position inappropriate for the task. "If you're breathing hard and your heart's pounding during a space walk, stop! You're doing something wrong. Take the time first to position and tether yourself properly, and put the work right here—in front of you. Going slowly, you'll still work faster than you can taking the brute force approach." EVA was a ballet, he said. "You've got to think about your moves like a dancer, only you'll be balancing on your hands, not your feet. If you're doing it right, you will move around in free fall with just the tips of your fingers."

He gave the Hairballs a few other practical tips. "Why wait until orbit to start getting used to zero-g? I start my adaptation during quarantine." To simulate the headward fluid migration that occurs in free fall, "Put the foot of your bed up on cinder blocks, so you sleep head down all night." Story claimed that the resulting congestion and headaches were a good thing; they gave your body a several-day jump on coping with the same symptoms in flight. Most of the Hairballs countered that space adaptation syndrome sounded bad enough as it was; why give yourself an extra week of misery? Story did admit that his strategy had its faults: it was tough getting a good night's sleep in quarantine when your body kept sliding into the headboard.

He ended the talk by challenging us: "Become a creature of space. Take advantage of the opportunities offered by zero-g." How? "Think outside the box. When you're using the urine hose on the commode, there's no reason to maintain a one-g orientation. Try going to the bathroom in a heads-down orientation. I guarantee you the rest of the crew will get a kick seeing your feet floating above the compartment door."

AS ASCANS OUR ATTENTION was focused intensely on the shuttle, but in late 1990 Dave Finney, an FCOD pilot and engineer, began briefing us every Monday morning on NASA's next human spaceflight project: the Space Station. Proposed by President Ronald Reagan in 1984, the *Freedom* space station was supposed to be in orbit by 1994, furnishing American, European, and Japanese researchers with a spacious orbital laboratory for materials and biological research. But Reagan's overall cost estimate of $8 billion had long since been exceeded by contractor overruns and lax NASA oversight of the project. Congress had put a budget ceiling on the Freedom program, whittling away its size and capability, and the station seemed no closer to achieving orbit than it had six years earlier. The first launch of Freedom components on the shuttle seemed so far down the road to us Hairballs that we paid little attention to the project's halting progress. I had no inkling that I would one day become intimately involved in its planning and construction.

In August 1990 Brandenstein was predicting that congressional budget pressure would force a redesign of the Space Station in an attempt to get it established in orbit more quickly. By October of that year, the cuts had materialized, reducing the station's annual budget from $2.4 billion to $1.9 billion. NASA was grappling with several options to try to get under the cost ceiling. They included creating a slimmed-down, less expensive version of *Freedom*, or coming up with a new, simpler design for a station: cheaper, but with reduced science capability.

There was no clear winner. By November the agency was wrestling with a host of tough questions about the station's future. What was the best way to assemble the station in orbit? Was the huge "erectable truss" (the station's backbone) too complicated to build in orbit? Could we simplify the design to enable a simple Skylab-type station to be quickly deployed in orbit? How could we counter congressional moves to make the station "man-tended," only periodically visited by astronauts? Where would we find the funds for the station's critical lifeboat, needed to place a permanent crew aboard?

In November 1990, facing a projected $7 billion shortfall in the Freedom budget over the next six years, NASA considered shrinking the size of station modules and substituting a "mini-Lab" and "mini-Hab" (habitation module). It appeared that just a few months after we arrived, the station that the Hairballs had seen in mock-up in Building 9 would be significantly downsized. It certainly wasn't a project that generated much enthusiasm among us.

A new wrinkle in the station story emerged at the end of November. John Fabian, a veteran of the 1978 astronaut class, briefed us on his visit to Moscow to discuss a proposed cooperative space venture with the Russians. His remarks sparked a lively discussion about the wisdom of working with our former enemies and space rivals. Fabian challenged the gathered astronauts: "Is it wise to allow the former USSR to collapse? Perhaps both countries can benefit by working together in space."

By early 1991 NASA had produced a smaller and less costly station design, but it failed to impress congressional critics. NASA administrator and former astronaut Dick Truly was quoted as saying that the new version looks "too much the same"; although easier to assemble, the new design did not save enough money.

In February 1991 NASA had come up with yet another design that called for one less solar array, a ground-assembled truss (as opposed to an astronaut-built structure), a shortened Lab and Hab module, a lifeboat and life-support systems for four crew members, and sufficient supplies for ninety-day crew stays in orbit. Although there was not much change in external appearance, it reduced up-front costs and greatly reduced the orbital assembly required. John Young liked the idea because much of the station's testing could now be done on the ground before launch.

Our suspicion of all these constantly revised plans was evident when Mike Foale, a member of the 1987 astronaut class helping with the station effort, got up at the Monday meeting on June 3, 1991. Popping a tape in the VCR, he showed us a computer-generated view from the station's cupola, a greenhouse-style structure that would give astronauts their most expansive look at the outpost's exterior. The "camera" panned across the windows, showing the spacecraft, the black sky above, and finally the blue planet below. The computer imagery of Earth's surface was decidedly blurry, and someone piped up sarcastically: "Is the problem that the glass isn't of optical quality?" Foale didn't miss a beat: "No, that's how poorly your eyes will be working by the time you get to fly on the space station."

Foale's comment reflected the mixed reaction astronauts had about the station. NASA saw it as its institutional future, "the next logical step" in the von Braun scheme of exploring the solar system. Yet the astronauts knew that weak management and congressional manipulation had gutted the original concept for a research and exploration stepping-stone into Earth orbit and beyond. Its costs were spiraling ever upward; the NASA field centers, especially JSC, didn't trust the station managers in Reston, Virginia; and little if any hardware was being built. In the Astronaut Office, real spaceflight meant "space shuttle."

A space station might very well be the next destination for the shuttle, but for now, *Freedom* existed only as a stack of NASA briefing documents. The Hairballs would believe in a space station when, and only when, it moved off the drawing board and out to the launch pad. In 1991 no one was holding their breath waiting for that to happen. I frankly favored bypassing the station for a push toward President George H. W. Bush's long-term goal of returning to the Moon and reaching for Mars.

The shuttle program, of course, was not immune to problems. In February 1991 astronaut Loren Shriver noted that photos of the external tank taken from the orbiter *Columbia* just after main engine cutoff showed dinner-plate–sized pieces torn from the tank's foam insulation. Apparently a faulty application of the foam caused air trapped beneath to pop chunks of insulation loose. The worry was that the foam could impact and potentially damage orbiter tiles. We were told that a new method for applying the spray-on foam insulation would be implemented at the Michoud factory to address the problem. The Hairballs put this news into the ever-growing category of stuff we were too busy to worry about.

By late spring of 1991 our year of Ascan training was nearly over. We had completed the required field trips, the hands-on survival training, the history lectures, the space food demonstration, the zero-gravity ride on the Vomit Comet, scuba qualification, the workbooks, geography lessons, hundreds of hours in the simulators, and dozens of hours of T-38 flying. We twenty-three Hairballs were technically qualified to handle a crew assignment, a challenge we all relished. Each of us wondered, When? More often we wondered, Who?

The author and Tammy Jernigan in front of the jammed airlock hatch on STS-80. (NASA)

The author and his wife, Liz, with the shuttle *Endeavour* at Launch Pad 39A, Kennedy Space Center, Florida, August 1994.

Astronaut preparation involves frequent
flights on the STA, or Shuttle Training Aircraft
(lower aircraft), and smaller T-38 Talon, seen
here over White Sands, New Mexico. (NASA)

The crew of the STS-59, Space Radar Lab 1, which orbited earth 183 times in April 1994, taking detailed radar images of Earth's ecosphere, oceans, and geology. From left: Godwin, Chilton, Jones, Apt, Gutierrez, and Clifford. (NASA)

Astronauts on the STS-59 Red Shift—Kevin Chilton (left), Linda Godwin, and Sid Gutierrez—relax off-duty in *Endeavour*'s middeck as they try to corral the printout of the day's updated flight plan from Mission Control. (NASA)

The author's wife, Liz, with children Annie and Bryce on the Launch Control Center roof at the Kennedy Space center just after liftoff of STS-59, *Endeavour*, April 9, 1994. (Author collection, photo by Mary Ellen Weber)

The STS-68 *Endeavour* crew in orbit on Space Radar Lab 2, October 1994. From left: Bursch, Wilcutt, Baker, Smith, Wisoff, and Jones. Flags of SRL partners—the U.S., Germany, and Italy—line the sleep compartments. (NASA)

Space Radar Lab 2 operating in *Endeavour's* payload bay. The SIR-C (marked JPL) and X-SAR (narrow hinged panel at right) radar panels stretch 40 feet to the rear of the payload bay. The MAPS instrument (marked LaRC) points earthward in right foreground. (NASA)

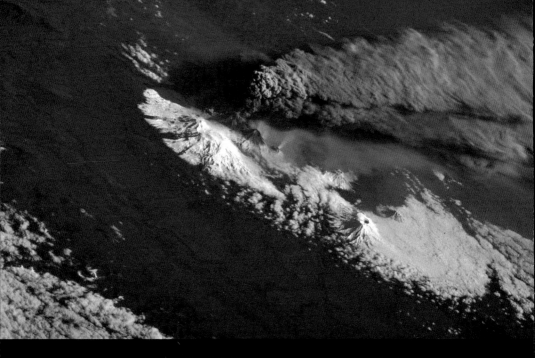

The STS-68 crew documented the Kamchatkan volcano Kliuchevskoi's eruption plume, which rose ten miles into the atmosphere as the jet stream carried dust and ash far out over the Pacific. (NASA)

BOTTOM RIGHT: The STS-80 crew at the pad during their countdown rehearsal: (from the left) Musgrave, Jones, Cockrell, Jernigan, and Rominger. (NASA).

The author prepares for a 1996 underwater training session in the WETF, which simulated through neutral bouyancy the working conditions of a spacewalk. (NASA)

*Columbia's* robot arm holds the Wake Shield
Facility satellite above the payload bay and
Earth's sunlit horizon, just prior to berthing.
(NASA)

*Columbia*, lit by xenon searchlights and trailing wingtip vortices in the humid morning air, glides in for a landing at Runway 33 at the Kennedy Space Center, Florida, on December 7, 1996, after the longest shuttle mission in history. (NASA)

BOTTOM RIGHT: The STS-98 crew heads for the launch pad and *Atlantis*, February 7, 2001. From the left: Curbeam, Jones, Ivins, Polansky, and Cockrell. (NASA)

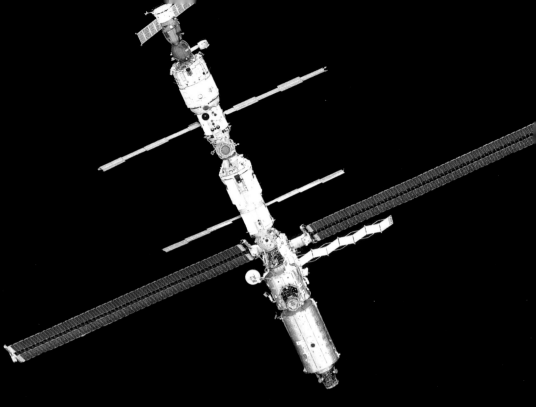

The International Space Station after *Atlantis* undocking,
February 16, 2001. From top to bottom: Soyuz, Service
Module (*Zvezda*), FGB (*Zarya*), PMA-1, Node I (*Unity*), the
Lab (*Destiny*), and PMA-2. (NASA)

*Atlantis* approaches for docking with the ISS, February 9, 2001. The *Destiny* Lab module rides in the aft payload bay. *Atlantis*'s airlock and docking port are just aft of the cabin's twin overhead windows. (NASA)

LEFT: Shadows from *Atlantis*'s launch plume fall toward the face of the full Moon, February 7, 2001. *Atlantis*, thirty miles up, is the white dot at the plume's tip. (NASA)

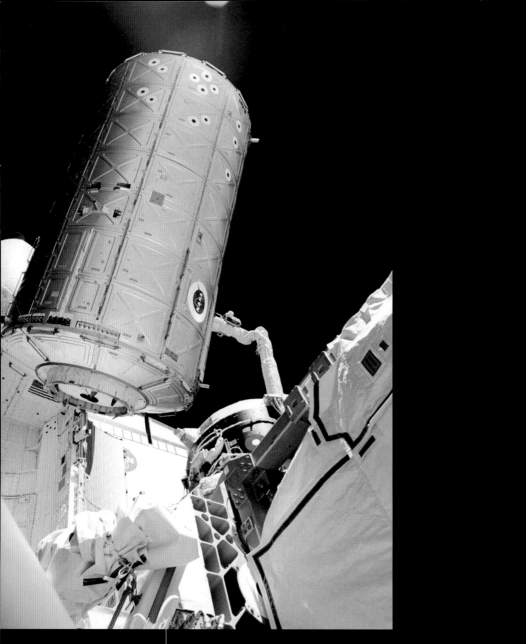

The *Destiny* Lab emerges from the payload
bay as Marsha Ivins executes the "Big Flip,"

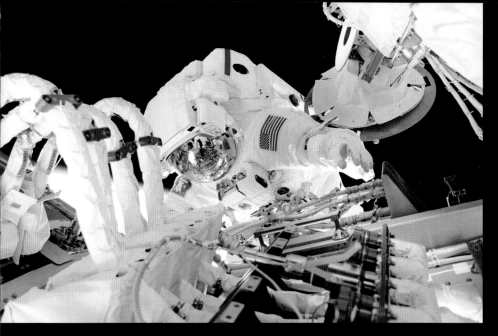

The author connects electrical lines from ISS to the *Destiny* Lab, February 10, 2001. Ammonia coolant lines extend to the Lab at left. (NASA photo by Bob Curbeam)

ISS solar arrays stretch across black space behind the author during his second spacewalk on STS-98, February 12, 2001. (NASA photo by Marsha Ivins)

The STS-98 crew members return to their families at Ellington Field, Houston, February 21, 2001. From left: crew members Curbeam, Jones, Cockrell, Polansky, and Johnson Space Center Director George Abbey. (NASA)

# 4

# COUNTDOWN
# TO FIRST FLIGHT

*What kind of man would live where there is no daring?...If*
*one took no chances, one would not fly at all.*

CHARLES LINDBERGH,
THE WARTIME JOURNALS OF CHARLES A. LINDBERGH, 1970

By early 1991, five years after the loss of *Challenger*, the Space Shuttle
program was finally hitting its stride. Six flights were scheduled for
the year, and the agency had announced plans to send astronauts to
repair the faulty Hubble Telescope. By summer, three orbiters and their
crews had successfully executed their missions. If NASA could maintain
this flight rate, the Hairballs might soon be snagging seats in the rotation.

Like most of my classmates, I wondered where I stood in the peck-
ing order. Dan Brandenstein explained his astronaut personnel policy
to the Hairballs during a briefing in March 1990. It was a rare session
with the chief. He began by telling us tongue-in-cheek that he considered us "the salvation of the Office" because of the manpower we would
add to an astronaut corps stretched to the breaking point by the increasing flight rate and an expanding array of technical commitments. Brandenstein simply didn't have enough people. He had been "holding his

finger in the hole in the dike," waiting to throw us into the breach. All the technical jobs we would be assigned were important, the chief said, and by performing well we could not only help the Astronaut Office but also impress the guy who made crew assignments—him. He would be looking for teamwork, initiative, the ability to learn, and any original contributions we might make to the program. Brandenstein reminded us that our technical jobs not only provided the Office with insight into the shuttle program but were also a vehicle to boost morale and dedication at JSC and throughout NASA. The astronauts hoped that people would work more carefully knowing that mission success or even a crew member's life depended on their actions. We were, he said, the point of the spear thrown into space by the much larger NASA team.

Then Brandenstein turned to flight assignments, our favorite subject. He expected to start assigning Hairballs to flights in early 1992, with some of us actually flying in mid-to-late 1993. We could expect our first, second, and third missions to come in relatively short order, with mission specialists first, then pilots as slots opened up. We could expect a break in flight assignments after flying three times (that was precisely the pattern my flight career would take). His choices would be influenced by a mix of our office performance and the specific mission requirements. The chief's closing advice to us was practical and positive: keep your ego deflated, remember how visible your position is, and serve as a good ambassador for the manned spaceflight program and NASA.

Brandenstein moved quickly after our briefing to plug us into the holes in that leaking dike. By late March he had turned us over to the branch chiefs. I was assigned to Dave Hilmers, chief of the Mission Development branch, which assessed the feasibility of proposed shuttle payloads from an operational and safety perspective and provided astronaut operations advice to shuttle experiment designers.

Dave assigned me to work with the developers of the small science payloads flown in the orbiter's middeck lockers. The work involved attending the designers' technical meetings, reviewing any safety problems the experiments might present, and helping the science "customer" train the crew's payload operators. I enjoyed the assignment because it put me in direct contact with a variety of experimenters and introduced me to the kind of scientific work I would no doubt be performing in orbit as a mission specialist. My immediate boss was Bonnie Dunbar, an astronaut since 1980 and a two-time shuttle flier. Bonnie had a reputation for being a bit prickly and aloof, but in the few months before she went off to train for her next flight I found her encouraging, open, and easy to work with.

While the Hairballs focused on our new job assignments, we couldn't help but notice that the revamped space station *Freedom* seemed to be going nowhere. The Monday morning status reports rarely contained good news. On July 8, we heard that NASA's deputy administrator, J. R. Thompson, had kicked off a study of extending the shuttle's orbital endurance to an astonishing 270 days (a typical shuttle flight then lasted anywhere from 5 to 9 days). The idea was that modifying the orbiter to stay aloft for nine months would be less expensive than building a separate space station, and it would give the United States a long-duration space research capability sooner than the Freedom program. This long-duration orbiter could also serve as a substitute if the Space Station was canceled outright.

A reliable source of pithy commentary about NASA's doings was John Young, officially the technical advisor to the director of the Johnson Space Center but still a fixture at our weekly Astronaut Office meetings. The man who had been an astronaut for nearly thirty years was incredulous at Thompson's plan for a nine-month orbiter: "Two-hundred-seventy days! You might as well just train everybody to eat little white pills instead of food because there won't be enough room in there to pack it." And the long-term effects of microgravity on the guys who would have to land the shuttle worried him. "Do you know what you're goin' to have to do?" he asked the room at large. "You're going to have to freeze-dry the pilot and commander!"

I didn't pay much attention to these concerns, although I was curious about how engineers might modify an orbiter to achieve such long flights. My immediate objective was shedding the "Ascan" label and getting a shuttle mission assignment. The Hairballs had nearly completed a full year as Ascans. How would the Office officially recognize the formal completion of our training? Like the rest of the class, I eagerly awaited the award of the silver lapel pin designed by Mercury and Gemini astronauts, which would mark our official qualification for a flight assignment. A gold version would follow after our first flight.

Monday, July 15, 1991, marked the end of our first year in Houston. Bud Ream, who had traveled and flown with all of us over the past year, stood up that morning to address the astronauts crowded into the conference room. Bud was a likable pilot and had periodically briefed the Office on our training performance and the highlights of the Hairballs' field trips.

"Ladies and gentlemen," he said, "I'd like to announce that tomorrow the Ascans will formally finish their year of training. They've worked hard and learned much in these last twelve months. Having reviewed their progress with Jim Voss, Merri Sanchez, and their instructors, I'm pleased to tell you that the Hairballs are fully trained and ready, willing, and able to take on a flight assignment."

In the conference room there was dead silence. No applause, no murmurs of admiration from the crowd, and significantly, not a peep of agreement from Brandenstein. Bud had inadvertently trespassed on the chief's authority—that of formally recognizing the new class, a necessary prerequisite to crew assignments. Bud sat down, embarrassed but unsure why his announcement had fallen flat with the veterans. The Hairballs were just as perplexed. Then Curt Brown, the only remaining unassigned member of the 1987 class, spoke up in the same bright tone Bud had used: "I'd just like to remind everyone that I've been here for *four* years, and I'm *still* ready, willing, and able to take on a flight assignment." The comeback brought down the house and defused the tension in the room caused by Bud's little speech.

By the end of the Ascan year, my family had settled comfortably into Clear Lake and the JSC community. I traveled less often than I had earlier in the year. Liz felt better about my schedule and the local lifestyle. We had finally moved into our new home nearly three months after arriving in Houston, but the added wait just heightened our happiness with the new neighborhood. I was ten minutes and two traffic lights from the JSC back gate, and our quiet cul-de-sac was close to schools, St. Bernadette's Catholic Church, shopping, and the homes of other astronauts.

The twenty-three Ascans had become very close over the course of the year. We had traveled the country and spent hundreds of hours together in T-38s and simulators. We had camped together in the wilderness, hiked the rugged terrain of New Mexico on geology field trips, and endured mind-numbing briefings from well-meaning but sleep-inducing NASA officials. We became intimately familiar with the Hairball personalities, caricatured humorously by classmate Susan Helms in the Hairballs' irreverent log of our training experiences. We helped each other move, shared rides to Ellington, and caught lunch together in the

# The 24 Types of ASCANS

cafeteria. And we tried to ease each other over the rough spots in life that inevitably hit some harder than others. Though friends, we still wondered which one of us would be first to win a slot on a flight crew.

Most of the Hairballs kept a mental scorecard of the shuttle missions scheduled a year or more down the road that had yet to have a crew assigned, particularly those in late 1992 and beyond. By now we knew the

Office personalities fairly well, and we often discussed who might be in or out of the rotation. A commander who had flown twice in three years was due for a rest; a pilot who had flown just once wouldn't become a commander without the requisite second mission as a right-seater. (In dual-pilot aircraft like the B-52, for example, the left seat is the pilot's seat; the right, the copilot's. But no astronaut wanted to be referred to as a copilot. As far back as the mid-1960s, in Gemini, NASA's first two-person spacecraft, the solution was to call the left-seat occupant the commander; next to him, then, sat the pilot.) A further complication was that a few veterans had developed difficult reputations and were unlikely to be at the top of the chief's list. At any given time, we guessed, Brandenstein might be working up assignments for three or perhaps four missions. Applying such factors and eliminating those astronauts already assigned or just returned from a flight, we could narrow the field to twenty or thirty colleagues.

Crew announcements are always a major event at the Astronaut Office, and when my secretary, Kiki Chaput, told me of an "all-hands" meeting one day late in August 1991, speculation resumed in earnest. Every astronaut available showed up in the conference room for Brandenstein's announcement, especially the Hairballs, anticipating the day when we would be participants—not spectators—in this drama. Brandenstein thanked everyone for coming and quickly got to the point: he had *eight* new shuttle crews to announce. The room murmured with excitement as the chief placed each new crew list on the overhead projector: STS-50 . . . STS-46 . . . STS-47 . . . STS-52, and then STS-53. Scheduled to fly in late 1992, this new crew comprised Dave Walker, Bob Cabana, Guy Bluford, Jim Voss, and—Rich Clifford, a Hairball! Rich was one of the funniest and best-liked of our bunch, known to the Hairball women as one of the "BA-BA's" (Bad-Ass Boy Astronauts), an irreverent group that included Leroy Chiao and Charlie Precourt. BA-BA or no, Rich was headed for orbit.

Brandenstein turned to STS-54 . . . with Hairball Susan Helms joining four other crewmates! And STS-55 would fly with a crew that included Bernard Harris, the third Hairball mission specialist named that day. Rich, Susan, and Bernard were the first to crack the assignment barrier. As our jubilant class crowded around to congratulate them, we basked in the glow of their success, for their selection opened the floodgates for the rest of us.

* * *

WITHOUT A CREW ASSIGNMENT, most of my contacts with veteran astronauts were limited to my office mates and the handful of colleagues in the Mission Development branch. The surest way to talk to and learn from the rest was to go flying. As a mission specialist I needed to log an average of an hour a week in the high-performance Northrop T-38 Talon. The T-38 is a sleek, sculpted needle of an aircraft, with tandem seats and two afterburning turbofans that give it high maneuverability and a top speed just over Mach 1. NASA used its thirty or so Talons for spaceflight proficiency training. Forty years after its adoption by NASA, the T-38 still serves as one of the most useful (and fun) tools for preparing astronauts for space.

For the mission specialists new to flying, the backseat of the Talon served as a superb classroom for the shuttle, where decision making and crew coordination were at a premium. The T-38 introduced the novices to the fast-paced world of the cockpit, where they dealt with challenges of shifting weather, changing flight plans, communications foul-ups, and most important, the routine occurrence of the unpredictable.

Practicing crew coordination was the main benefit, but the T-38 backseat also offered mission specialists frequent opportunities to get some stick time. Although in the Air Force I had flown the T-38 solo, I was content with NASA's policy of reserving the front seat for shuttle pilots and a few active-duty military pilot mission specialists. By splitting the approaches and cockpit chores with my front-seater, I usually snagged at least half the flight on the controls.

The T-38 admittedly offers an unbeatable convenience factor. Need to attend a meeting at the Cape? By commercial airline, you would drive to Houston's Hobby or Bush Intercontinental Airport, take the parking shuttle, check in, fly to Orlando, rent a car, and drive to the Cape. Total time, five hours minimum. Taking a T-38, you would drive to Ellington in fifteen minutes, park nearly on the flight line, take thirty minutes to plan the flight, walk to the jet with your parachute and briefcase, fly to the Cape, and arrive at your meeting twenty minutes after landing. Total time, three hours, tops. And on the way, the T-38 offered ninety minutes of stick time at 39,000 feet. The choice was easy.

Often I found myself tagging along with a shuttle pilot to a meeting or shuttle training aircraft session at the Kennedy Space Center. After long hours spent in payload safety reviews or a half-day dealing with malfunctions in the motion base shuttle simulator, the Cape run was a welcome relief. I'd meet the pilot at Ellington, check the weather and takeoff data, file the flight plan, and fifteen minutes later the afterburners would be thumping us back in our seats on takeoff. Heading south toward Galveston, we would pop up through haze and fluffy cumulus into the clear blue atmosphere above the Gulf coast. Steering first toward Leeville on the lower Mississippi delta, south of New Orleans, for the next hour we would track steadily eastward across the Gulf of Mexico. The cockpit was cool and comfortable at cruise altitude; chatting through our oxygen masks, we would catch up on office news while winging along just below Mach 1. Landfall would come over Tampa on the Florida west coast, and air traffic control would start us down on a 100-mile glide toward the shuttle landing facility (SLF) at KSC. Skirting Orlando at 20,000 feet, I could dip a wing and just see the distinctive shape of one of my old B-52Ds, serving as a monument to the SAC crews who had once pulled alert at McCoy Air Force Base, its runways long-since absorbed by Orlando International.

From fifty miles out, if the weather was clear, we could see the prominent sugar cube of the Vehicle Assembly Building on Merritt Island and then the white slash of the SLF itself just to the north. Our trip from Houston would end with a sweeping left turn past the runway, descending steeply to level out a thousand feet over Launch Complex 39. For an Ascan, nothing made the heart beat faster than the sight of a shuttle poised on the pad, the brown bulk of the fuel tank flanked by the slim white shafts of the solid rocket boosters. A minute later, we would flare over the SLF's blazing white concrete, gently touch down, and open our canopies to the humid but refreshing Cape breeze. Debarking at the spaceport, I'd grab an RC Cola and a Moon Pie and privately exult at being part of all this.

For the pilot astronauts, the nimble jets were perfect for honing their flying skills. The front-seaters could perform aerobatics out over the Gulf, fly precision instrument approaches, and team up for some exhilarating formation flying. The T-38s also served as the astronauts' commuter jets, getting the pilots out to NASA's operating bases at El Paso and the Cape, where they did some *real* flying.

The shuttle pilots trained almost weekly in one of four Gulfstream II business jets heavily modified to replicate the handling qualities of a gliding orbiter. Based at El Paso and periodically at KSC or Edwards, the twin-engine, twelve-seat aircraft puts the shuttle pilot and an instructor in a simulated orbiter cockpit and provides the visual cues, feel, and performance of a space shuttle on final descent. Using the drag from extended landing gear and thrust reversers, the Shuttle Training Aircraft (STA) descends steeply toward the runway with a glide ratio of 5 to 1: it glides just 5 feet forward for every foot of altitude lost (the B-52's glide ratio is 20 to 1). Computers give the stick the stiff feel and lagging control response of the larger, heavier orbiter; throw in real winds and turbulence, and the STA mimics a shuttle approach with heart-pounding authenticity. In a single ninety-minute training flight, a shuttle pilot might sweat out as many as ten practice shuttle landings. Prior to their first spaceflight, shuttle pilots have flown approximately five hundred mock shuttle approaches in the STA; commanders have racked up nearly a thousand. And that's not counting the many landings each pilot will have seen in the Shuttle Mission Simulator.

I was sometimes lucky enough to grab the backseat to El Paso or the Cape with John Young, who had flown two Gemini spacecraft, been to the Moon twice, and commanded the shuttle twice, including the maiden voyage of *Columbia*, the very first shuttle mission. The shared flying chores on a trip with John weren't too demanding: a couple of fingers on the control stick, an occasional nudge of the throttles, and every few minutes a radio call and the twist of a knob to tune a new frequency. That left plenty of time for conversation: "John, what was it like driving the lunar rover on *Apollo 16*? How steep a slope could you and Charlie Duke handle in that thing?" Or, "How many g's did the Titan II's first stage give you and Gus on *Gemini 3*?" Surprisingly, John preferred the Moon's one-sixth-g surface gravity to orbital free fall, saying it was a relief to be able to pour water into a cup and have it stay there. John's long experience and familiarity with space center management were always good for an impromptu lesson in the technical challenges of spaceflight or the latest in NASA politics.

After spending a couple of hours at the Cape "kicking the tires on the orbiter," as John put it, we would boom out of the shuttle runway and usually be home in time for a late supper. Although the El Paso run couldn't rival the Cape's space hardware, launch pads, and tempting Atlantic beaches, it had its own laid-back satisfactions. While John was out flying the STA at the shuttle's White Sands alternate landing field, I would grab a late Tex-Mex lunch at Grigg's restaurant or pick up a few bags of fresh tortillas from the factory just down the road from the NASA hangar. After passing an hour or so doing work brought from Houston, I would hear the radio announce that the STA crew was inbound from their approaches. John would step off the STA and have a cup of coffee, and I would have the flight plan filed for our return leg. We would walk fifty yards out to our T-38, perform the preflight inspection, crank engines, and be airborne ten minutes later. Rocketing up over the Rio Grande in a climbing left turn to 17,000 feet, we would turn our backs on an orange sun disappearing over the desert horizon. Out over the vast emptiness of west Texas at 37,000 feet, an indigo sky would start to bloom stars, while to our right, the dark shoulders of the Davis Mountains sank into the blue haze of the distant Big Bend country. Turning down the cockpit lights, we would take turns identifying the constellations winking into view in a sky even more brilliant than I had seen as an astronomer on Hawaii's Big Island. "Almost as many stars as there are in orbit, old buddy," Young would comment, then get back to some choice observations about "the damn troubles with the space station" or other meaty topics of the day.

Young was famous for his frank internal memos on how the space agency could do a better job, often targeting NASA managers, headquarters bureaucrats, mercenary contractors, or shortsighted members of Congress. At a Monday meeting in June 1991, Carl Walz briefed the Astronaut Office on options to reduce the number of shuttle "cue cards"—cardboard checklists velcroed around the cockpit as handy references for the crew. The cards were multiplying like rabbits; they were both expensive to produce and tended to get lost in free fall. An astronaut suggested reducing the clutter by converting some cue cards to decals on the orbiter instrument panels. Carl responded that if we later wanted to change a cockpit decal, the shuttle's prime contractor, Rockwell International, would charge the government roughly $10,000 to alter the orbiter's engineering drawings. This was too much for Young, who indignantly observed, "That's the problem with this whole goddamn program—Rockwell!" Amid the ensuing laughter, he added a quick, "No offense."

When the room quieted, someone suggested we could eliminate at least one cue card, the one showing space walkers where to attach their chest electrodes, by tattooing the locations onto the astronaut's body.

Again John's voice rose above the laughter: "I've got tattoos for eight damn electrodes on my body right this minute, and the bastards never used a one of 'em." Young's jokes were far outnumbered by his thoughtful comments on everything from piloting technique to the right approach to astronaut and shuttle safety. He had lost close friends and colleagues in the *Apollo 1* fire and had witnessed the tragedy of *Challenger*'s loss firsthand as chief of the Astronaut Office. Most of us regarded Young as the conscience of the astronaut corps.

Brandenstein awarded the Hairballs our silver astronaut pins in August, with our spouses and the many veterans in town for the Astronaut Office's reunion dinner in attendance. As our second Christmas in Houston approached, the calendar for the first half of 1993 showed three missions with crews as yet unassigned; a few Hairballs would likely fill some of those open seats.

Kiki tracked me down one morning in February 1992 with a message from Don Puddy's secretary. Could I get over to Building 1 for a brief meeting at 11:00 a.m.? There was no hint of what he wanted; there was also no other answer except "I'll be there."

I was shown into Puddy's large corner office; it was my first visit to the rarefied atmosphere of the eighth floor. Linda Godwin, a physicist who with her crew had deployed the Gamma Ray Observatory on STS-37 the previous spring, sat with Puddy at the small conference table. Although nervous, I tried to keep my voice cool and pleasant as we shook hands. I vaguely remembered that Linda had been assigned earlier to work on long-range planning for one of next year's shuttle missions, but I couldn't recall which one. After some happy talk, Puddy soon got to the point of my summons:

"Tom, you might know that Linda's been working with the folks at JPL [Jet Propulsion Laboratory] on the plans for STS-60. Linda, why don't you give Tom a quick summary of the mission?" Linda, a former payloads flight controller whom I knew mainly from staff meetings in the Mission Development branch, gave us a thumbnail sketch of the "Spaceborne Imaging Radar-C," a powerful synthetic aperture radar whose two predecessors had flown on the shuttle back in the 1980s. This latest instrument was much more capable and versatile, and its shuttle flight would be a proof-of-concept for an eventual permanent orbital observatory that could image Earth's changing surface, round-the-clock, on a global scale.

After her technical introduction, Puddy picked up the story. "We put Linda on the job almost two years before launch because so much early development work needs to be done on the crew's role in aiding the radar science investigations from orbit. How would you like to join her as the science deputy on the mission?"

Flushed with excitement and smiling with genuine pleasure, I babbled something about how I would enjoy using my planetary science background to help with the radar remote sensing technology we would be using. I added that I would be more than delighted to work with Linda on—what was the mission again?

"You'll be flying with Linda on STS-60. You two get things rolling; we'll name the rest of the crew later for Space Radar Laboratory 1—SRL-1. Its launch date is September 30, 1993."

# 5

# Seeing Earth in a New Way

*To see the Earth as it truly is...*
*Is to see ourselves as riders on the Earth together.*
*Brothers on that bright loveliness in the eternal cold.*
Archibald MacLeish, after Apollo 8, 1968

Riding the elevator down from Puddy's office with Linda after the meeting, I enjoyed a giddy hint of weightlessness. I had experienced free fall a few dozen times by then, both in Air Force trainers and, of course, on NASA's zero-g training aircraft, the famed KC-135 Vomit Comet. I still felt a bit of that floating sensation when I got back to my office and called Liz, eager to share with her the good news and the few details I knew about the mission. By the time I got home that evening, she had whipped up a celebratory dinner of steaks and my favorite chocolate cake, which she and Annie had decorated with the astronaut corps emblem, lots of sprinkles, and a big frosting inscription: "Hurray for Daddy!" We ate in the dining room, a venue usually reserved for holidays or entertaining. This was a very special day, we told Annie and Bryce. Your daddy will be going on a trip into space.

The next day, Brandenstein held an all-hands meeting to announce new crew assignments. Along with me, he had tapped Hairballs Janice

Voss and Charlie Precourt for missions in 1993. Amid the hubbub of handshakes and congratulations once the chief had finished, one could discern the distinctive chest-thump-and-cough Hairball "salute."

## CREW ASSIGNMENTS ANNOUNCED FOR FUTURE SHUTTLE MISSIONS

*From NASA Press Release 91-25, February 21, 1992*

Thomas D. Jones, Ph.D., will be a mission specialist on the Space Radar Laboratory-01 flight, STS-60, in late 1993. This is the first flight for Jones . . . a member of the 1990 astronaut class. Linda M. Godwin, Ph.D., was assigned in August 1991 as mission Payload Commander. SRL-01 will acquire radar images of the Earth's surface for making maps and interpreting geological features and resources studies.

Additional crew members on . . . STS-60 will be named at a later date.

Between our other job commitments, Linda and I quickly got to-gether to go over the basics of the mission. But first things first: armed with her shirt size, I drove over to Baybrook Mall and bought a couple of polo shirts in brilliant teal. Within a couple of days we both sported these "training" shirts, our names and "SRL-1" embroidered on the left breast. As a rookie, I had lusted after one of these for nearly two years, and now I had wearable proof that I had joined the ranks of assigned crews.

Properly clothed, I could now focus on the mission details. There are no bad shuttle missions, but SRL was an especially good assignment. Linda and I would be flying on a "Mission to Planet Earth," applying so-phisticated planetary remote sensing techniques to our own world. Sim-ply put, SRL-1 was a powerful radar camera that would create highly detailed digital images by recording the echoes of radar energy beamed at Earth from the shuttle. The mission was designed to prove the worth of space-based radar in detecting and monitoring natural and human-caused changes on Earth's surface. The prototype radar system was a joint effort by NASA's Jet Propulsion Lab and the German and Italian space agencies. SRL-1 would be a nine-day mission, with the radar oper-ating around the clock to capture as much Earth imagery as possible.

Rather than scan the planet indiscriminately, the radar science team would focus most of its efforts on "supersites" around the globe where many problems in Earth science could be addressed in daily radar passes over the region.

Linda had been working with the JSC mission manager and the JPL science team to develop the crew operations concept for the flight. As payload commander, she was responsible for helping mission designers come up with well-defined duties for the crew and for ensuring that her fellow astronauts were sufficiently trained to get the best science out of the mission. My remote sensing and planetary geology training, I hoped, would help us accomplish that. As a scientist, I had used an infrared telescope to detect the presence of water-bearing minerals on asteroids; in the same way, reflected radar energy could help us extract detailed information on the geology, hydrology, and vegetation cover of Earth's surface.

Linda and I quickly scheduled a visit to JPL, in the Los Angeles suburb of Pasadena, to learn about the mission from the engineers and scientists who had conceived it. As a pilot and a scientist, I thought I knew a fair bit about radar already. I discovered I had much to learn.

Space-borne radar had already been used successfully for both scientific and intelligence-gathering purposes. The last three Apollo crews scanned the lunar surface from orbit with radar, documenting the Moon's rugged terrain and probing as deeply as half a mile beneath the battered surface to investigate the structure of the ancient lunar crust. Both American and Soviet reconnaissance satellites used radar to track naval vessels and image military installations from orbit; space shuttle *Atlantis* launched the first in the Lacrosse series of sophisticated radar reconnaissance satellites on the classified STS-27 mission in December 1988.

The main virtue of imaging radar in planetary studies is that it's an *active* remote sensing technique. Conventional imaging or reconnaissance satellites rely on sunlight to illuminate the scene for their sensors; thus clouds or darkness interfere with their ability to see the surface. Radar, in contrast, carries its own light source. The radar antenna beams microwave energy at the surface, then catches the reflected radio echo. (Microwaves are just another form of light, with wavelengths typically measured in centimeters.) Scanning the Earth with radar is like lighting up a dark room with a flashbulb. And because radio waves can pierce most clouds and rain (the waves are much longer than the tiny water droplets in a cloud and all but the biggest raindrops), bad weather doesn't interfere with the ability of the radar antenna to "hear" the echoes of the transmitted energy. A space-borne radar is an all-weather, round-the-clock monitoring system (exactly why it's so well suited to defense reconnaissance efforts).

Synthetic aperture radar (SAR) technology uses the motion of a spacecraft or aircraft to electronically synthesize an antenna much larger than its physical size, yielding higher imaging resolution. The concept was soon applied to a planet that seemed tailor-made for its use: Venus. Shrouded in perpetual sulfuric acid clouds, Earth's sunward neighbor thwarted nearly all efforts at telescopic or robotic imaging of its surface; only a few stark images of barren lava plains had been returned from Soviet landers before they succumbed to Venus's furnace-like 900°F temperatures. In the late 1970s the US *Pioneer Venus* orbiter used a radar altimeter to create a crude global terrain map of Venus. A decade later planetary scientists were ready to investigate the planet with the higher-resolution SAR technology.

On May 4, 1989, the space shuttle *Atlantis* deployed the *Magellan* spacecraft into low Earth orbit; its inertial upper stage then fired it off on a fifteen-month journey to Venus. The probe spent the next four years creating a global radar map of the planet's surface, revealing details as small as 100 meters across.

*Magellan*'s creators at JPL had also been using orbiting radar to study our own planet, first examining Earth's oceans with *SeaSat* in 1978. JPL's radar team followed up with more advanced imagers on two early shuttle flights: STS-2 in 1981 and STS-41G in 1984. Mounted in the shuttle payload bay, these two radars scanned long strips across Earth's oceans and continents, proving the science potential of SAR technology. One of the most exciting and intriguing results was the discovery of buried stream channels beneath the sands of the Sahara Desert, the so-called radar rivers. The third radar in the series, Spaceborne Imaging Radar-C (SIR-C, pronounced "sir-see"), was put on hold by the loss of *Challenger*, awaiting a slot on a future shuttle flight.

The cutting-edge SAR system on the Space Radar Lab transmitted in three different radar frequencies tuned to obtain maximum information about the complex surface of our changing world. Echoes in these different radar "colors" could be combined to construct Earth images of unprecedented detail. The new technique promised a dramatic increase in the data we could return, the equivalent of replacing a black-and-white newspaper photograph with a color, high-definition image. A successful Space Radar Lab mission could give us a fresh, highly detailed snapshot of Earth's varying terrains: oceans, forests, lava plains, deserts, cities, croplands, volcanoes, glaciers, earthquake faults, even archaeological sites.

The Space Radar Lab mission was an international effort spearheaded by NASA, with the German and Italian space agencies contributing important elements. Complementing JPL's SIR-C radar, the German Deutschen Zentrum für Luft-und Raumfahrt space research agency was contributing the X-Band Synthetic Aperture Radar (X-SAR, pronounced "ex-sar"), and the Italian space agency was building the ground-control equipment.

Much more sophisticated than *Magellan* or previous radar satellites, SRL was the proof-of-concept for a permanent radar observatory. The shuttle offered an attractive science platform for this shakedown test, providing power, pointing the payload bay antennas at the science targets, and transmitting and recording radar science data. The orbiter would also return the radar to Earth for repair, adjustment, and reflight. The crew would not only operate the shuttle but would save further expense by simplifying the data-recording system and providing on-orbit repairs or adjustments. Avoiding the construction of a satellite to perform these functions would save hundreds of millions of dollars. SRL could then be refurbished and reflown to take another in-depth look at our changing planet.

On our first JPL visit, Linda and I walked over to the Spacecraft Assembly Facility with Mike Sander, the JPL project manager, and Ed Caro, SRL's chief engineer. In the high-ceilinged clean room where other planetary spacecraft like *Voyager* and *Galileo* had been built, aluminum forgings and beams lay stacked on the gleaming white floor. These were the building blocks of the SIR-C antenna, a promising yet still unfinished instrument. Like this radar, our new shuttle crew would have to be assembled and tested before we were ready to play the demanding role assigned us by the science and flight control teams.

The JPL scientists and engineers made it clear that their chances of success in orbit would increase if the crew became active members of the SRL science team. To perform on that level, we had a lot to learn, including radar remote sensing theory, the SRL mission design, and many aspects of the dozen or so scientific disciplines involved in the SRL experiments. Over the course of the next year and a half, we would also undertake the equivalent of a master's degree program in terrestrial geography and geology.

The eighteen months we had to prepare for our September 1993 launch seemed scarcely enough. But then NASA gave us some breathing room. Three months after my assignment, shuttle program managers moved up the Hubble Space Telescope repair mission on the flight schedule, shifting SRL-1 from its autumn 1993 slot to spring 1994. I remember my frustration at the resulting six-month slip: Wasn't our science mission just as important as Hubble's? But Linda reacted with professional patience: she had waited six years for her first flight, and delays were just part of the astronaut business. I had been granted the privilege of someday flying in space. Some uncertainty in the actual delivery date was part of the bargain.

The SRL science team comprised some fifty investigators from thirteen different countries, including Australia, Brazil, Germany, Italy, and Japan. They would set up experiments during the SRL mission on every continent except Antarctica (although the surrounding Southern Ocean would be heavily studied). The JPL radar science leads, Mike Kobrick and Ellen Stofan, introduced us to as many science team members and their investigations as possible. But for detailed insight into how the radar data would be used, there was no substitute for getting out of the classroom and visiting some of those researchers.

Our visits to radar science supersites began in the spring of 1992. Linda and I traveled to Mammoth Mountain, California, to learn about remote sensing of the water content of the Sierra Nevada snowpack; to Kerang, Australia, to see how radar could measure saltwater intrusion into prime farmland; to Chickasha, Oklahoma, for an introduction to remote sensing of soil moisture; to Michigan's Upper Peninsula, for measurements of forest biomass; and to Hawaii's Volcanoes National Park, where we hiked the lava flows with geologists using radar to assess eruption hazards. While in the field, we also developed an easy working relationship with our JPL counterparts, whose advice would be instrumental in dealing with any SRL problems in orbit.

Although the travel did deliver excellent science training, my frequent trips involved long absences from home, which put additional stress on Liz, who was left parenting two small children alone. An August 1992 trip to Virginia's Langley Research Center wound up generating a perfect storm of domestic repercussions. Langley was home to the SRL's Measurement of Air Pollution from Satellites (MAPS) instrument, which used an infrared detector to determine the amount of carbon monoxide in the atmosphere below. MAPS would locate the sources and track the movements of this important pollutant produced by automobiles, factories, forest fires, and agricultural biomass burning.

While the MAPS experimenters briefed us on the instrument's purpose, design, and operation, Hurricane Andrew bore down on the southern tip of Florida. On August 24, Andrew slammed into Homestead, a small town south of Miami, with winds of 145 miles per hour. The hurricane plowed across the Everglades from east to west and headed into the Gulf of Mexico, taking ominous aim at the upper Texas coast. My position was a bit precarious: I was in Virginia with Linda, while Liz, Annie, and Bryce were in low-lying Houston with a hurricane just two days from landfall.

I went straight to the Langley weather forecasters and got the latest storm track predictions. In a phone call home the next day I tried to reassure Liz. "The NASA weather guys say that Andrew's going to hook north into Louisiana. You won't see a trace of the storm in Clear Lake."

"Then why," she replied, "did the schools cancel classes today and tomorrow so families can evacuate? Why is all the milk, bread, peanut butter, and bottled water gone from the supermarket? And why are Brian and Carol down the street nailing plywood over their windows?"

Hmmm. The storm, still two days out, was nevertheless making itself felt. I could imagine Liz's displeasure at having to load Annie and Bryce, Vesta the cat, the photo albums and other mementos into the station wagon and hit the road to Dallas alone. I promised her I would be on a plane home the next morning if the storm continued toward Houston.

As I had hoped, Andrew turned north at last and drenched the Louisiana swamps. But the episode was a perfect example of the astronaut wife's lament: "They're never home when you need them." Focused on my work in tidewater Virginia, it had been too easy for me to dismiss the storm; next time I would head home if a hurricane so much as winked in Houston's direction.

\* \* \*

FROM THE RESULTS OF the two previous shuttle radar flights, SIR-A and SIR-B, Linda and I learned that the crew could provide crucial information from orbit to help researchers interpret the radar imagery. Environmental conditions such as rain, thunderstorms, dust, and standing water could alter the radar echo and confuse the image analysis. At their supersites the SRL teams planned to record local conditions, including wind velocity, relative humidity, temperature, soil moisture, and the presence of rain or severe storms. This "ground truth" would be used to calibrate and correct the orbital data, extending the usefulness of radar imaging to much larger areas.

Although ground truth was available at the supersites, few of the hundreds of other radar targets around the globe would have investigators present to validate the orbital measurements. This was where our crew came in. From orbit, we planned to radio our visual observations of each target to the Payload Operations Control Center (POCC), just across the hall from Mission Control in JSC's Building 30. Using an array of fourteen cameras, the crew would record video and shoot thousands of still photos of the targets for later comparison to the radar imagery. Our visual observations and science photography would serve as the ground truth for our global mission.

JPL had initially suggested that we take notes or make sketches to document our orbital observations. From her STS-37 experience Linda was fully aware of the difficulty of putting pen to paper in free fall and instead suggested that we carry sensitive microphones in the cabin. They would enable us to dictate our verbal reports directly onto the video recorder capturing the view from a camera aligned with our radar scan. We would thus have a visual and audio record of each observation. Linda also programmed our laptops to accept the brief summaries we planned to type up following each data take. The POCC could retrieve those files daily by radio and distribute them to the science team.

ABOUT 1:30 A.M. ON Monday, November 2, my sister, Nancy, called from Baltimore with serious news. The night before, Dad had suffered a heart attack and collapsed at a church hall in our hometown. He was in intensive care. "You'd better make plans to get home," she warned. Her voice was full of worry.

While in the simulator on Tuesday morning, I received another call from my brother Ken. "Tom, you'd better get up here. Dad hasn't regained consciousness; he's in a deep coma."

"I've been working on plans, Ken, but I'm tied up all day," I said. "I'll have the plane reservations this afternoon."

He was insistent. "Tom, Dad's really in a bad way. They suspect serious brain damage. We might lose him any time. This might be your only chance to say good-bye. And Mom needs you up here. You need to get home now."

Ron Grabe, the simulator crew commander for the day, assured me that the rest of my colleagues could cover the cockpit duties. By early evening I joined my mom and three siblings in a sad vigil at Dad's bedside. His heart attack had caused irreparable brain damage. He could not recover.

Four days later, with Dad still lying unconscious, I flew home to Houston. Stepping off the plane at Ellington, I found Liz and the kids waiting for me outside the small terminal. Her expression told me the bad news. Three-year-old Bryce squeezed himself against my leg: "Daddy, I'm sorry your dad died," he said. Annie threw her arms around my neck and hugged me as I slumped to the curb.

I thought back to the previous January when I'd given Dad and Mom a tour of the Cape launch facilities. We had ridden the launch pad elevator to the White Room, checked out the crew quarters, and stood beneath the black-tiled belly of an orbiter. My mom had told the tile technicians to be sure they were doing a good job! I flew home in a T-38 that evening, but the folks had stayed around to see the STS-42 crew launch on their International Microgravity Laboratory mission. Mom told me that Dad had been as excited as a little kid at the launch, and a month later he was on top of the world at my assignment to SRL-1. His death hit me hard. I took some solace in knowing that when my space shuttle jolted skyward in little more than a year, his reassuring presence would be with me in the cabin.

THE REST OF OUR SRL crew wouldn't be assigned until nine months to a year prior to launch—enough time to complete the usual shuttle training syllabus and learn the skills they would need for radar operations. As Linda and I surveyed the Astronaut Office, we agreed there were only a handful of candidates for command of the mission. Most of the shuttle commanders had either already been assigned or had just returned from a mission; only a few of the available pilots had racked up the requisite right-seat experience and were ready to move up to command.

At the Office's 1992 Christmas party in Galveston, Liz and I found ourselves in the buffet line with Sid and Marianne Gutierrez. Sid had given me my Mission Control tour as an "Ass-Ho" (Astronaut Hopeful) back in 1989, and we had since trained or flown the T-38 together half a dozen times. A 1973 Air Force Academy graduate, an experienced fighter and test pilot, and a member of the 1985 astronaut class, Sid had first flown in space in June 1991 as pilot of STS-40. Seven years in the Astronaut Office meant he was eligible to command STS-59 (SRL's new mission number). While filling his plate with roast beef and Gulf shrimp, Sid expressed more than a little interest in the SRL mission profile: a long flight (planned for nine days), a high-inclination orbit, which would offer extensive views of Earth, and a payload that heavily engaged the crew in the science. Sid wasn't sure he would get his Christmas wish, but as he and Marianne headed for their table, he offered the hope that "maybe we'll have the chance to work more closely together" in the coming year.

That would suit me fine. Given our shared Zoomie background and Astronaut Office work, I knew we would have no problems working together. On our T-38 sorties he had been generous in sharing both his Office and orbit experiences; I had learned much from him about how the astronaut corps had changed since *Challenger* and how it was run today.

By March 1993 Linda and I had been training for SRL for over a year. Early in the month I got a call at home one evening from Rich Clifford, back only a few months from his first flight on STS-53 in December. He asked if I had heard the news. "What?" I asked, crossing my fingers. "We're flying together on STS-59, TJ!" he exulted. I'm sure he could hear me grin right through the phone line. But who else was on the crew? "Sid's the commander, Chili's the pilot, and Jay Apt is our other MS." "Chili" was Kevin Chilton, who had been my office mate for over a year, and though I didn't know Jay very well, I knew he was highly regarded technically. This could be a very good crew.

## STS-62 AND STS-59 SPACE SHUTTLE CREW ASSIGNMENTS ANNOUNCED

*From NASA Press Release 93-42, March 5, 1993*

NASA today named the crews of STS-62 and STS-59, two space shuttle missions scheduled for launch in early 1994. USAF Colonel Sidney M. Gutierrez will command the STS-59 Space Radar Laboratory mission aboard *Atlantis*. Other crew members are USAF Colonel Kevin P. Chilton as Pilot, and mission specialists Jay Apt, Ph.D., and Michael R. "Rich" Clifford, USA Lt. Colonel. Previously announced crew members are Linda M. Godwin, Ph.D., named Payload Commander in August 1991 and Thomas D. Jones, Ph.D., named mission specialist in February 1992.

The Space Radar Laboratory, STS-59, will take radar images of the Earth's surface for Earth system sciences studies including geology, geography, hydrology, oceanography, agronomy, and botany; gather data for future radar system design including the Earth Observing System, and take measurements of the global distribution of carbon monoxide in the troposphere.

I had barely known Kevin Chilton at the Zoo; he had been a year ahead of me, and I didn't run into him again until 1990. A Distinguished Graduate of the Academy and then a fighter pilot in the F-4 and F-15, Kevin graduated first in his 1984 test pilot class at Edwards. Coming to NASA three years later with the first class of astronauts after *Challenger*, Kevin had been the chief astronaut's personal choice as pilot on the first flight of *Endeavour*, the replacement orbiter for *Challenger*.

Since the accident, the Office had assigned two astronauts to escort the crew's families to the Cape, ensuring that the spouses would have instant, experienced help if, God forbid, they faced another "shuttle contingency." I had served as one of the family escorts for Kevin's STS-49 crew, flying with their spouses and children to the Cape for their launch in May 1992. Steve Oswald, the senior escort, and I were happy to do anything for the families: pick them up for the ride to Ellington, load baggage on planes and buses, and drive them from their Cocoa Beach hotel to the Cape for daily visits with their spouses. Most important was our presence with the families at liftoff on the roof of the launch control center. I got to know Kevin's wife, Cathy, a flight test engineer herself, during our stay at the Cape; their first daughter, Madison, was just two at the time, and Mary Cate was an infant. At the very least I could competently handle any diaper emergencies.

Lifting off into evening twilight, *Endeavour*'s brilliant trail of flame pierced a thin layer of clouds and lit it up from within, like lightning flashing inside a thunderhead. For an instant, the rippling exhaust and thin cloud deck glowed like a mushroom cloud against the darkening backdrop of the Atlantic. It was a spectacular send-off for the new orbiter and her crew, who were to chase down and repair the crippled *Intelsat VI* communications satellite.

During the exciting week that followed, I frequently joined the spouses (Kaye Akers, Jane Brandenstein, Cathy Chilton, Jeannie Hieb, Kaye Melnick, Steve Thornton, and Cheryl Thuot) in the glass-walled gallery at the rear of the flight control room. The crew rendezvoused flawlessly with the drifting satellite, and soon it was time for the spacewalkers, Pierre Thuot and Rick Hieb, to bring it aboard. But two EVAs in as many days failed to snag *Intelsat*: each time Thuot tried to nab the satellite, his custom-built capture bar nudged the spacecraft and sent it wobbling away. At each attempt, his wife, Cheryl, sat white-knuckled but stoic, frustrated at seeing Pierre just inches from the 4.5-ton spacecraft yet unable to complete the grapple. Brandenstein and Chilton brought *Endeavour* back for a third, final attempt, and this time Tom Akers squeezed into the airlock with Rick Hieb and Pierre to lend another pair of hands. It was the only three-man EVA in American spaceflight history; the trio barely fit into the cylindrical confines of *Endeavour*'s airlock.

Fidgeting in our viewing-room seats, we watched tensely as the space-walkers reached up in unison and finally snared the wayward *Intelsat* in their white-gloved hands. The flight control room exploded in cheers and relief. After Brandenstein's expert landing out at Edwards a few days later, the crew emerged from *Endeavour* and met their spouses near the runway for a jubilant reunion. A private dinner together at the Silver Saddle Ranch that evening furnished the perfect ending to a dramatic flight. I looked forward to the pleasure of flying with Kevin on his second mission.

Rich Clifford, of course, was my Hairball classmate and the first of our class to fly in space. A 1974 West Pointer, Rich had flown attack helicopters and then trained as an Army test pilot before joining JSC's vehicle integration and test team in 1987. With the VITT, he had helped certify and integrate the post-*Challenger* shuttle crew escape system. After his return from STS-53 late in 1992, Rich and I had trained in and flown NASA's Cessna Citation II business jet, leased by the agency to supplement the overworked T-38 fleet. The two of us rejoiced at being front-seaters again, and in a Citation to boot.

When we weren't sharing the Citation cockpit, Rich and I were regulars at the astronaut gym's racquetball court, often joined for a pickup game by Leroy Chiao, competing in a blitzkrieg round of "cut-throat." The mystery of how Rich consistently whipped me (a younger, faster player!) remains, but I was willing to take a licking if my fellow Hairball would show me the ropes in space.

Jay Apt rounded out the STS-59 crew. Jay earned his doctorate in physics from the Massachusetts Institute of Technology in 1976, performed planetary research at JPL in the early 1980s, and worked as a payload flight controller in Mission Control from 1982 to 1985. Joining the Astronaut Office with Linda in the 1985 class, he first flew in the spring of 1991, earning high marks for his two space walks on STS-37. While Linda operated the robot arm from inside the shuttle cabin, he and Jerry Ross had freed a stuck antenna on the Gamma Ray Observatory that threatened its successful deployment. On a second space walk, while testing space-suit mobility aids and a variety of tools for the future space station, Jay faced a hand-intensive series of tasks that caused so much wear and tear to his gloves that a metal strap inside one palm worked its way through the pressure bladder. When he removed the glove back inside, he was startled to find the fabric liner stained red with a bit of blood. Fortunately, oxygen leakage through the small puncture was negligible. During the round-the-clock science work on the STS-47 Spacelab-J flight in the fall of 1992, Jay served as MS-2, the flight engineer. With his scientific credentials, familiarity with JPL, and skill as a photographer, he would serve as a jack-of-all-trades mission specialist for SRL.

Our crew quickly met to hash out how we would tackle this mission. Linda, the payload commander, outlined what JPL and the JSC flight designers saw as our crew's three main responsibilities. First, we would operate *Endeavour* as a science platform, maneuvering repeatedly to point the twelve-ton SIR-C/X-SAR instrument at the planned science targets on the ground. Second, we would manage three high-speed digital recorders in the cockpit that would store the flood of imaging data streaming in from the three radar antennas. And third, we would obtain extensive photographic coverage of the science targets to aid in radar image interpretation.

Because SRL's radars operated around the clock, regardless of weather and lighting conditions below, our crew would split into two shifts of three crew members each. These Red and Blue teams would each pull a twelve-hour shift, with another four hours allotted for shift handover, housekeeping chores, meals, exercise, and whatever unplanned work might pop up.

Sid broke down the shift assignments for us. He needed the two pilots on the same sleep cycle, both fresh and wide awake for launch and landing days. Linda would join them, enabling Sid to get the payload commander's input on any experiment or orbiter problems without having to wake the opposite shift.

Jay, Rich, and I would comprise the Blue Shift. Because Jay had handled orbiter cockpit operations on his last flight, he would serve as shift supervisor, manage the orbiter, and oversee the cameras and film for science photography. As Linda's backup, I would handle SRL science operations. Rich would cover science duties, too, and back up Jay in the cockpit.

All of us would cross-train on the Radar Lab operation to cover the primary science mission. Linda would handle any robot arm operations, and Rich would serve as flight engineer for launch and landing. Finally, Sid designated Linda and me to handle any payload or orbiter emergencies that might occur outside the cabin. Jay had already performed two space walks, and Rich had trained in this role on his first flight. By putting Linda and me through the space-walking drills, Sid boosted our chances for an EVA on a future mission.

With a year before launch, the six of us embarked on a rapid-fire series of training trips to introduce the science goals of the Space Radar Lab. JPL hosted our four new crewmates at Mammoth Mountain, California (snowpack water content) and Death Valley (climate change and dune field morphology); a few weeks later, in May 1993, we joined researchers at the US Geological Survey in Flagstaff for a firsthand look at desert landforms, volcanic cinder cones and lava flows, and an active sandsheet in the valley of the Little Colorado River. Over dinner, Jack McCauley, Carol Breed, and Jerry Schaber, all veteran geologists, recounted their explorations of Egypt's western desert in a ground search for the radar rivers revealed by SIR-A and SIR-B. On the banks of one of these buried Saharan stream channels, they found hundreds of crude stone hand axes made by our distant ancestors nearly 200,000 years ago. These early humans of the Aechulian era, called *Homo erectus* because of their upright gait, used these primitive tools to hunt the herds of antelope and wildebeest that once roamed the grassy plains of the Sahara before climate change turned it into a scorching desert. The scientists brought out the Indiana Jones in me with their story of a Persian king who vanished with his entire army into a desert sandstorm. Because SRL's L-band radar waves could penetrate several meters into dry soils and sand, the Flagstaff team thought there was a slim chance that buried metal or ancient structures might turn up in radar images from the Sahara.

In the fall of 1993 Sid announced we were all going camping. A shuttle mission, he said, was very much like an extended campout. Living conditions were spartan, food was just a faint echo of home cooking, and privacy largely went by the boards. The key to success, he said, was good teamwork; all of us would benefit from learning how to work together under field conditions.

On Thursday, October 22, 1993, the six of us drove to Brazos Bend State Park, a fertile tract of wetlands and forest an hour south of Houston. Seated at a picnic table in camp, we reviewed our flight plan; our strategy for "post insertion," the first three hours after arriving in orbit; our individual tasks in orbit; and our plan for getting the orbiter prepared for reentry. Each of us took a turn as camp cook: Sid grilled steaks, I served up scrambled eggs and bacon, and Jay turned out a shrimp paella. We also got used to seeing each other right out of the sleeping bag, a twice-a-day occurrence on our dual-shift flight.

We were getting along well, for the most part. Linda and I were good friends by now, veterans of overseas expeditions to Australia, Germany, Hawaii, and Italy. (Liz and the crew were skeptical, of course, that we had done anything but sightsee and enjoy fine meals on those science training trips.) Rich was my classmate and mentor, sharing his many practical tips for efficient living aboard the shuttle. Kevin, my cheerful old office mate, was as sharp as they come, with a solid confidence gained on his first trip piloting *Endeavour*. Our commander was an easy-going leader who believed in letting each person do his job, only reluctantly stepping in with advice. I particularly liked Sid's inclusion of our families as vital members of our team.

Jay and I have since become good friends and colleagues, but we got off to a rocky start. I got the impression he would really have been happier running the Blue Shift by himself. He would often introduce an opinion with the phrase, "This is my third spaceflight"—a not-so-subtle reminder to his listeners that they were talking with a veteran astronaut.

His manner took a little getting used to. During our first Blue Shift meeting, Rich had shot me an amused glance when Jay briefed us on his "shift philosophy." He wanted our threesome to look professional in front of the instructors and Mission Control (as did we all). He correctly pointed out that we should always have a crewmate backstop us when flipping a critical switch or running a malfunction procedure. No arguments there. But he floored the two of us with his last proposal: "Until things are going well and we're feeling comfortable in the sim or on orbit," Jay said, "I'd like to handle any radio calls we make to MCC." In other words, Rich and I could leave the microphone to him.

Jay's reasoning was that by handling communications, he could filter out any embarrassing or confusing transmissions from our end until we had our act together. Rich and I both spluttered in protest—I had been flying for nearly twenty years, while Rich had gotten rave reviews for his performance on STS-53. It was ridiculous to suggest censoring our radio calls. We told Jay he would simply have to trust us, his crewmates. And that was tough for a man who had perfect confidence only in himself.

Jay politely backed off, but Rich and I couldn't help feeling we were somehow on probation. The exchange fed into *my* biggest anxiety: that I might slip up under the stresses of spaceflight. In our final six months of training, I worried most about making some sort of rookie mistake that would let my crewmates down. Despite the high fidelity of our simulations, the space environment and its stresses couldn't be replicated on Earth. The ultimate proof of one's skills waited out there on the launch pad.

# 6
# FINAL COUNT

*The only way to discover the limits of the possible is to go beyond them into the impossible.*

ARTHUR C. CLARKE, "TECHNOLOGY AND THE FUTURE," IN
*REPORT ON PLANET THREE*, 1972

I t was time to ride the merry-go-round. Enduring the thousand-horse-power centrifuge at San Antonio's Brooks Air Force Base is an exercise every rookie astronaut undergoes: a chance to check the fit of the orange launch and entry suit (LES) while experiencing the shuttle's ascent acceleration profile. As the suit technician strapped me into my seat on that cool morning of December 3, 1993, I remembered John Young's tales of the punishing g forces he had experienced during Gemini and Apollo missions. "It wasn't too bad . . . maybe 8 g's on entry, old buddy." We were four months from launch.

Inside the sealed cab, smaller than the interior of a compact car, I felt the centrifuge's twenty-foot arm lurch into motion at simulated liftoff. Staring at the g-meter, I sensed no rotary motion at all, just a heavy squeeze back into my seat as the "shuttle" powered upward. The acceleration grew, until two minutes into the run it seemed as though a giant hand had pinned me solidly into the seat. I was spinning at 18 revolutions per minute (rpm), sustaining 2.5 g's. This was kid stuff for fighter pilots, who endure eight or nine times the normal acceleration of gravity as they maneuver in a dogfight. High g forces can also make a pilot pass out as blood drains from the head under head-to-toe acceleration. The g's are easier to endure on a rocket launch, where astronauts take the acceleration front-to-back, supported evenly by their cockpit seats.

The centrifuge forces lightened considerably after the simulated burnout of the solid rocket boosters: at three minutes, I was pulling only 1.1 g's at 7.5 rpm. But the acceleration built steadily over the next four and a half minutes, until the cab was making a complete revolution every three seconds. My body now weighed 465 pounds, and my LES felt like a jacket of lead. The final minute of ascent ticked off the clock as I forced air into my lungs, waiting for main engine cutoff. At shutdown the sudden relief from the g's brought a momentary illusion of tumbling end-over-end. I wasn't weightless, but the instant loss of acceleration and noticeable feeling of lightness was just as disorienting as my initial experience on the Vomit Comet. Two centrifuge runs to "orbit" assured me that the physical gauntlet of liftoff and ascent were tolerable. Far more daunting were the mental challenges of the flight.

Back in Houston, we were fighting through a seemingly endless sequence of orbiter systems reviews, classroom lectures, and SST malfunction training. By six months before liftoff, we were in the midst of our high-fidelity SRL dress rehearsals in the Shuttle Mission Simulator.

The SMS was actually two simulators. The fixed base simulator, used for teaching orbit operations, replicated the flight deck and middeck of the shuttle cabin and presented the crew with convincing views of Earth and space out the ten cockpit windows. Thrusters thumped, klaxons wailed, and all cockpit instruments and switches functioned exactly as they did in orbit. Fifty feet away was the motion base simulator cockpit. This simulator, used to practice ascent and entry skills, housed only the four flight deck seats and forward cockpit displays. The motion base used hydraulic jacks and servo-motors to give the cockpit crew an accurate simulation of launch and landing sounds and vibrations. Every ascent run started with the cab rocked back 90 degrees, the crew lying on their backs as if they were about to rocket off the Cape's Pad 39.

Michelle Truly, our shuttle training team leader, and her cadre of a half-dozen orbiter systems instructors put us through two levels of simulator training: stand-alone and integrated. Our introductory lessons were stand-alone, with the crew rehearsing a portion of the mission on their own and the instructors both monitoring and deliberately sabotaging the works (once we had the normal operations down pat). The training team prepared scripts for each simulator session, with orbiter malfunctions inserted at the worst possible moment: a failed engine on the deorbit burn, a leaking landing gear hydraulic system, or a blown tire on touchdown forced the crew to deal with the trouble and still find a way to get safely back on the ground.

Stand-alone training lasted five to six months, with our crew completing a comprehensive series of progressively more difficult lessons. In the second phase, integrated training, our assigned simulation supervisor (Sim Sup) directed the training team to work not only with the crew but also with the flight controllers. With Mission Control plugged into the simulation via telemetry and voice links, astronauts and controllers handled the mission rehearsals together. Our crew of six teamed with several dozen flight controllers in MCC, while Michelle's instructors provided a steady stream of systems failures. Integrated simulations are a bit easier on the crew because the flight controllers can help diagnose and respond to malfunctions. But the downside is that they usually present subtler, more intricate failure scenarios, difficult to resolve and demanding close coordination between the crew and MCC. A slow thruster propellant leak, for example, might cripple the shuttle's ability to point the SRL radar for hours as controllers and the crew work to isolate the failure and restore attitude control.

The integrated simulations brought all elements of our mission training together, from orbiter flying skills to complex science and radar operations. Ascent or entry sims typically lasted less than half a day, but seven weeks before liftoff our culminating SRL orbit operations sim ran for a marathon thirty-two hours. Both Red and Blue Shifts took their respective turns in the fixed base, living by the flight plan as if we were actually in orbit. Out the aft flight deck windows, television screens fed us images of the sprawling SIR-C/X-SAR antenna filling the cargo bay. Above us, through another pair of television "windows," the Earth rolled sedately by; computer-generated but still-recognizable coastlines, rivers, and cities appeared, slid beneath our orbiter, then receded off the tail. The radio crackled with conversations between us and our capcom in MCC; on a second air-to-ground channel, we radioed mock Earth observations to our JPL colleagues in the payload operations control center. Between malfunctions, we prepared appetizing freeze-dried and thermo-stabilized meals in the middeck galley. At the end of each shift, we would grab some sleep at crew quarters, then return eight hours later to resume our duties. For me, the team's only rookie, it was all pretty exciting. I was becoming part of an astronaut crew whose next run through the flight plan would occur *off* the planet.

Our JSC Earth observations team strove to give us the array of geography skills we needed for the Radar Lab. By early 1994, after reviewing thousands of Earth images culled from previous missions, our crew could easily distinguish the Rann of Kutch from the Dzungarian Gates; tell Benghazi from Brindisi; and zero our cameras in on Uluru (Ayres Rock) in the vast rusty sands of Australia's Outback.

Using JPL predictions of our orbit's ground track, we cut and pasted dozens of military jet navigation charts into a set of four large-format atlases. The trace of our projected 160 orbits stitched a spider web of black ink across the continents. On these we plotted and labeled each of the 400-plus SRL science targets. A new laptop program written by JSC's in-house software team proved to be our best tool for keeping oriented over the Earth. Running on our brand-new IBM Thinkpads, the software displayed a moving map tracking *Endeavour*'s path across the planet.

Jay worked with our photo trainers to turn the rest of us into proficient amateur photographers. We trained with our Nikon, Hasselblad, and Linhof cameras both outdoors and in the simulator. The Linhof was a monster of an aerial reconnaissance camera, equipped with side handgrips and a lens barrel looking more like a mortar tube than a precision optical assembly. In the simulator, we practiced loading, aiming, and triggering this photographic arsenal, simultaneously using the orbiter's external video cameras to record the same swath of Earth being scanned by the radar. Each forty-five-minute daylight pass would require sharp geography skills and quick coordination between three camera-wielding astronauts; soon we were clicking away on the simulator flight deck like a mob of sidewalk paparazzi chasing down a movie star.

We didn't neglect our "downstairs" training, either. The middeck served as kitchen, science lab, storage closet, sleeping quarters, dressing room, and bathroom. We prepared space food in the galley, memorized the location of our equipment in the lockers, and practiced the detailed procedures needed to operate the waste collection system, using a fully outfitted and functional shuttle toilet.

\* \* \*

DURING THE YEAR PRIOR TO launch, much of our radar and science training had taken place at JPL, where we practiced with the science team and got some hands-on experience with our high-speed data recorders and their hefty magnetic tape cassettes. On each visit we noted the steady progress in assembling the radar antenna and electronics. In a massive rotating jig, JPL technicians on the Spacecraft Assembly Facility floor had completed the aluminum antenna structure, mounting on its flat surface the twelve-by-four-meter expanse of the SIR-C transmitter/receiver panels. The radar's power conditioners, control electronics, and data processors were mounted on a Spacelab pallet to be cradled in the cargo bay beneath the antenna structure. In late January the entire 23,000-pound antenna and pallet rolled east through Houston in a specially built trailer, the first leg of its journey to space.

During the first month of 1994, I noticed a distinct change in the pace and intensity of training. Nine months prior to launch, we had dropped our desk jobs for full-time mission preparation. But our daily schedule still granted us a couple of hours each day to keep up with routine desk work: mail, phone calls, reading, and even less serious aspects of the flight, like selecting from a catalog the shirts we would wear in orbit. Now our crew scheduler scrutinized our remaining training objectives and floored the accelerator.

Shuttle crews just back from orbit were unanimous in their dislike for the grueling pace of the final ten weeks of training. Work weeks routinely ran to sixty hours or more. Scheduled activity (STA flying, simulators, EVA training, Earth observation reviews, cabin stowage classes, payload procedures sessions) ran from 8:00 a.m. to 6:00 p.m., five or six days a week, spilling over into several evenings and weekends. A half-hour lunch was a luxury; afternoon workouts at the gym a distant memory. I usually made it home for dinner three or four nights a week, but it was hardly quality time for the rest of the family. Away from my desk all day, I shifted routine but necessary office work to my evenings at home. Facing Liz across the two desks in our study, I would wind up reading systems or payload manuals, sifting correspondence, or making notes in my crew notebook until bedtime. By mid-February I found the days rushing by in a blur; it was difficult to focus on anything but that day's schedule, and I rarely looked more than a day or two ahead. The STS-59 mission had taken over my life.

Linda and I had nearly completed our EVA training. The last in a series of four underwater runs, preparing us to deal with unplanned repairs outside, took place ten weeks prior to launch, in late January 1994. These sessions began at 7:15 a.m. in the divers' locker room with a brief physical performed by a flight surgeon. After a quick review of our tools and work objectives with our instructor, we would change into our water-cooled long underwear, then meet poolside for the laborious process of suit donning.

Lying on a mat next to the WETF's sparkling surface, I wriggled into the suit's trousers and boots, then shuffled over to the donning stand. My back to Linda, I squirmed upward through the hard upper torso of my suit. Once I had inserted my arms into place and popped my head up through the neck ring, my suit technician sealed the trousers onto the torso, locked gloves into place, and snugly seated the helmet. Immobilized by my 375-pound bulk, I stood patiently on the stand as the test conductor completed our life-support checks and cleared us into the pool. The hoist operator smoothly lifted the donning stand and lowered us carefully into the water.

Once released from the stand by our divers, we ran through airlock egress procedures, tool familiarization, and standard repair techniques for the orbiter's many cargo bay mechanisms. During one session, for example, we freed the robot arm's stuck shoulder brace; retracted the orbiter's high-gain dish antenna; manually stowed the thermal radiators on the payload bay doors; and cut jammed pushrods that were preventing those doors from closing.

Those clamshell doors *had* to close properly. Sixty feet long, they enclosed the cargo bay and added structural stiffness to the shuttle's fuselage. If we reentered the atmosphere with one door even slightly open, the searing plasma and hypersonic air loads would likely tear the shuttle apart. Redundantly powered motors ensured the doors were reliable, but on that proverbial bad day, perhaps the only way to get safely home would be to secure them manually.

But a pair of astronauts couldn't hope to close them using brute force, because an unpowered door, though weightless, still had tremendous inertia. Instead, Linda and I would hook a line to a door edge, then haul the door to the ready-to-latch position with a hand-powered winch.

That was half the job. The doors were held shut by thirty-two motor-driven latches that clamped the doors to each other and pulled them home against the forward and aft bulkheads. The latches were simple and reliable, but Linda and I could apply a manual backup in case of a serious failure. On our final WETF training run, we worked from inside the closed cargo bay to apply a set of clamps specially designed to seal the doors together.

This "doors and latches" class took place under the metal overcast of two payload bay door mock-ups. In the half-lit gloom, divers handed us the heavy stainless-steel door clamps. These latch tools would be weightless in orbit, but steel doesn't float very well in water; each weighed at least ten pounds. If successful in nestling the jaws of the clamp into position, we would ratchet the doors shut with a jackscrew handle similar to that on an automobile jack. But swinging that handle wasn't the problem. It was getting my body, and thus the clamp, into proper position. As Story Musgrave had told us, body position was everything: without it, finesse was impossible, and brute force was a poor substitute.

Our bulky suits prevented easy access to the tight angle between door and bulkhead, making the job frustrating and exhausting. After juggling a heavy clamp at arm's length for fifteen or thirty minutes, my fingers, hands, and forearms would scream with fatigue. Although the safety divers could support the tool while I rested (it was *supposed* to be weightless, after all), placing the clamp correctly depended on endurance and determination. By the time I had succeeded in installing a pair, even "max cooling" from my Liquid Cooling and Ventilation Garment (LCVG) couldn't keep up with the heat load. I wrestled the last clamp into place as sweat stung my eyes; the helmet prevented even a quick wipe of my brow.

What little elation I felt at finishing this last lesson was outweighed by the fresh realization that some physical challenges in space could put me near the limits of my ability. I barely had the strength to do the job in training—would adrenaline carry me through a real EVA? I wondered again whether I had the mental as well as physical stamina to overcome the surprises I might face in orbit.

To complete our space-walk training, we ran through three six-hour suit-up rehearsals in Building 9's shuttle mock-up, donning the space suit in the cramped confines of the orbiter's airlock. Kevin Chilton, who had assisted with STS-49's four space walks, was our in-cabin coordinator. When Linda and I, wearing just our long johns, joined Kevin to wriggle into our space suits in an airlock only slightly larger than a bathroom shower enclosure, we became close friends by necessity. The mission would bring us closer still.

My graduation exercise on March 2, 1994, five weeks before liftoff, was a dress rehearsal in my Class 1 space suit, the actual rig I would fly with on *Endeavour.* The test took place in Building 7's vacuum chamber, a copy of the shuttle's airlock that could be pumped down to high vacuum. Linda and Kevin helped button me up in the suit, suspended from a weight-relieving harness in the cylindrical airlock. With helmet and gloves locked in place, I was now breathing pure oxygen. The four-hour prebreathe would flush nitrogen from my blood and reduce my risk of getting the bends. I settled in to watch a Sylvester Stallone movie (*Cliffhanger*) and do a little reading in Stephen Sears's Civil War history *To the Gates of Richmond* (although it was a little tough flipping the pages with my sausage-fingered gloves). Soothed by the steady rush of oxygen through the helmet, I eventually dozed off until the bustle and chatter of the test team woke me for the trip to vacuum.

Prebreathe complete, we ran through the final steps of the depressurization checklist. Inner hatch closed, I double-checked the suit indications, then opened the depress valve. Air whistled out of the airlock into the much larger adjacent chamber, taken to a near-vacuum earlier by powerful pumps. As the pressure dropped to 5.0 psi, I paused for a last suit check. I noticed a surgical glove, tied off and hung from the ceiling by the test team, had swelled grotesquely to the size of a cantaloupe. Suit status good: down to vacuum we go. I rotated the valve open again. As my "altitude" rose toward the equivalent of 400,000 feet, nearly seventy-six miles up, my suit pressure gauge settled reassuringly at 4.3 psi. "Outside," the air pressure was below 0.5 psi, and the pan of water on the ledge in front of me boiled furiously in the near-vacuum, its edges rimmed with ice created by the frenzied evaporation of the liquid. In this hostile environment, the pressure equivalent of low Earth orbit, only the fourteen thin layers of my suit separated me from rapid decompression and death.

The exercise was designed to instill confidence in the suit's integrity and function. The test conductor, who annoyingly addressed me as "Subject" through our entire conversation, had me run through a series of suit checks as I removed my orbiter umbilical, which furnished me with electricity, oxygen, and cooling water, and switched to battery

power. Now I was completely dependent on the backpack for life support; in a sense, I was the occupant of a small but fully self-contained
spacecraft. After discussing our recovery plan, the conductor directed
me to switch off the battery. With a flick of the chest-mounted switch,
the noise of fan, water pump, and radio disappeared into silence. I remembered the tag line from the sci-fi movie *Alien:* "In space, no one can
hear you scream." *That was certainly the case right now.* But of course,
there was nothing to yell about. I stood, snug and safe, inside what was
essentially a thick-skinned balloon. A battery failure in orbit would end
a space walk, but I would have ample time—at least thirty minutes—to
make my way back to the ship's airlock and plug into the orbiter's power
supply.

Once powered up again, the test conductor walked me through the
emergency procedure for dealing with a failure of the fan motor, which
circulated both oxygen and cooling water through the suit. To prevent a
buildup in the helmet of carbon dioxide, heat, and humidity, I would
have to vent oxygen from the suit long enough to make my way back to
the airlock. Reaching up to the left edge of my faceplate, I deliberately
opened my helmet purge valve. This was serious stuff: hunching down in
the suit an inch or so, I could look right through the spaghetti-sized hole
into emptiness. The whistle of escaping gas nearly overwhelmed the radio as oxygen streamed directly from my helmet into the vacuum outside, to be replaced by fresh, cool gas from the supply tank. When I
flipped the valve closed again a minute later, I was satisfied that the suit
would do its job.

TWO TRIPS TO THE CAPE in February and March underlined the fact
that launch day was fast approaching. The Crew Equipment Interface
Test brought the six of us to the Kennedy Space Center for two days to
explore our spaceship's work and living spaces. With the ship still undergoing final preparations in the hangar, we dressed head-to-toe in
"bunny suits" to avoid contaminating the crew cabin with clothing
fibers, stray hairs, or skin particles.

Crawling aboard *Endeavour* on February 25, 1994, was an important
but routine exercise for the rest of the crew, but for me it was like a day-
long pass to Disney World's Tomorrowland. Secretly giddy at being inside a real orbiter cabin, I spent hours checking out flight deck control

panels, SRL switch panels, payload recorders, and the life-support machinery beneath the middeck floor. Out in the payload bay, nearly filled by the massive radar antenna and its avionics pallet, Linda and I examined our tools, tethers, and access to the SRL mechanical systems we might have to repair outside. I flew back to Houston anxious to take on the last month of training. On March 19, 1994, *Endeavour* rolled to the pad.

Our preflight press conference in Houston drew the reporters' usual questions about the STS-59 mission plan and our crew's activities in orbit. One journalist observed that I had worked as an engineer for the CIA and was now about to fly with a sophisticated new tool for taking a detailed look at Earth. Would I care to comment on how my CIA work had prepared me for this mission? I thought a moment about how to handle this. "One thing I do remember very well about my work for the Agency," I said, smiling, "is the severe penalty they promised for divulging classified information." The reporters laughed with me—no more questions on *that* topic.

My crew returned to the Cape a few days later for our culminating dress rehearsal: the Terminal Countdown Demonstration Test. With launch just two weeks away, we spent two full days familiarizing ourselves with KSC's safety and launch control procedures, ending with a fully suited mock countdown out at Pad 39A.

In one exercise the six of us took turns driving KSC's lime-green M113 rescue vehicle, a twelve-ton armored personnel carrier designed to protect us from a looming shuttle explosion while en route to helicopter evacuation sites a safe distance from the pad. Grinning from the driver's hatch while working the twin joysticks, I hurtled along dirt roads through the marshes surrounding the pad. A mere twitch of the wrist would put me into the ditches lining our evacuation route, and I was anxious to avoid the fate of a fellow Hairball, Ellen Ochoa, whose inadvertant amphibious exploits in the M113 had caused the KSC rescue folks to rename a certain roadside pond in her honor.

Early the next morning we suited up and rode out to the pad in the "Astrovan," an Airstream motor home. It was just a rehearsal, but the excitement for me was real: this was the way launch day would play out. Strap-in at the pad was uneventful (we had practiced it in Houston), and I took my middeck seat to Linda Godwin's right as my brain wrestled with the strange cabin orientation. With rocket ship *Endeavour* poised on its tail, the middeck was rotated 90 degrees nose-up from the horizontal orientation of the simulator. I lay on my back, knees bent as though sitting in a chair, feet pointing up at the white lockers that now formed our "ceiling." Two hours of working through the countdown with the launch control team by radio brought us to T-zero, the point of liftoff.

But there were no propellants in the big orange-brown ET, and no igniters installed in the boosters to touch off their rubbery solid fuel. Instead of the simulated main engine ignition, we heard the blare of a cockpit master alarm. Three miles away, the launch control center ordered a "Mode 1 egress," an emergency exit from the orbiter. I hurriedly unplugged radio and oxygen lines, unstrapped the parachute, and crawled after Linda through the open hatch to the White Room. Pausing momentarily as I exited, she led the way across the swing arm and around the pad elevators to the landward side of the launch tower. Breathing hard on emergency oxygen, I glanced through my sealed helmet visor at the dizzying view from 195 feet above the flat Florida scrub, then clumsily vaulted into the escape basket with Linda. She slapped the paddle that would have sent us zinging down a quarter-mile of cable toward the landing area just short of the blast bunker. Only a safety chain kept us from an eye-watering ride down the wire at speeds approaching 55 miles per hour. NASA, with no hint of irony, considers this too risky a ride for astronauts to practice.

The dress rehearsal complete, I was quiet on the T-38 ride back home, musing that my next scheduled departure from the Cape would be on a more exotic form of transport: space shuttle *Endeavour*. I knew that my life and those of Liz, Annie, and Bryce would turn on the outcome of that upcoming morning's events. The two possible futures were radically different. Launch was set for April 7, 1994.

The remainder of the training syllabus included a few refresher sessions in the motion base simulator. To mark the formal completion of STS-59 sim training, we invited our spouses over to experience a practice ride to orbit. As the wives took their turns in the right rear flight deck seat, Michelle Truly and her team inserted the usual ascent failures so that the crew could display its finely honed skills.

I plugged a headset in at the instructor station and listened as Sid and the flight deck crew prepared for another ascent. The motion base cab rocked to the vertical, lurched with the jolt of ignition, and rumbled "off" the pad. Sid, Kevin, and Rich stayed on top of each malfunction as *Endeavour* continued its simulated ascent to 120 nautical miles.

Just like clockwork, the instructors took an engine down just four minutes into the flight. Rich read off the checklist steps as Sid and Kevin smoothly threw the proper switches, and *Endeavour* shifted its trajectory to arc across the North Atlantic to the emergency strip at Zaragoza, Spain. Suddenly they felt a lurch and a series of jolts in the cockpit accompanied by a master alarm. Sid thumbed the Backup Flight Software switch on his control stick, cutting in the last-ditch flight control computer, and—the instruments froze, all three computer screens displayed big green X's, and the hydraulics brought the cab to a halt. The instructors pursed their lips, checked displays, and shook their heads worriedly at Michelle.

It was an LOC—a loss of control. *Endeavour* had suddenly tumbled out of controlled flight, veering irretrievably from its precise abort attitude. Both autopilot and the crew's manual control inputs had been unable to recover. Had the flight been real, orbiter and crew would both be gone. It wasn't quite the demonstration we had in mind.

The instructor and sim teams later traced the LOC to a simulation software glitch, clearing the flight software from suspicion. But even before the incident, Liz understood that the shuttle was never more than a few seconds from potential disaster during ascent or entry. With less than a month to go, her understandable worries about the impending flight grew more insistent. Added to our discussions about wills, life insurance, and the LOC incident was our selection of a casualty assistance officer, my friend and classmate, Carl Walz. He had the explicit responsibility of helping Liz if I didn't make it back from STS-59. She also knew from my own experiences that family escorts Charlie Precourt and Kevin Kregel were not going to the Cape just to schlep baggage for the spouses. If the worst happened, they were charged with smoothing the path for Liz, the kids, and the other families, getting them safely away from the launch site and back to Houston, while NASA literally picked up the pieces.

To be sure, Liz had come to understand over the past four years the amazing yet risky characteristics of the shuttle as flying machine. A year earlier I had recorded in a taped diary this note about one of an ongoing series of problems with the orbiter:

**FEB. 11, 1993:** What I wanted to talk about was this safety issue we discovered just this past couple of days . . . that the tip seals on several of the high pressure oxygen turbopumps on *Columbia* coming up for launch here had not been inspected at the proper interval. . . . [W]e weren't sure of the status of the tip seals in those pumps. Because of that doubt, all three of those turbopumps have to be changed out, and that's going to affect the [STS-55] launch date by about two weeks. It appears that several other shuttles in the past year have flown without proper inspection as well. That leads to worry about what other items we're missing that are of critical importance and that [might] rise up and bite us [with] an engine failure. Engines are the most catastrophic element in the system. So that report on the news last night about how NASA may have missed inspections for the past year sure made Liz's ears prick up and ask me what was really going on here. She's worried about what else we are missing in the system, and quite frankly so am I. Our office awaits the explanation about how that procedural happenstance occurred.

I believed, as did she, that the shuttle program continually took a hard look at itself and didn't flinch from finding and correcting technical hazards. Of the dedication of the people who worked on the shuttle, she and I had no doubts. What worried us was the possibility that NASA couldn't anticipate every possible shuttle vulnerability. What new combination of technical and human failings lurked out there, waiting to surprise us from a totally unexpected direction?

*Challenger* had disintegrated just eight years earlier. How many more missions could we safely fly before statistics caught up with NASA? For months, Liz had seen a familiar tide of worries crest and then recede: concern for my safety, what life as a widow would be like, and how she would raise two children without a father. She had no prospect of banishing any of these black possibilities until she got through *Endeavour*'s launch.

I was praying hard enough for the two of us. I had my own cadre of saints lined up to put in a good word with God: St. Thomas More, for the grace of clear thinking; St. Joseph, patron of families; St. Christopher, the traditional patron of the traveler; the Holy Family, to help me keep my cool under pressure. It also comforted me to know my father and my grandparents were able to present my needs directly to God. I knew He was listening, and that was reassurance enough.

A month before launch, seven-year-old Annie selected a little workbook titled *Great Disasters* from her teacher's list of school reading assignments. Annie mentioned to Liz that the book included a section about the *Challenger* accident. When Liz called the school the next day, Annie's teacher was apologetic, but the damage had been done. Two weeks later, Liz was brushing Annie's hair and said, "Well, Daddy's launching in two weeks. Are you excited?" Annie thought for a moment. "No." "Why not?" Liz asked lightly. Annie turned to look up at her mother: "It might blow up."

Thursday, March 31, 1994: The long, relentless months of training were over. Medical quarantine was about to begin. Late that evening I tucked in the kids with the promise of seeing them in Florida. Kissing Liz good-bye for the moment, I threw my duffle bag and briefcase in the car and drove the ten minutes over to JSC. Sid and the rest of the crew were just arriving at our temporary quarters in a quiet corner of the space center next to the astronaut gym. For the next week we would prepare for launch while isolated from the day-to-day family and office encounters that might expose us to a risk of infection. Although our spouses, given a quick medical exam by the flight surgeons, could visit every day, our children, conveyors of whatever germs might be circulating in the schools, were off-limits until we returned to Earth.

AS THE SRL SCIENCE team moved to Houston to man the Payload Operations Control Center consoles or were deployed to science supersites around the globe, the STS-59 crew began the shift to an orbit-based work and sleep cycle. The pilots and Linda on the Red Shift would have to wake at 2:00 a.m. for our early morning launch on April 7; they would be fresh for ascent and their first twelve-hour shift on orbit. The Blue Shift, by contrast, would launch near the close of our new sixteen-hour workday. We would need to turn in just five hours after launch in order to wake and relieve the Red Shift seven hours later.

To get our schedules in sync, we all followed a light therapy protocol designed by Flight Medicine. Bright light, natural or artificial, would reinforce our waking periods. Subdued room lighting and eight hours of complete darkness while sleeping would help lock in the new sleep cycle. For the Blue Shift, it started with "the slam": Rich, Jay, and I stayed awake all that first night, finally turning in exhausted around 8:00 a.m. In this new routine, we would rise at 4:00 p.m., then go outside or work under banks of brilliant fluorescent lights to reinforce our bodies' awake cues. The fluorescents were bright enough to wash out television screens and bring on an involuntary squint. By contrast, we slept in light-tight, pitch-dark rooms, with the Red Shift quartered down another hallway to minimize disturbances.

Our spouses visited daily around six for dinner. While they and the Red Shift enjoyed an evening meal prepared by the JSC space nutrition staff, we on the just-awakened Blue Shift tucked into pancakes and omelets. The dual menu was just another way of shifting to and maintaining the new sleep pattern. By the fourth day in quarantine, the light protocol had worked its magic, and I felt rested, refreshed, and comfortable with my new circadian rhythm.

Quarantine is a godsend to shuttle crews. Free of formal classes, we were able to get in some vital last-minute studying, review dozens of small but important mission details with our crewmates, complete the personal notebooks we would take to orbit with us, and get in a daily workout.

Early on Monday morning, April 4, our sleep-shifted crew climbed into our T-38s and taxied past our training team, who had come out to see us off. With afterburners stabbing electric blue in the growing light of dawn, we climbed out over the Gulf, Venus's fading brilliance beckoning us toward the sunrise. Slicing down ninety minutes later in a long glide toward the Cape, our formation crossed the Banana River and leveled just south of the Vehicle Assembly Building. In the lead jet, Sid led our four Talons straight toward the beach. Approaching Pad 39A, he banked left, the other three jets strung out above and to our right. I looked down. Wow! That's my rocket! *Endeavour,* still cocooned in the pad's gray service structure, was nearly ready to go. The hair stood up on the back of my neck as I stared past the wingtip at this impressive display of American technology: On Thursday I'd be riding that machine. What would it be like? Would I be ready?

Our escorts accompanied our families from Ellington on NASA 2, a venerable Grumman Gulfstream turboprop. I was asleep when they landed that afternoon after a three-hour flight, touching down at the Cape's old Skid Strip, where Gemini and Apollo astronauts had once arrived for their trailblazing missions. Settled in at their Cocoa Beach hotel, the wives joined us at crew quarters that evening for a relaxed arrival dinner. We greeted them with the news that shuttle managers had postponed Thursday's launch attempt by twenty-four hours. Technicians examining a shuttle main engine at Rocketdyne, the California SSME manufacturer, had found liquid oxygen guide vanes in the turbopump preburner whose critical dimensions were out of tolerance. A thinned or deformed vane could break off in the oxidizer flow, shattering the turbopump blades downstream. The launch delay would give technicians time to examine the corresponding vanes in *Endeavour*'s three main engines.

The extra day in quarantine was fortuitous: despite our medical isolation, Kevin came down with a nasty virus that evening, a potential threat to the launch. The docs moved him down the hall to an examination room for observation. Banned from the crew quarters proper lest he infect the rest of us, Kevin was now doubly isolated while the crew underwent final suit checks and reviewed flight plans and crew procedures. It was a race between Chili's recuperative powers and the ticking countdown clock.

The Cape is not a bad place to be quarantined. Around our conference-room table we watched *The Milagro Beanfield War*, one of Sid's favorite movies, and Rich and I had plenty of time during the graveyard shift to work up a sweat in the astronauts' Apollo-era racquetball court an eighth of a mile down the deserted corridors of the Operations and Checkout (O&C) building.

Living in the O&C was a daily reminder of the frenetic days of the Moon race. Others were close at hand. Taking a break from studying one day, I jumped into one of the crew's rented LeBaron convertibles and made a quick pilgrimage to the historic launch sites down Cape Canaveral's old ICBM Road, south of the shuttle pads. Near the Cape lighthouse was Alan Shepard and Gus Grissom's Complex 5, its Redstone blockhouse now preserved by the Air Force Space and Missile Museum. Turning back north, I found Launch Complex 14, site of John Glenn's historic ride to orbit on *Mercury-Atlas* 6 in 1962. A mile north, at Pad 19, its rusting flame bucket staining the concrete orange, Grissom and John Young had inaugurated the Gemini program in March 1965. Their silvery Titan II booster had been built near my home in Essex, Maryland, and I still remembered glimpsing a pair of the imposing rockets on a Cub Scout field trip. Most poignant of the old pads, however, was Launch Complex 34, its massive blockhouse just a quarter mile from the Atlantic surf. Out on the pad apron, abandoned these past twenty-five years, stood the launch pedestal for a Saturn IB rocket. Wally Schirra's crew had left Earth from here in 1968 on *Apollo 7*, but on this same pad in January 1967, Grissom, White, and Chaffee had met death in the inferno of the *Apollo 1* spacecraft. Alone with the sea, the wind, and whatever ghosts lingered still, I stood beneath the launch ring and thought about those heroes of my boyhood.

North of Pad 34, past the Air Force's Titan pads, is the Kennedy Space Center Conference Center, known to the astronauts and their families as the "Beach House." A remnant of Cape Canaveral's prerock-etry days, the weathered bungalow stands just behind the dunes south-east of the shuttle pads. Almost hidden from the road behind scrubby palmettos, it was the perfect spot for a picnic lunch or dinner, either in-doors or out on the deserted beach. Although patrolled by the base's se-curity police, Liz and I could still walk along the surf, deep in conversation, thankful for each other's company. Our visits were rich with emotion and suffused with the tensions of the approaching launch.

Along with time alone, we welcomed the chance to socialize with the rest of the Blue Shift. Surprisingly, much of our conversation with Rich and Jay and their spouses, Nancy and E.B., centered on our guest arrangements for the launch. Our invited guests, including parents, sib-lings, and hundreds of relatives and friends from all over the country, were now arriving; inevitably there were problems.

For launch, each astronaut was allotted seventy-five "car passes," each good for a single vehicle admission to the NASA Causeway span-ning the Banana River some five miles south of the pad. In addition, our extended families and a very few close friends would take buses out to a row of bleachers three miles west of the pad. As our guests checked in, our veteran crew secretary, Carolyn Morris, who was endowed with saintly patience and courtesy, forwarded any questions to us. At the Beach House, Liz and I sorted through the guest problems. How silly it seemed to be spending our last few hours together unraveling the mys-teries of launch morning for our guests. But this was a way to thank the many who had honored us by coming to Florida.

Wednesday dawned with clear skies and mild breezes, perfect for breakfast with our spouses and close family members, all of whom had to pass a NASA physical. Rich, Jay, and I led the way out to the beach just before sunrise, with Sid and Linda following. Sadly, Chili was still recuperating back at crew quarters under the ministrations of the flight surgeons. The sun was just emerging above the tranquil Atlantic when our guests spilled out of the NASA bus. There were introductions and hugs all around. Crew and guests mingled and enjoyed breakfast burritos, barbecue, and even a few beers (it was supper, after all, for the Blue Shift). I made sure none of my family missed the view of *Endeavour* looming above the sandy thickets a few miles away from the Beach House deck.

There were a few questions about turbopumps, but most of the talk centered on my crewmates and what each would be doing in orbit. After breakfast, Sid invited the families out onto the beach to pray together. Astronauts, their friends, and families gathered in a circle, the surf lapping near our feet. Father Tom Bevan, who in Baltimore thirty years before had trained me as an altar boy and later officiated at my wedding, led us in reflection. The morning sun warmed our faces as he reminded us how God had formed out of nothingness the universe, our planet, its oceans, its continents. In two days we would leave His Earth on that rocket just over the dunes and see His handiwork in a new way.

We were not just detached explorers of a static, ancient universe, Father Tom noted, but active participants in a Creation that continues to renew itself. I looked past the priest's thoughtful face, the soft morning light playing on the sands. Just offshore, pelicans skimmed the surf, and beneath those waves a surging ocean teemed with life. Around me stood my mother, my brothers, and the families of my crew. All were evidence of the ongoing vigor of birth, of life, of Creation. Our impending voyage represented the leading edge of humanity's expansion into the cosmos. Joining hands, we asked God for a successful voyage and our crew's safe return. Father Tom's simple words on the beach left me with a strong sense of peace—one that carried me through the difficult parting with family. Worries dropped away, and those last few days at crew quarters were spiritually calm.

With Chili finally well enough to rejoin the crew, the six of us met our spouses on Thursday, the day before liftoff, for a tour of the launch pad. The couples fell silent at seeing *Endeavour* poised on the mobile launch platform. As we posed for a last crew portrait, the ET and boosters soared 185 feet above us, spires of a cathedral to be hurled at the stars.

I saw the kids one last time that night near the pad, where the families and guests arrived after sunset for a view of *Endeavour,* radiant now in the intersecting beams of xenon floodlights. I talked briefly to Annie and Bryce across the roadway separating us according to quarantine rules. Annie burst into tears when she realized I couldn't come near enough for a hug. My brother and sister-in-law comforted her, though, and it was an upbeat crew and crowd who gathered to see the dazzling shuttle. After half an hour, our Blue Shift said our good-byes and drove off, the spouses following in a van driven by our escorts. At the Beach House, Liz and I had a final few minutes together. There wasn't time for a long walk on the beach, but we discussed the coming morning and our plans for a vacation alone when all this was over. We kissed, reluctant to release our hold on these moments together. It was good-bye.

HIGH WINDS KEPT *ENDEAVOUR* nailed to the ground the next morning, Friday. Lying for a fifth hour on my parachute, back stiff and aching, I could hear the flight deck gang of Sid, Kevin, Rich, and Jay report only blue skies above. But strong crosswinds gusted across the shuttle landing facility, the three-mile-long runway north of the Vehicle Assembly Building. The winds would play havoc with an emergency landing back at KSC if we had to abort for a serious failure immediately after launch. Kevin asked over the intercom which of our patron saints would have the inside track to God today. "St. Theresa is the patron of aviators," I responded. "Or there's St. Joseph of Cupertino, the astronauts' patron saint. Maybe he can get the wind machine turned off."

"What's St. Joseph of Capistrano got to do with astronauts?" Sid asked.

"Cupertino. He used to levitate over the altar of his monastery chapel," I answered. "He discovered zero-g three hundred years ago."

Sid recommended we try them both. "Couldn't hurt," said Kevin. "Even with the way my back is aching, I still don't want a scrub. It'd take them an hour to get us out of these seats. The quickest relief is just eight and a half minutes away—straight up."

But it wasn't to be. With the wind still gusting out of limits more than three hours after the scheduled launch time, Bob Sieck, the launch director, reluctantly ordered a scrub. We would try again at dawn tomorrow, an hour earlier than today's attempt, when the clouds and the winds might be more cooperative.

It was a quiet ride back to crew quarters for our regular sleep shift, followed by a quick phone call to Liz. Rich, Jay, and I spent another long night studying while the Red Shift caught their last eight hours of sleep. Finally, after breakfast and suit-up, it was time for the clumsy walk in our orange LESs down the hallway, the crew quarters staff waving and blowing kisses to us as we shuffled to the elevator. We told the staff that we hoped we wouldn't be seeing them again for a few days. The feeling was mutual, they assured us.

As the Astrovan rounded the bend in the crawlerway, I rose in a half-crouch, straining to get a predawn glimpse of *Endeavour* caught in the brilliant glare of the xenons. "No more scrubs," I thought. This would be the day.

I was thirty-nine when I stepped out on the pad with the rest of the crew, but I gazed up at *Endeavour* with a child's amazement. The orbiter reared above us, gleaming against the night sky. Moving to the edge of the flame trench for a clearer view, I shivered with excitement at the sight of my now-ready spaceship. The shuttle's clean arrowhead profile seemed about to leap upward, as if it too were intent on a launch this morning.

Helmet bags in hand, we rode the elevator to the 195-foot level. Twenty-three men had passed this way en route to the Moon. We weren't going that far, but we hoped to get a closer look at Earth than they did.

I waited with Rich and Jay at the railing near the swing arm as Linda and the pilots strapped in. We stared down at the twin pillars of the boosters, their nozzles suspended over a red plastic tarpaulin holding a foot of water to help muffle the blast of booster ignition. Wisps of ice fog fumed around the supercooled liquid oxygen and hydrogen lines from the ET to the orbiter's belly. A trail of vapor streamed white against the dark sky from the "beany cap" vent hood over the external tank's nose. In the steady hiss of escaping nitrogen, purging the cargo and engine bays, the orbiter seemed to signal its eagerness to leave. *Endeavour* was alive.

Just before my turn was called from the White Room, I acted on another of Rich's valuable tips. From a small handset next to the swing arm, I dialed a number I had scrawled on my palm. The phone rang in the launch control center office of Norm Carlson, one of the shuttle launch managers, where Liz and the kids would be arriving shortly. From the picture windows fronting his desk, the crew's families could watch the shuttle on the pad and follow the countdown until it was time to walk up to the roof at L-9 (launch minus nine) minutes. Carlson's secretary, Nora Ross, whom I had met several times during my own escort visits, took down my message for Liz. It was brief but never more heartfelt. Thanking Nora for the favor, I hung up and turned to board my spaceship.

In the White Room, a pair of pad technicians helped me into my parachute harness. Donning my white communications cap with its built-in headphones, I snapped on the chinstrap, shook hands with the techs, and crawled into *Endeavour*, my home for the next ten days.

Troy Stewart, our lead suit technician, was waiting for me as I squeezed past Linda's seat, projecting oddly out of the "wall" to my left. Troy, a vigorous and supremely competent man who had strapped in Apollo crews before the shuttle flew, expertly guided me into the lightweight seat. As he clipped me into the parachute, I took in the view of my little universe. Even with our training exercises and the countdown rehearsal, the cabin was still disorienting. The forward wall of lockers formed a low "ceiling" just eighteen inches above me. To my right was the gunmetal stack of four sleep bunks. Behind my head the curved metal housing for the escape pole arced across the middeck toward the hatch. Beneath me was the airlock's closed inner hatch, protected for the moment by temporary "floor" panels. Satisfied with his final checks, Troy withdrew from the cabin, and the hatch crew outside swung the heavy circular door closed, engaging the latches with a solid "clack." Cinched snugly into her seat by the galley, Linda shot me an encouraging smile. With less than two hours until launch, we were alone.

Over the intercom, we heard the pad crew clear off and head for their sandbagged checkpoint a mile and a half back up the crawlerway. This was a prudent move, given that we were sitting atop a half-million gallons of explosive liquid propellant and 2.5 million pounds of solid rocket fuel. Inside *Endeavour*, we talked quietly on the intercom, tossing around a few nervous jokes. The background fan and machinery noises in the middeck seemed louder than usual, but they gave me the privacy to record a few notes as the countdown proceeded.

**APRIL 9, 1994. 5:00 A.M.** We're a little worried...in the back of our minds [lurks] the fact that if we don't launch today, there is a range conflict that will scrub us until the 23rd of April, almost two weeks from now. Want to get off today, God willing. I've got all the saints pulling for me.

**6:19 A.M.** I'm getting a sore back, but it will all be worth it if we can get off here this morning. It's hard to breathe. It's hard to take a deep breath [because of the parachute harness], or inflate your lungs, but if you just stay still and don't breathe too hard it's not bad. It's going to be interesting to see how 3 g's feel after being on my back for two hours.

I'm thinking of everybody on the bleachers and on the causeway. It's great of them to come out here, and I really appreciate their support and their love and can't wait to get back and see them all and tell them all about it. I don't want to wind up getting out of here today on a scrub and then going home.

T-20 minutes and holding. About 35 minutes before launch time. The countdown is proceeding normally. We can only wonder about the weather. The winds are strong. If they get too far away from the runway direction, then we're going to have a crosswind problem just like yesterday.

T-9 minutes and holding. We don't have any words on the crosswinds, but we do know the sky looks clear outside, and we're ready to go down here. Thinking about Liz and Annie and Bryce going up on the roof soon....

As the count resumed at T-9 minutes, the spouses, children, and escorts began the walk to the roof. Liz had found that she could smile brightly and assure everyone she was doing fine, but she couldn't ignore her body's response to the building stress. She had had trouble sleeping for weeks, and walking out on the roof the day before, her chest had felt painfully tight. Stepping onto the roof for the second time, the chest pain didn't reappear.

Walking with Nora toward the rooftop railing, Liz looked up. This first sight of the shuttle on the pad caused her knees to buckle. Nora asked quickly, "Are you all right?" "I'm fine, but *that's* never happened before," Liz answered, surprised by her legs' sudden betrayal. She looked around at the scene on the roof, drinking in the details. Closest to the rail were the families, astronauts in their blue flight suits, and other assigned escorts. Liz noticed Trudy Davis, the lead NASA administrative escort, appraising her with a practiced eye. Standing discreetly at the rear was a knot of security and medical personnel.

> In the firing room, everyone says they're "Go." Bob Sieck has just given us a go for launch. Inside of nine minutes, the countdown is proceeding, and we're all strapped in and ready to go. It looks like there are no constraints to launch this morning. We're getting to try it.
>
> Starting to get a few butterflies. A little hollow pit in the stomach, but otherwise looking forward to the experience. Kevin is getting ready to start the APUs; he's going to crank them up pretty soon. I've got to put this recorder away . . . we'll talk to you after orbit.

Just inside five minutes to go, Kevin started the auxiliary power units (APUs), three hydrazine-powered turbines that drive hydraulic pumps to move orbiter aero control surfaces and main engines. The initial whine of the APUs carried up to us through the orbiter structure. The time for jokes and small talk was over. Outside the cabin, *Endeavour* was rousing herself.

# PART TWO

# SHUTTLE

# 7
# EARTH AND SKY

*The surface of the Earth is the shore of the cosmic ocean.*
*From it we have learned most of what we know.*

CARL SAGAN, *COSMOS*, 1980

*E*ndeavour crew is go for launch!"

Sid's ebullient statement to Bob Sieck in the launch control center (LCC) summed up how all of us in the cockpit felt. With the astronauts and launch team ready to proceed, the orbiter test conductor
(OTC) took up the count.

"*Endeavour*, close and lock your visors, initiate $O_2$ flow, and 'Vaya
con Díos!'"

"Thanks a lot, Mark, and we'll see you in about ten days."

Shivering with excitement, I hurriedly closed my clear helmet visor,
snapping the bail into place to seal the faceplate. At my left knee I
flipped a silver lever on the suit to start the flow of ship's oxygen. Sid
checked us in on the intercom: "CDR's up," and we all answered in turn.
"PLT, MS-One . . . MS-Two . . . MS-Three . . ."

"Four," I chimed in, tail-end Charlie. With adrenaline flowing freely,
I tried to slow my breathing in the close confines of the helmet. I tugged
my shoulder straps a little tighter.

"Vaya con Díos," Rich said. "That was a nice send-off."

"Want me to translate that for you, Sid?" Kevin teased our commander, a native of Albuquerque.

"Nope, I think I got that one. The real translation is 'God be with you as you go.'"

The butterflies were in full flight now. Could these be the last few moments of my life? I toyed with that one for a second or two, then discarded the idea. I didn't believe it. Even years ago as a student pilot, I had no interest in calculating risks. I saw no point in worrying about events beyond my control. I just focused on doing my best.

The butterflies weren't due to the imminent hazards of launch but to worry that I might fail in front of my colleagues. These were the rookie jitters. Well, I told myself, it's too late to worry about what else you can do to get ready. *Endeavour* was waiting; let's see what this machine can do. Having imagined riding a rocket for thirty years, I was eager for the physical experience. I had pulled 6 or 7 g's in a T-38, been weightless in the KC-135, hammered through turbulence at low level in the B-52. It was time to find out if *Endeavour* would live up to the centrifuge or Cabana's T-38 simulation.

One minute to go. Linda gave me a thumbs-up and a smile. Fingers tingled with excitement now. "This is *great!*" Chili said, echoing my thoughts. While the flight deck team was "hot mike," the intercom picking up their every word, Linda and I had to key our microphone buttons to be heard. The middeck crew has few duties during launch, so I kept quiet and listened. For the last minute of the count, we heard only Mark Paxton, the OTC, sitting at his console three miles away.

"T-minus-thirty-one," Sid said quietly. Beneath me, I heard the whine of the three auxiliary power units far aft in the engine bay and the faint buzz of the cabin fan beneath the middeck floor. The ground launch sequencer (GLS) computer was in charge for the last few seconds of the count.

Launch Control: "GLS is go for auto sequence start."

*Jesus, Mary, and Joseph, be with me now.*

"Fifteen seconds." On the flight deck the twin attitude indicator spheres snapped smartly to the correct launch attitude. "Nav init," confirmed Sid: the orbiter guidance system incorporated its final update, ready to put us on track for space. Things happened in a rush now.

"We have a go for main engine start." On the LCC roof, Liz later told me, she heard the final call over the loudspeakers and pulled Annie and Bryce close to her, an arm around each diminutive shoulder. We had watched one launch together, but this one had an entirely different feel. Now *I* was sitting on the rocket out there.

*How can you do this to me?* thought my mother, trembling in the viewing stands a couple of miles north up the Banana River. Distrustful even of airplanes (she has never flown), the countdown had put her in uncharted territory despite a steady diet of tranquilizers. She was sure something terrible was about to happen—*and me without your father!* Mom remembered the awful pictures of Christa McAuliffe's parents, their expressions caught on film as *Challenger* exploded, shock and grief playing across their faces as they grappled with the enormity of what they were seeing. She prayed to Mary, the Blessed Mother: *Make him come home and be a garbage man!*

T-6.6 seconds, and tons of water flooded the pad and cascaded into the flame trench in a Niagara designed to cushion the shock waves of ignition. *Endeavour's* computers commanded the start sequence for the three main engines. Far aft in the orbiter's engine bay, turbopumps spooled up and the dentist-drill whine carried to us in the cabin.

A rumble shook the stack from ten stories below as the main engines coughed fire and shivered their way up to full power. Six seconds tumbled by as the entire stack rattled with the barely restrained fury of a million pounds of thrust. "Three at a hundred!" said Kevin over the rumble as the engines shouldered *Endeavour* sideways, steel booster casings flexing under the load. Still bolted to the pad, the twin solid rocket boosters took up the strain, then sprang back to the vertical. The computers raced through engine checks and marked the time: zero.

Wham! Twin boosters ignited and instantly added six million pounds of thrust to the fight against gravity. Gravity lost—just as explosive charges shattered the eight nuts holding the SRBs to the pad. *Endeavour* jolted into the air, sending a crash-bang wallop through the cabin.

A giant mallet hammered my seat from below, and I felt the upward surge. Swiveling booster nozzles whipsawed us left, right, forward, and back, striving for the vertical. It was all I could do to plant a shaky finger on my kneeboard's digital stopwatch and stab it into life, my sole duty during ascent. *The clock is running!*

Seven million pounds of thrust shot us past the top of the launch tower, the entire stack already powering upward at more than 100 miles per hour. Chili called, "Tower clear!" amid a nasty shaking delivered by the turbulent exhaust ripping from the booster nozzles. Each SRB greedily consumed five tons of propellant per second, burning at a scorching 5,000°F.

On the shaking middeck, I felt the cabin whirl as the ship pirouetted around its five rocket engines, picking up our orbital track inclined 57 degrees to the equator. "Roll program, Houston," Sid called, his voice confident amid the vibration. We were "heads down" as we arced over the Atlantic. Although hanging upside down in the seat, I felt only a slight tug headward; most of the acceleration came straight back through my chest. Already at 2.5 g's, I imagined my dad standing next to me, his reassuring hand on my shoulder. *Be not afraid!*

Down on the roof of the LCC, Liz followed the golden trail of flame against dark clouds now backlit by a rising sun. Happy at the moment of launch, she was now subdued as she followed the orbiter's racing ascent. Annie and Bryce snapped pictures with giddy excitement as the thunder of *Endeavour*'s passage crackled through the surrounding air. As the Vehicle Assembly Building's steel walls rippled under the pounding shock waves, my wife failed to notice the very roof shaking beneath her.

I held on as a thin howl from outside penetrated the cabin walls. As we accelerated through the lower atmosphere, I shrank from the mental image of a fiery tornado, half as hot as the Sun, burning not twenty feet away. Just the cabin wall and two thin steel SRB casings stood between me and, well, vaporization. *Think about something else.*

The main engines throttled back to reduce the loads on the stack as we encountered maximum dynamic pressure (Max Q), the peak period of aerodynamic stress. Forty-five seconds—"Mach 1!" from Chili; we had punched through the sound barrier.

"Mach 2, four sixty-two [knots]," Chili noted again as we rattled our way upward. "A little shock coming off the nose there."

The crackle and thunder still rattled the Banana River grandstands, shaking my mom with worry. "Is that normal?" she asked no one in particular. *If my chest is shaking, what's it like for . . . ?* She had yet to bring herself to even look at the rising space shuttle.

Safely past Max Q, the computers commanded full thrust again, and *Endeavour* responded with a will. The thin banshee howl rose to a full-throated roar of unrestrained power. With the engines now at 104 percent power, pumping out over seven million pounds of thrust, I was squeezed firmly back into my seat. *Endeavour* demanded my attention: *Let me show you what it really means to ride a rocket!*

Capcom Ken Cockrell, "Taco" to everyone, called from Houston: "*Endeavour*, go at throttle-up." "Roger, go at throttle-up," Sid replied, and to us: "Feel how we're accelerating!" *How could you miss it?* Taco's throttle-up call triggered the same reflexive thought in every astronaut: this was the point where we lost *Challenger.*

Rich called me from upstairs: "Mach 2 at 1:19, TJ!" Eighty seconds into the flight, *Endeavour* had burned over two-and-a-quarter-million pounds of propellant; the stack now weighed only half of what it did at liftoff.

"Three at one-oh-four," Chili exulted. *Endeavor's* engines were back at 104 percent power. "Good-lookin' motors!"

"We're with you!" is all I could manage under the two-and-a-half-g load. Rich was confident, enjoying himself. "What a beautiful ride."

*Aaahhhh!* The thrust lessened as the boosters hurled their final tons of solid fuel aft into our billowing exhaust. I still felt as if someone were trying to drag me up off the seat through my shoulder straps. Close to two minutes since liftoff.

"$P_c$ less than fifty," Sid noted as the SRB chamber pressure tailed off, close to burnout. I braced for separation a few heartbeats later, a solid, metallic *klong!* as explosive bolts sheared the boosters from the tank under the brief thrust from separation motors. The two 150-foot casings, streaming a shower of orange sparks, fell lazily back toward the Atlantic.

"Okay, we're off the SRBs!" Sid confirmed with obvious satisfaction. "$O_2$ off, visors up!" Automatically I glanced over at Linda, and we grinned with the same relieved stress. No words were necessary. I reached out to my left, and our gloves clasped in a squeeze of encouragement and thanksgiving. I flipped the faceplate upward and sucked in a breath of cabin air. *Thank you, Lord!*

"Some kinda ride, TJ!" called Rich over the intercom, and Linda and I both whooped loud enough to echo all the way to the flight deck. Our friends upstairs laughed right back. Thirty miles up at 2,700 miles per hour: "Wow!" was about all I was capable of saying.

"*Endeavour*, performance nominal," called Taco from Houston. He confirmed that the boosters had done their job during first stage. The main engines continued to push us upward, drinking fuel and oxidizer from the ET at a rate that would empty a big backyard swimming pool in twenty-five seconds.

The howl from outside was gone. We were just a little over 1 g. Ever lighter, we accelerated with almost no noticeable vibration. "Boy, this is electric drive . . . real smooth," noted Sid.

Kevin chimed in: "I got the horizon and I see space out there, boys . . . and ladies! Forty-five nautical miles. What a sight!" The sky outside the windows was completely black, yet sunlight streamed in from the east through Sid's side window. "Tom, you're almost an astronaut," he called down. "Yeah, it's been a hell of a ride!" I said. "It gets better!" Kevin assured me.

Fifty miles high now, and Sid announced with grace and a bit of formality, "Congratulations, Tom, you're now an astronaut!" My friends echoed his welcome words. "Thank you," I called back, and considered the idea. I had just begun this mission and done little but hang on for the ride, but I had just achieved one of the biggest goals of my life.

The g's started to build. Jay called that we were passing Mach 7.2. "And the families are happy," Rich said to all of us, mirroring our thoughts of wives and kids back on the LCC roof, perhaps relaxing just a bit with the smooth progress of ascent. More than a hundred miles behind me, Liz knelt between Annie and Bryce as they followed the brilliant white dot of *Endeavour* against the brightening eastern sky.

Rich spoke up, "Boy, this is a smooth ride."

"*Columbia* was nothing like this," Sid answered.

Although squeezed ever-deeper into my parachute, I felt no trace of vibration. Only a thin metallic hum through the seat betrayed the fact that our three main engines were still rocketing along at 104 percent.

"It's not just *Columbia*," Jay added. "We had a Spacelab on *Endeavour* last time and it was rocking the whole way."

"It's that Golden Gate Bridge we're carrying," answered Rich, referring to the massive SRL antenna. "Mach 10 at five minutes."

Sid broke in just after Mach 11: "Got the sun in the left window. That's great. Good morning, Mr. Sun!"

The altimeter tape showed us barely climbing, nudging 2 g's as the computers calculated the most efficient way to convert remaining fuel into the required velocity and altitude. *Endeavour* stayed at full thrust, angling down slightly to build up speed, tearing along at a pace that would nearly double our velocity in the final two minutes.

"Mach 17 . . . seven minutes. . . . Pulling some g's now," Rich said with a grunt of effort as the g-meter ticked upward toward three times the force of gravity. "There's that 300-pound gorilla," Sid joked. My suit and I weighed nearly 700 pounds, and I had to work to force the air into my lungs.

"Three g's at 7:35," said Kevin. Checking the engine gauges: "Three at a hundred." The four computers throttled back the engines to below 100 percent thrust, keeping the orbiter's lightweight structure below 3 g's. A very long minute remained until we reached orbital velocity.

"Mach 21." Rich grunted out. Nine seconds later, "Mach 22," this time from Sid. "Mach 23; thirty seconds to go," called Kevin. "Mach 24. Here comes the bug." On the backup trajectory display, a small cursor marched across the center screen from right to left: our velocity closing steadily in on the cutoff target. The guidance calculations showed we were just a dozen seconds away from the computer-issued shutdown signal. I sat silent, listening, enduring the g's as I watched the clock tick toward eight-and-a-half minutes since liftoff. The engines were back at idle now; we were heading northeast across the Atlantic at nearly 17,000 miles per hour.

The four general-purpose computers (GPCs) commanded *Cutoff!* precisely on time, and the main engines thundered into silence. We coasted above the atmosphere, going from three g's to free fall in under a second. The pressure on my chest was gone. I felt light under my straps. *This must be weightlessness.*

"Welcome to space!" Rich exclaimed, and our tumbled congratulations echoed through the cockpit even as the big tank departed with a bang and a flurry of thruster firings. Our cutoff velocity was near-perfect. "Welcome to orbit, TJ!" Rich called downstairs again. "I love it!" is all I could answer.

I grinned over at Linda. Had I not been strapped in, I would have kissed her and everyone else on the crew. I conducted my first orbital experiment: tugging off my left glove, I released it to spin lazily, drifting inches from my face—proof we were in orbit.

After cutoff you have to hit the ground running (while trying to avoid bouncing off the ceiling, which I did all too often). Unstrapping, I broke out cameras and passed them from the middeck to Linda and Jay upstairs, who photographed our jettisoned ET to document any foam insulation it may have shed during launch. I considered this just a pursuit of engineering trivia at the time. Back on the middeck, I hustled about, passing needed equipment to the upstairs gang, busy transforming the orbiter from rocket ship to orbiting science platform. I raced to finish our setup work before our first Blue Shift sleep period began five hours after launch.

Although the sensation of free fall was strange, I found it was easy to mentally choose an "up" direction: toward the middeck's ceiling. I had no idea where Earth and true "down" really were. But I suffered no disorientation. The dozens of hours I had spent in the simulators and JSC mock-ups made the middeck wholly familiar: I knew the switch locations, the storage locker layout and contents, and just what to do next. So far, I was coping with space—*inside Endeavour*. I still hadn't had time to look out a window.

I was getting very warm in my bulky LES, and I was relieved to have Linda help me out of that rig when she came back downstairs from photographing the ET. Worn in case of a North Atlantic bailout, my winterweight long johns were soaked in perspiration, and I floated quickly to the fat, flexible fresh-air hose, aptly called the elephant trunk, to bathe in a blast of cold air. In shorts and polo shirt, I joined Linda in freeing our colleagues from their own suits, stowing their ascent gear, and activating middeck systems. First was the WCS (waste collection system, or toilet), followed by critical circuit breakers, the galley, and plumbing for the orbiter's fresh- and waste-water systems. The physical activity and time pressure of post insertion always made it one of the busiest phases of my shuttle missions.

All the while, my body was undergoing a rapid adaptation to free fall. Fluid normally pulled toward my legs and lower abdomen by gravity migrated toward my chest and head. My sinuses began to close, and the WCS shaving mirror reflected my puffy face, flushed as if I had spent a long afternoon in the sun. The headward fluid shift also fooled my arterial sensors into concluding I had too much blood volume, causing the kidneys to kick into overdrive. The most crucial change, I think, was neurological: my inner-ear balance organs, lacking a clear "down" acceleration, began transmitting a stream of confusing impulses to my brain.

These effects rapidly transformed my initial feelings of euphoria. Within an hour I felt a wave of nausea, sensing it just in time to pluck my plastic airsickness bag from a pocket. As the wave crested, I doubled over in an involuntary spasm, caroming across the middeck, eyes tightly closed. For several minutes I drifted, feeling completely miserable. I recovered quickly and was soon back at work, but two more episodes followed, leaving me worried about my long-term effectiveness. It was time for Doctor Rich to make a house call.

At least it would be a fellow Hairball needling me. I braced against a sleep station as Rich stuck me with a hypo full of Phenergan, our anti-nausea drug of choice.

Turns out that in space, they *can* hear you scream. The effect of the shot, however, was nearly instantaneous. My stomach queasiness and general malaise vanished within a minute. "No charge, TJ," Rich said sympathetically as he applied a Band-Aid and administered a slap of encouragement, thoughtfully avoiding my bruised posterior. A noticeable soreness persisted there for days.

Rich left a stream of satisfied patients in his wake. Between my bouts with space adaptation syndrome, I floated about, delivering checklists and equipment upstairs and stowing much of our launch gear in lockers, sleep stations, or the airlock. Unused to free fall, I endured a fair number of bumps and collisions before learning to slow down and move using only finger power. Drifting past the side hatch window, a sliver of light from outside drew my attention. Incredibly, I realized that I had yet to look back at Earth. Palms cupped around my eyes, I pressed close to the window.

We were in darkness. A faint glow—starlight reflecting from the clouds—lit the planet below. As *Endeavour* rode toward sunrise, I gasped. Between heaven and Earth was a vision of pure beauty, the robin's-egg-blue of the atmosphere backlighting the darkened horizon. That luminous rim of Earth heralding the coming dawn triggered a flood of emotion in me: memories of the thirty years I had dreamed of flying in space and the two decades spent working toward that goal. Now those thoughts gave way to gratitude that God had granted my wish. I fought back unexpected tears, said a silent prayer of thanks, and turned reluctantly back to work.

By now, three hours into the mission, Linda was leading Sid and Kevin into the activation procedures for the Radar Lab. While they worked upstairs, Jay, Rich, and I took another two hours to outfit the middeck for life in orbit. We packed away parachutes and helmets, folded and stowed the middeck seats, strapped the escape pole against the ceiling, and set up our cabin air cleaner. But the day had been a long one for the Blue Shift, and I was grateful to slip at last into my coffin-like bunk. I slid the bunk door closed, snapped off the light, and was instantly asleep.

> **MISSION ELAPSED TIME:** Zero days, thirteen hours, and twenty-one minutes (MET 0/13:21). Had breakfast on my first full flight day. Mexican scrambled eggs and a sausage patty, orange juice, even some coffee with artificial sweetener. It was all great. Feeling good this morning: just a little head congestion.
>
> . . . We're just getting a marvelous sunrise over East Africa. The nearly new moon is just above the horizon, with just a thin sliver of a crescent left. It was almost immersed in the delicate blue glow of the horizon. Now the sun is coming up ever stronger.

After just four hours of sleep, the Blue Shift had taken over Radar Lab operations. Rich, Jay, and I were on our own on the flight deck for our first full day in orbit. The usual routine was sixteen hours awake followed by eight hours of sleep. Our shift would normally begin after an hour or so of "postsleep" activity: time to clean up, grab breakfast, and take a short handover briefing from Sid, Linda, and Kevin. Each of our nine work days would be full of radar and photography work, focusing on the more than 400 science targets the SRL team had selected for radar examination. After twelve hours, we would turn operations over to the Red Shift again, taking another three hours for dinner and routine housekeeping before heading off to sleep.

Orbital mechanics determined the pace of our efforts. At an altitude of 120 nautical miles, our ship completed one orbit in just under ninety minutes; sunrises and sunsets alternated every forty-five minutes. Just as we had trained in the simulator, we focused on our three main responsibilities: orbiter pointing, data recording, and photography. Our cockpit bustled with SRL science photography during daylight; night passes were somewhat calmer.

Pointing *Endeavour* to align the radar with our earthbound targets proved to be the most demanding challenge for the crew: we would make a record 495 maneuvers during the mission. Each had to occur precisely on schedule to cover the targets and obtain the highest-quality radar images; each required us to tell *Endeavour*'s computers exactly when and how to fire the forty-four thrusters that controlled our attitude in space. Consulting the flight plan and the printed attitude timeline up-dated by fax each day, we would key the planned maneuver into the com-puter, double-check the start time, and execute. The GPCs would then fire the thrusters at the appropriate time to get us in position for the data take. For large changes in attitude, the big primary thrusters thudded and spat flame, each pivoting the ship with 870 pounds of force. We could see smaller corrections through the rear windows as the tiny vernier rockets with just twenty-four pounds of thrust flickered against the star-filled sky.

Accuracy was all-important: MCC monitored our keystrokes from the ground and usually gave a verbal confirmation that we had loaded the maneuver correctly. Rich, Jay, and I took turns handling the entries and managing the autopilot, generally about once every forty-five min-utes. My fingers never touched the shuttle's control joysticks; all SRL maneuvering was done via the computers, which initiated each thruster firing, slewed the ship around, and stopped us at the proper attitude.

The flat, rectangular SRL antenna looked "up" from the cargo bay, so to point the antenna panels at the ground, *Endeavour* flew upside down with respect to the Earth. We usually flew with a constant 26-degree roll bias from the vertical, aiming the radars out to the left or right of our ground track to image a swath of Earth anywhere from nine to fifty-six miles wide and hundreds of miles long. There was no spinning radar dish out in the payload bay: instead the SIR-C radar beams were steered elec-tronically by the phased-array antenna panels, eighteen each for L-band and C-band, respectively. The German-Italian X-SAR antenna, a nar-row 12.0-by-0.4-meter strip along the upper edge of the SIR-C array, was tilted mechanically by ground command to match SIR-C's radar beam; the X-SAR antenna was the only moving part of SRL we could see.

Another high-priority but low-tech job was recording the precious imaging data. The digital echoes from the three separate wavelengths (designated L-, C-, and X-band in radar terminology) transmitted by SIR-C and X-SAR produced 225 million bits of data (megabits) per sec-ond, far exceeding the orbiter's maximum transmission rate to the ground of 50 megabits per second. To capture this flood of radar im-agery, we carried 166 tape cassettes, each slightly larger than a VCR car-tridge and capable of storing 50 billion bytes (gigabytes) of data. The Radar Lab would fill a tape about every thirty minutes. Each planned tape change was noted on our science timeline faxed up to us daily by the POCC. We set timers on board to make sure we didn't miss a tape swap for the three payload high-rate recorders (PHRRs). The job was about as difficult as inserting a tape and changing the channel on a VCR, but a missed tape would result in irretrievably lost data. We couldn't afford a mistake.

Our most rewarding job was science photography. During daylight passes, Rich, Jay, and I reviewed upcoming radar targets, referenced the kind of imagery required by the science team, and armed ourselves with up to four 70mm Hasselblads and two large-format Linhof cameras. When a target such as California's Death Valley came over the horizon, we had about thirty seconds to acquire it visually, take a light-meter reading, and position ourselves for a clear view through the windows. Snapping madly away from two overhead and a pilot's side window, we would document the weather, surface, and vegetation state on film, then type the general conditions and specific film magazines used into our laptop log. We would also slew the payload bay's color TV camera to match the radar's downward view of the Earth, recording the environmental conditions below on videotape. We looked like a crowd of tourists mobbing the bus windows at a Grand Canyon overlook, but we had more than 10,000 frames of film to expose and were determined to bring back an extensive optical record to compare with the radar images.

The view from the overhead and aft windows was stunning, like staring up through the roof of a greenhouse at a planet filling half the sky. A glance "up" from work on switches and computers at the scenery rolling by just 138 miles below always made me draw a sharp breath. I couldn't screen out the sheer emotional impact of seeing our planet's subtle beauty and glorious complexity. I could never grow bored with this view.

MET 0/15:56. . . . about 9 o'clock on a Saturday night back in Houston [launch day]. We are in a night pass over the Pacific. I'm up here on the flight deck by myself. We've been steadily taking pictures, using all the cameras. X-SAR and SIR-C all seem to be working very well, so we're happy, really happy, about that. The recorders are working beautifully, and I feel pretty good. I'm still a little tentative when it comes to maneuvering in weightlessness.

It's really sort of unreal. It feels like you're in the simulator, except the views are stupendous. We just had a spectacular sunrise a while back with a crescent new moon rising just before the sun.

The night side of our orbit brought new vistas. Jay would darken the cockpit so that we could take long exposures with our Nikons, trying to capture some of the constellations, meteors, and glittering city lights.

**MET 1/00:58.** We've just finished a night pass down the southern portion of the orbit where we saw the Aurora Australis: green fingers forming and pulsing a little and advancing and retreating on the margin.... It's in sort of a large sweeping arc around the South Pole. Receding from the aurora, we could see the jets firing off the tail and the aurora glowing green on the horizon. While I was watching the aurora I was listening to the sound track from *Bladerunner*.... Some nose jets firing: Bump-bump, big thump-thump ... somebody is kicking the orbiter! Linda, Kevin, and Sid are all up on the flight deck getting their SRL ops done, and they're fully into the swing of it now.

I just want to tell Liz, Annie, and Bryce that I'm thinking of them right now.... Maybe they're just getting up and getting ready to go to church. It's Sunday here in space as well, and Kevin has brought along some Communion....

With the rhythm of the radar governing our lives, we fell into a demanding but comfortable routine. The intriguing litany of science targets became reassuringly familiar as we made repeated passes over Earth's varied surface. Soon I was adept at finding Mount Everest, or the gray smudges of frozen cities along the trans-Siberian railroad. After our daily twelve-hour shift, we had time for a relaxed dinner, a few chores, and some recreational Earth viewing before bed.

Our sleeping accommodations were quite cushy by space standards. *Endeavour* carried four sleep stations stacked along the starboard wall of the middeck. Our crew "hot-bunked" in the top three, using the bottom unit as an extra storage locker. Each station had a sleeping bag, reading light, fresh-air vent, and sliding door panel that kept out light and muffled the sounds of fans, pumps, and the inevitable noises from the on-duty shift.

**MET 1/01:24.** I'm getting ready to go to sleep in my bunk. Actually very cozy in here. I can imagine I'm just standing up in this slim little box, but it feels more natural to just imagine myself lying down, floating here. The ceiling is only about six inches above my face. Only about two feet wide, but the "mattress" is very soft. There's a pillow under my head. I don't even need to strap my head down like I did last night. [I felt comfortable with my head just floating off the pillow.] I know I'm just going to drift right off to sleep (I've had a long day with only four hours sleep the night before, but I feel remarkably well for all that).

I woke up hungry on my third day in orbit. The shuttles don't carry a refrigerator or freezer, which would consume too much valuable power, so all food must tolerate room temperature storage. At a tasting session back at JSC, the crew had sampled almost every item on the shuttle's extensive menu. We chose our individual menus, which were then approved by a nutritionist. At mealtime I'd slide a food package into the galley's rehydration station, puncturing the plastic pouch with a blunt needle that delivered the desired amount of hot or cold water. Military-style "Meals Ready to Eat" (MREs) were even simpler to prepare: the foil pouches emerged piping hot ten minutes after being tossed into our small convection oven. The galley also dispensed water to fill our drink bags, dissolving the powdered drink mix or coffee inside the silver foil pouch. Insert a plastic straw, give a quick shake, and presto—a chilled fruit drink, hot chocolate, or delicious Kona coffee.

A spoon sufficed for all meals, as the food was sticky enough to make the trip from package to mouth without much trouble. With reasonable care, even soups could be eaten this way, and I quickly learned how to dine neatly in free fall.

**MET 1/00:58.** For dinner tonight I had green beans with mushrooms (good), chicken with noodles (good), tapioca pudding, raspberry yogurt, and a couple of fluid bags of orange drink with artificial sweetener. Those were all very tasty; it was an excellent meal. My appetite is fully back. Had a couple of Tastykakes today, too.

Although all of us found space cuisine agreeable, the galley itself was giving us some trouble. Its rehydration needle was delivering a liberal dose of air bubbles with each shot of hot or cold water. Both Sid and Kevin ingested enough air with their food to cause stomach problems, and without the downward tug of gravity, trying to burp out the excess air was surprisingly difficult. The gas pains ruined their appetite for food and drink, and if the problems continued, dehydration could potentially force an early end to the mission. Sid started to weigh our options even as MCC tried to develop a repair method. Something had to be done.

Acting on a tip from the flight surgeons, Kevin managed to get bubble-free water by twirling a big water bag around his wrist, flinging the denser water to the outlet end of the bag while leaving most of the bubbles behind. We admired his ingenuity, but after a couple of days, the moist interior of this thick plastic bag, designed for use inside our space suits, began turning an algal shade of light green. Although *his* appetite was slowly returning, the sight of his water bag did little to improve those of the rest of the crew.

Rich, Sid, and Kevin finally determined that the galley's water injection needle was punching too large a hole in the food packages, allowing air to be drawn in as well. The solution suggested by Mission Control was to seal that gap with improvised rubber washers; adding water slowly to the food packs also minimized the air intake. By the fourth flight day the galley was back in business.

A daily highlight aboard was the arrival of a batch of e-mail from the families; our answering messages went down once a day to Mission Control with our Earth science logs.

LETTER HOME, FLIGHT DAY 3:
*Dear Family:*

*I pulled out all your pictures and gazed at them longingly. It seems I'm so far away while I'm whirling around the globe. I've seen some gorgeous sunrises and sunsets, the aurora, the stars especially bright, and some terrific scenery over Asia. My thoughts and longings are for all of you, and I am looking forward to our reunion in another week. . . . I love you and miss you.*

By the fourth day of the flight those feelings of loneliness passed, along with my stuffy nose. My spinal vertebrae, no longer compressed by my weight, had stretched about an inch and a half; the resulting backache had vanished after a few nights of rest. The radar work was a pleasure, particularly our Earth observations and the radio interactions with our JPL colleagues, our link to the science team around the globe. The SRL people were ecstatic about the radars' performance, detecting features as small as sixty-six feet across. We filled tape after tape with high-quality imagery. MCC faxed a few samples of the pictures up to us, along with data from the MAPS carbon monoxide experiment, which corresponded well with the smoke and fires we were seeing from orbit. The shuttle radar passes, pollution numbers, and ground truth measurements were stitching together a detailed portrait of Earth, exceeding the best hopes of SRL's designers and investigators. Aside from our galley problem, *Endeavour* had experienced only a minor glitch or two.

LETTER FROM HOME, FLIGHT DAY 4:

*From Annie: Hey Dad, how's the flight going? I've been getting those messages of yours. Mom, Bryce, and Vesta [our cat] are waiting to see you again. I almost cried when you said in one of your messages that you were looking at our surprise pictures longingly. That is what made me almost cry. Love, Annie*

*From Bryce: I hope you had a good flight. My favorite lizards are iguanas. I like my ambulance that you got me, and I love you, Daddy. Mostly at Sweet Mesquite, the people give me too much ice cream.*

LETTER HOME, FLIGHT DAY 6:

*Annie, sorry I'm making you cry, but I understand how much you miss me. I feel the same way about you, too, up here. Bryce, why do you like iguanas all of a sudden? I'm flying over some islands where really big ones grow.*

Although I loved the quirky agility granted by free fall, the thrill of helping operate a spaceship, and the satisfaction of looking down on places like the Galápagos Islands, perhaps the best part of this adventure was sharing the space experience with my crewmates. In fact, I regretted how little we saw of the Red Shift gang, our time together limited to the hour or two of overlap at the beginning and end of our workdays. When work permitted, we would gather on the flight deck to share a snack or trade stories and new discoveries. Sid also called a daily crew meeting to review the latest news from MCC, check on everyone's health, brainstorm solutions to any technical problems, and keep both shifts on the same wavelength.

*Endeavour's* cabin had become a familiar and reasonably comfortable home. Six or seven hours of sleep would get me through the following workday with ample energy. The day's flight plan would be faxed aboard by MCC first thing in the morning, along with the latest updates to our maneuver and science timelines. Jay, Rich, and I would review those over breakfast, then hit the flight deck for our twelve-hour shift.

During the forty-five-minute orbital "nights," when darkness made science photography impossible, Jay, Rich, and I took turns squeezing in a workout on the ship's exercise bike. With my shoes clipped into the pedals, I'd ride the cycle ergometer much like a recumbent exercise bike, cranking away for thirty minutes or more. With a Walkman drifting above my head on its headphone "tether," I found it easy to pedal 8,000 miles in half an hour. No need for a seat: a cloth belt held me loosely against the padded backrest, my body floating comfortably a few inches above the middeck floor. The strangest thing about exercising was the free-fall behavior of perspiration. With no convection to carry heat away and circulate cooler air around the body, evaporation was much less efficient. Sweat coated every inch of my arms, chest, and face as my body strained to keep cool. The only solution was to wipe down frequently with a towel, also using the elephant trunk to blast a stream of cold air at my face and torso.

LETTER HOME, FLIGHT DAY 5:

*I ate well and even exercised on the bicycle today. It took an hour and fifteen minutes to exercise and get cleaned up again. Wait til I tell you the gruesome details of living and keeping clean up here!*

Bathing was easily if not luxuriously done with a washcloth, hot water, and rinseless soap; if Linda happened to be awake, we would just ask her to avoid our middeck locker room for a few minutes.

The few inconveniences of life in orbit were easily forgotten when we had time to share a meal, compare e-mails from home, or spend a quiet half-hour looking at Earth together. On the middeck we greatly expanded the scientific body of knowledge on the behavior of food in free fall. My trademark "flying saucer sandwich" was an irradiated chicken breast glued between two tortillas with picante sauce, spun lazily across the cabin to a lucky diner. Rich carefully dispensed water or orange juice from a drink bag to create tennis ball–sized globules, examining their undulating surfaces as they drifted slowly across the cabin. We would take turns sucking up the blobs with a straw before they could collide disastrously with any of several real science experiments running on the middeck.

During one crew dinner, I fashioned a blowgun from the plastic tube left over from a roll of fax paper, firing Whopper malted milk balls at my crewmates, impromptu targets in a cosmic shooting gallery. Playing the carom from a ricocheting Whopper and snaring it in midair took real skill. Rich helped us explore the mental dimensions of free fall, coaxing us to sit, dangling our legs inside the circular opening of the airlock hatch. Facing each other, with a little effort we could imagine ourselves sitting relaxed around the rim of an orbital hot tub. If I focused on my friends' faces and not the inconvenient orientation of the middeck walls and floor, it was easy to make the mental transformation to a new "vertical." In free fall, "up" and "down" were choices, not externally imposed truths.

During our twelve hours on Blue Shift duty, we were so focused on maneuvers, Earth observations, and tape changes that we often lost track of our crewmates' activities downstairs. Once they handed over the flight deck to Jay, Rich, and me, the Red Shift was out-of-sight, out-of-mind. I would wish my friends a good night's sleep, turn back to work, and promptly forget about them.

So it was with some surprise one morning that I heard Rich whispering to me from the interdeck access opening in the flight deck floor: "Hey, TJ! Get down here and see this!" I floated down headfirst through the square opening and nearly laughed out loud in amazement. Chili was asleep but nowhere near his bunk, drifting across the middeck with eyes closed, arms and legs in the fetal position. We guessed he had been typing a note home on the laptop and, exhausted, just dozed off. Rich and I gently corralled our pilot. Kevin murmured for an instant but never opened his eyes as we quietly tucked him into his bunk and slid the panel closed, wishing him sweet dreams of Cathy, Madison, Mary Cate, and Megan.

In addition to the daily e-mails uplinked from home, each of us had an opportunity during the mission to speak directly with our spouses and children. Piped into a conference room near Mission Control, the radio-only link (the TV channel was tied up sending radar imagery) offered Liz and the kids a ten-minute conversation with me.

Through the American Radio Relay League, each of us had also arranged by prior coordination to contact our homes via the small Shuttle Amateur Radio Experiment (SAREX) set we had aboard. *Endeavour* was in darkness, passing northeast across the Pacific, when I transmitted our ham radio call letters into the ether through the small antenna temporarily mounted in Sid's side window. A strong signal brought the voice of a Hawaiian ham operator aboard; he immediately dialed Liz, and we had about five precious minutes of conversation, Hawaii to Houston, before the islands disappeared over the horizon and I lost the wonderful sound of her voice. I wished I could touch her, and for a moment I was keenly aware of the daunting challenge still ahead in getting safely home on a spaceship 120 miles up, circling the Earth at Mach 25.

Just two days after launch, I used SAREX to honor my dad's memory with a call back to Baltimore. He had been a counselor at Deep Creek Middle School for more than twenty-five years, and a crowd of students had gotten up at four in the morning to be there when my call came in. Over the ham radio set up in the cafeteria, we had a good ten minutes of questions and answers about life in space as *Endeavour* sailed up the East Coast in the predawn darkness. Some of Dad's old colleagues shared with the students and me some good memories of the man who had so encouraged my quest to fly in space.

Our mission's rare combination of low altitude and high orbital inclination gave us an extraordinary view of our planet, carrying us from the latitude of the Aleutians to South America's Cape Horn. Our orbital vantage point, half as high as that of a typical shuttle mission, revealed an impressive swath of the globe. Each ninety-minute revolution produced both day and nighttime views of Earth's wonders.

I've been stunned by the beauty of the Earth, and how beautiful the sunrises and sunsets are. It's not dramatically different from being in a high-flying airplane.... It marvelously appears to me that the Earth is unrolling in front of us, and scrolling by. It's just an amazing way to see the planet....

**MET 3/16:30.** We're on a night pass over the South Atlantic, and Jay's snapping away at Sirius and Orion setting over the tail, getting some thruster glow. And we've been looking at the setting Moon and Venus in the blue band on the horizon. I just saw a meteor below us off the coast of South America. We're seeing an ionized gas trail behind it—something to see! Jay has pointed out the noctilucent clouds. [Saw] a couple of more meteors, and . . . the atmosphere and sky are incredible. You can almost see a thumb-width into the atmosphere between the star going into what you perceive as the top of the glow of the atmosphere, and the surface itself. Great stellar occultation theory being demonstrated here.

**MET 4/03:50.** We had a great aurora pass in the far South Pacific, south of New Zealand. At times we were flying right through the long, thin streamers of the aurora, projecting straight up through the atmosphere, a very ghostly pale yellow green. . . . So we could see these long streamers going up above us, but at times we flew right over the long shimmering arcs of the aurora. We could see the shimmering curtain below us, and when we flew over the top of it, it would become edge-on to us, and we could look straight down on this line of the aurora. Just a fantastic sort of ghostly sight.

**MET 6/06:14.** A day full of another set of sights and wonders. Saw some marvelous aurora . . . swirling all around the shuttle and below us with big streamers of green going up off the atmosphere and into the sky . . . some pulsating curtains and rippling ribbons of light.

**MET 10/20:24:57.** OK, here comes . . . I see Newport News, and James and York River and now the night lights go up to Richmond. There's Fredericksburg. OK, here's Washington, D.C. Oh! Just the whole Dulles airport area is laid out. I can see where my house used to be. . . . And there's Baltimore, Middle River . . . and—oh!—and we're coming out right now [directly above] New York City. If I look back over my head out the tail I can see the entire Midwest string [of cities]. You can see Manhattan Island distinctly. Brooklyn! There is the Hudson going up towards West Point. . . . There is Long Island Sound. Oh, my! Boston is a little bit cloudy. Just an incredible night pass.

. . . Now the eggshell blue light of the sunrise is coating the horizon. The payload bay is now going bluish white as we come up out of the darkness. Across Nova Scotia now, and Labrador, and still no sunshine visible. I can still see the stars. No, not for long. Here comes the orange of the sun. Boom! Sunrise! Now the payload bay is pink-orange, yellow, going to white and it will soon be brilliant. Fantastic!

I thanked God each night before falling asleep for these glorious views of Earth and for the success of our mission thus far. I asked for the continued safety of our crew and a joyful reunion with our families. I was conscious of the special gift of each day in space, aware of the unique privilege I had been granted. And I remembered Father Tom Bevan's words on the beach back in Florida. When Sunday rolled around again, two weeks after Easter, it seemed particularly appropriate to share our thanks and thoughts on what we had seen. Sid, Kevin, and I—all Catholics—gathered on the flight deck one orbital night for a short Communion service.

Kevin, a Eucharistic minister, carried the Blessed Sacrament with him, the hosts protected within a simple golden pyx. The three of us thanked God for the views of His universe, for good companions, and for the success granted our crew so far. Then Kevin shared the Body of Christ with Sid and me, and we floated weightless on the flight deck, silently reflecting on this moment of peace and true communion with Christ.

As we meditated quietly in the darkened cockpit, a dazzling white light burst through space and into the cabin. Pure radiance from the risen sun streamed through *Endeavour*'s forward cockpit windows and bathed us in its warmth. What else could this be but a sign?—God's gentle affirmation of our union with Him. Drifting parallel to the floor, I rolled away from my crewmates, embarrassed at my reaction to that singular sunrise. Through tears I looked instead through the overhead windows at the Pacific far below, the dawn painting its surface with a rich, limitless blue.

"Look at that," I called out almost unconsciously to my friends. From the living water below, we drank in hues unmatched by the palette of any human artist. After a moment, Kevin said simply, "It's the blue of the Virgin's veil, Tom." He was right. He had found the perfect way to express the vision we were seeing out the window.

*   *   *

LETTER HOME, FLIGHT DAY 9:
> *The whole crew is awake now and the cabin atmosphere is light and full of laughter. Rich is a great guy to fly with, believe me. We are enjoying it together. No big news; just another productive day in which I didn't make too many mistakes.*

On the tenth day of the flight, we deactivated the Space Radar Lab. The science team was ecstatic: in 939 data takes, totaling 65 hours of radar operation, SRL had scanned 5.4 percent of Earth's surface, filling 166 tape cassettes with 47 trillion bits of radar imagery. Printed out on paper, those radar images would fill 20,000 volumes of an encyclopedia documenting the changes, both natural and man-made, affecting our planet's surface. The MAPS experiment had logged 211 hours of observations on Earth's carbon monoxide abundance, fulfilling 100 percent of its mission objectives. Through nine days of radar operation, we had had only two minor failures: one of SIR-C's eighteen C-band antenna panels, and the cockpit's Number 3 payload recorder. SRL-1 had proven an unqualified success.

Now we had to get all that data—and ourselves—home. From low Earth orbit, circling the planet at five miles per second, I had paused during the flight to reflect that *Endeavour* would have to lose 98 percent of its present velocity to get us onto final approach to the Cape runway. The shuttle would do that via a sizzling encounter with atmospheric friction.

The flight controllers' efficient use of *Endeavour*'s electrical power had given us an extra day in space for the Radar Lab's imaging operations. The Blue Shift spent most of that tenth day getting ready to come home: powering down the Radar Lab, stowing loose equipment, and preparing the cabin for entry. Just hours after the Red Shift awoke, the six of us suited up and prepared for our deorbit rocket firing planned for about ten days, five hours MET.

After our long last shift preparing for reentry, I was tired and anxious to head home. But the Cape weather was too cloudy and windy for a safe landing. Next morning, I recorded this note onboard:

MET 10/21:30. . . . we were within a half hour of the deorbit burn when they [flight controllers] decided to stop and wave off for one more opportunity to see if the weather improved in Florida. . . .

[T]he weather was getting worse, so we sat there in our suits for about another orbit waiting for the [conditions] to improve. When they did not, we were forced to back out of the deorbit prep procedures and get ready for a one-day extension.

It was rather disappointing... knowing that we were going to cancel the landing attempt. [We were] not able to get out of the suits and do anything until they gave us the official word from the Cape via the capcom, Ken Cockrell.... We got out of our suits, had a meal, and went right to bed.

I think we'll definitely land somewhere today, hopefully Florida where our families are, but if not there, then Edwards AFB, because the weather should be good there. They want us on the ground today.

That was our hope, too. Our scientific work in orbit was complete. More important, we were scraping the bottom of our pantry barrel. But for me, the real incentive for entry was my family: I was lonesome for Liz and the kids. I wanted to get back to them, to describe and share this adventure face-to-face.

After the gift of an extra twenty-four hours in orbit, I watched from my flight deck seat (as planned, Jay and I swapped seats for entry) as our two orbital maneuvering system engines flared into life behind us. Burning for more than two minutes above the Indian Ocean, the rocket firing reduced our velocity by 160 mph, just enough to drop us back into the upper fringes of the atmosphere as we raced toward a Pacific dawn. Still thwarted by low clouds and gusty winds at the Cape, our destination was Edwards.

Sitting just to Rich's right, I followed along in the entry checklist as Sid, Kevin, and Rich monitored *Endeavour*'s autopilot-guided trajectory. The first traces of the rarefied upper atmosphere whipped silently past our spaceship, kindling outside the thin tendrils of the glowing plasma that would eventually envelop us during our hypersonic descent. With atmospheric friction slowing us almost imperceptibly, I felt my body settle gently into my seat. A semblance of gravity returned to our inner ears, equipment, and spacecraft.

*Endeavour* fell swiftly, soon dropping into the heart of the entry heating regime, seared by outside temperatures approaching 3,000°F. Angling my kneeboard mirror in my lap, I stared mesmerized out the overhead window as the flickering plasma coalesced into a brilliant white-and-yellow comet tail. To any observers on ships or isolated islands below, we must have appeared as a brilliant meteor racing across the predawn skies. As *Endeavour* endured a near-constant 1.5 g's of deceleration, it was all my weakened muscles could do to heft a camcorder to eye level and record portions of our forty-five-minute fall toward California.

Hurtling into sunlight well off the coast, I glimpsed a glittering blue ocean and fluffy clouds through Sid's windows; they appeared startlingly close and our velocity alarmingly high. Over Kevin's shoulder, I caught a glimpse of the snowcapped Sierras as we whipped over the California beach at Mach 5. Los Angeles was off to Kevin's right, the green of the central valley streaking by on our left. "Eighty-five thousand feet, Mach 3. Here comes the roll!" called Sid. "Tally-ho on Edwards!" Our commander had the runway in sight.

I strained to keep my heavy helmet from tugging my head down onto my chest. We were falling like a brick now, trading altitude for airspeed as *Endeavour* raced toward the dry lakebeds surrounding Edwards. Sid and Kevin were cool and confident. They had practiced this together a hundred times in the simulator, and it showed.

"Mach 1.6, 62,000 feet," Kevin called. *Endeavour* began to buck and vibrate, a shaking nearly as violent as that from the boosters eleven days ago. "Mach 1, there's the Mach buffet," confirmed Sid. Desert slid past the cockpit windows as Sid, right hand gripping the joystick, took over manual control.

"Starting to come left," he told us.

"Making the turn nicely," Kevin confirmed. "I see you correcting up, nice correction. . . . What a flying machine!" he exulted.

"Runway in sight," Rich chimed in.

"Tally-ho on runway two-two," Sid confirmed. Our commander rolled the orbiter out on final at 12,000 feet, plunging nose down at 300 knots toward his aim point.

"Out of 7,000! Now through 5,000, good radars," Kevin called, his soothing patter to Sid narrating for the rest of us the accelerating pace of events in the cockpit. As we slowed to approach velocity, Sid was dead on: *Endeavour* was headed straight for the end of the runway.

"Two thousand feet, preflare next."

"Preflare, arm the gear," Sid commanded, beginning his gradual transition to the landing attitude.

Kevin rapped out the altitude calls: "1,000 . . . 800 . . . 700 . . . 600 . . . 500 . . ."

"Gear." Sid called.

Kevin punched the button. "—Gear's moving," he confirmed. His steady litany of altitudes and airspeeds paced *Endeavour*'s rapid descent as Sid brought us skimming over the dry lakebed. Timing his final flare just over the runway's approach end, he focused outside now on the "ball-bar" glide path indicator, bleeding away airspeed and settling the last dozen feet.

"Touchdown, 215!" Kevin confirmed.

"Shucks!" Sid chided himself for not shaving a few more knots off the airspeed before the main gear touched the concrete. But it was a gentle landing, one barely felt by the rest of the crew.

When Sid halted *Endeavour* on the runway center line with a final squeeze of the brakes, the cockpit erupted with handshakes and congratulations. The six of us were giddy with relief at being down, and exuberant at having pulled off the mission with our friends in MCC and on the science team. Linda's "Great landing, Sid!" was echoed by all of us. Voice tinged with justifiable pride, Sid keyed the mike to MCC. "Houston, *Endeavour*. Wheels stop!"

"Roger, wheels stop, *Endeavour*," answered Taco. "Sid, your Radar Lab has provided an unprecedented view of Earth, and you and your crew have been a joy to work with."

Sid matched the gracious call with his own: "Thanks, Ken, and thanks to the payload folks. All their years of hard work have paid off."

Rich looked over at me with the biggest grin I'd seen on the entire flight. My elation matched his smile. We shook hands, then opened our checklists to help the pilots through the postlanding cockpit checks. We were back on the home planet.

# 8
# THE ONLY MAN AVAILABLE

*Do you deserve this, this fantastic experience? Have you earned this in some way? . . . the answer to that is no.*
RUSSELL SCHWEICKART, *APOLLO 9*, WRITING IN 1988

rawling through the orbiter hatch, I practically fell into the waiting arms of the flight surgeon and suit technicians. After eleven days in space, my launch and entry suit felt like lead, but at least I had managed to make it down the ladder from the flight deck. Inside the transporter van, I peeled off my sweaty long underwear and changed into a clean flight suit.

Edwards Air Force Base wasn't the prime landing site, so formalities were kept to a minimum. After resting and sipping cold water for about thirty minutes, we all felt well enough to follow Sid outside for the traditional orbiter walk-around inspection. Still radiating heat from her Mach 25 passage through the atmosphere, *Endeavour* was in superb shape. Our impromptu host at Edwards was Rich's old STS-53 commander, Dave Walker; he became my friend for life by bringing me a cheeseburger and fries from the base Burger King while I grabbed a quick but luxurious shower.

Less than four hours after touchdown, the flight surgeons bundled the six of us, along with our exposed film magazines, into a Gulfstream II for the run back to Ellington. Technicians extracted the precious SRL data tapes from *Endeavour*'s cabin and hurried them off to JPL for image processing. I had been up nearly twenty hours by then, and I desperately wanted some sleep. Catnapping on the floor of the Shuttle Training Aircraft, I wrestled with my reintroduction to gravity. The feeling of extreme heaviness had passed within an hour or so, but I had boarded the STA with an uncertain gait. A few hours before I had been well adapted to free fall. Now, exhausted and back in one g, my sense of balance was shot. I needed my eyes in order to stand erect.

Taxiing to a stop late that evening on the floodlit tarmac in front of NASA's Hangar 276, we wobbled down the steps of the STA to meet not our spouses but the new JSC director, Carolyn Huntoon. Protocol satisfied, we turned at last to greet our families. As I looked for Liz, two torpedoes named Annie and Bryce hit me so hard that they nearly knocked me back to Edwards. As I held on to those two, there was Liz: it was so good to hold her in my arms again after two weeks of tenuous e-mail and radio contact.

After wading through a joyous welcome from our Houston friends, NASA colleagues, and SRL teammates, a NASA van drove us off the Ellington flight line under police escort. Home at last, I collapsed into bed, uneasy under the unfamiliar pressures of mattress, sheets, and pillows. Still tired, I rose the next morning to a normal Earth breakfast (food on a plate—what a concept!) and terrestrial realities like taking out the trash and catching up on mail. All of us on the crew savored the day off with our families . . . except Jay, who headed into the photo lab as soon as the first of our 12,000 Earth shots began emerging from the processors.

The rest of us followed the next day, and during the following week the STS-59 crew became regulars over at Building 8, scrolling through the thousands of frames of Earth imagery and in-cabin photography. We culled the top seventy-five or so for public release, then ran a gauntlet of NASA debriefing sessions. I had been forewarned by Rich and Linda, but I was still surprised by the heavy postlanding workload: except for that first day off, our schedule was almost as busy as during our training days, full of debriefings, the writing of our crew report, and the inevitable Astronaut Office paperwork. But it was already difficult to focus on the details of Space Radar Lab 1—after all, there were less than four months until my next liftoff. T-zero was set for August 18, 1994.

I recalled my surprise the previous summer when I met with Chief

Astronaut Robert "Hoot" Gibson. I was still eight months from launch on SRL-1, yet he had just invited me to fly again on the *second* flight of the Space Radar Lab. SRL-2 was scheduled for liftoff less than four months after the first Radar Lab landed. Hoot laughed at my startled reaction, but he wasn't kidding. What else could I say but yes?

### PAYLOAD COMMANDERS NAMED FOR FUTURE SHUTTLE MISSIONS

*From NASA Press Release 93-140, August 3, 1993*

Astronauts Tamara E. Jernigan, Ph.D., Thomas David Jones, Ph.D., James S. Voss, and Ellen Ochoa, Ph.D. have been named payload commanders on upcoming Space Shuttle missions.

. . . Jones, 38, is Payload Commander on the STS-68 Space Radar Laboratory-2 (SRL-2) mission scheduled for late 1994 aboard Atlantis. SRL-2 will take radar images of the Earth's surface for Earth system science studies, including geology, geography, hydrology, oceanography, agronomy, and botany.

Jones has a doctorate in planetary sciences and is a mission specialist on SRL-1 in the spring of 1994.

That was eight months ago. Now it was early May. The ink was barely dry on our SRL-1 crew report, and I couldn't imagine spaceflight without Jay, Kevin, Linda, Rich, and Sid. But my new colleagues, though polite, would only reluctantly share my allegiance with another crew. Even while I was in orbit on SRL-1, they had dropped subtle hints:

Dr. Jones—

*Great job! We've really enjoyed monitoring the mission.*
*Press coverage has been very positive. Can't wait to join you in*
*August.*
*Best of luck. Don't forget you start sims with us next week.*
*Your STS-68 Associates*

After a quick postflight getaway to Hawaii with Liz, I moved out of the STS-59 office and carted my stuff down the hall to join the STS-68 crew. It was a bittersweet experience leaving my first shuttle colleagues behind, but I knew how important it was to become an effective member of the new crew. I'd need every minute of the four months remaining until launch to achieve that goal.

Why had Hoot assigned me to SRL-2? JPL had lobbied the Astronaut Office to have its own radar experts fly on both SRL missions. They argued that a pair of payload specialists dedicated to just these two flights would increase the prospects for a successful mission. When Flight Crew Operations maintained that career mission specialists could handle the job, JPL requested that one or more SRL crew members fly again on the second mission to take advantage of hard-won experience.

Meeting JPL's request wasn't a simple matter for Hoot. Sid planned to retire from the astronaut corps after STS-59, and Kevin had now flown twice in two years and was due both a rest and a promotion to commander. Jay, who had chalked up three missions in three years, was also due for a breather. Similarly, Rich's two flights in less than three years meant that he and his family deserved to see him at a desk job for a spell. Finally, Linda had been tapped by Hoot to be the next deputy chief of the Astronaut Office, effective immediately. With my experience as her science backup on SRL-1, I was a logical choice for the job, but I was also the "only man available."

Assigned as payload commander in August 1993, I had to wait until October before I learned who my crewmates would be. The newly named STS-68 crew couldn't have pleased me more: there would be *three* other Hairballs on the flight to keep me company.

Our commander, Navy Captain Mike Baker, a Californian born in 1953, was a twice-flown veteran and would lead the Red Shift on SRL-2. (I'd met Bakes, as he was known, and his wife, Deidre, when I had served a year earlier as family escort for his STS-52 mission.) Terry Wilcutt, our pilot, a forty-four-year-old Marine test pilot and former Kentucky math teacher, was a Hairball making his first flight; he had drawn a great mission to kick off his spaceflight career. Jeff Wisoff, thirty-six, a laser physicist from tidewater Virginia, was an irreverent yet savvy bachelor who had been teaching at Rice University when selected for our 1990 class. On this, his second mission, he would round out the Red Shift and lead the pilots in Radar Lab science operations.

I was joined on the STS-68 Blue Shift by fellow Hairball Dan Bursch, thirty-seven. Dan, a 1979 Annapolis grad and flight test engineer, had just returned in the early autumn of 1993 from STS-51. His recent space experience made him a great choice to manage orbiter operations on the Blue Shift. Steve Smith, thirty-five, was a Stanford electrical engineer and veteran MCC payload officer, the first member of the 1992 astronaut class ("The Hogs") to be assigned. Steve was smart, outgoing, and indefatigably cheerful.

My new crew must have wondered how I was going to handle the six months of divided attention and the pressures of my first flight. After all, when Hoot assigned me to SRL-2, I had no track record in space. I might prove either incompetent or discover spaceflight was not my cup of tea. After training with me for eighteen months, I suppose Linda may have reassured the chief about my abilities. At least the STS-68 crew knew that my flying SRL-1 just four months earlier would be the best possible training for the second Radar Lab.

This second Space Radar Lab flight was largely a continuation of the work of SRL-1 but with some ambitious new wrinkles added. JPL had always planned to fly the new instrument at least two and preferably three times, testing the radars' ability to monitor seasonal variations in rain forests, croplands, wetlands, ocean currents, sea ice, soil moisture, glaciers, and snow cover. Similarly, the MAPS team wanted to track seasonal changes in carbon monoxide production, watching for shifts across the globe in industrial pollution and large-scale biomass burning. JPL would have preferred six months between missions to capture a full seasonal swing, but competing demands on the shuttle schedule moved SRL-2 up to late summer.

I put in an intense hundred days training with my new crew, applying the lessons learned from SRL-1 to our orbit operation simulations. Before I knew it, it was time to go:

### LAUNCH DIARY, AUGUST 18, 1994:

**5:52 A.M.** When you're lying on your back, your head gets a little bit stuffy and your feet tend to get a little tingly unless you keep moving them around. Of course they're up in the air a little bit higher than your body so they have some poor blood circulation. It's hard to find a good comfortable position for your feet. My back feels fine although there's a limit to how long you can lie here on your back [on the hard parachute]. Head's propped up at just the right angle . . . you keep moving your arms and legs around so that you don't get cramps. But the back's good and we only have about another 50 minutes to go. . . .

We are at 6:16 a.m. and just about 40 minutes until launch. They're still working outside the hatch but that seems normal. And we're getting a little bit of light out on the horizon. Things are still cooking down on the count. Feel optimistic down here. Jeff and I are playing rocks, scissors, paper—just like my Dad used to do. . . .

O.K. it's 15 minutes to launch . . . have our gloves on. We're in the T-9 minute hold. Just polled the launch team at 6:41—everyone is Go. . . . *Everyone* is Go. SRO [Supervisor, Range Operations] just gave us a "range clear to launch." Launch Director is ready to proceed. Payloads are Go. Engineering is Go. Safety and Quality is Go. . . . Weather has no constraints to launch. Bob Sieck [is] talking to MMT [Mission Management Team] now. Loren Shriver [former astronaut and head of the MMT] says we're clear to launch. Launch Director is talking to us, and Mike says we're ready—"Thanks for getting the vehicle ready for Mission to Planet Earth." "Go to proceed" from Bob Sieck, and we're getting ready to come out of the 9 minute hold. Countdown clock will pickup in one minute. . . .

Coming out of 9 minutes things are going to happen really fast. . . . Butterflies are starting to come—or maybe it's because I haven't eaten very much today. But we're feeling good . . . saw the White Room go away just now. Seven minutes to go. Terry's doing an APU pre-start. OK, we're 5 minutes till launch. . . . Definitely some butterflies. APU start is coming on; Terry's doing it . . . and we can feel the vehicle shaking as the APUs crank up. We're at 4 minutes now. Here comes the vibration of the vehicle: flight control surfaces moving, engines are cycling now with the hydraulics. OK, we're about . . . three-and-a-half minutes to go . . . two-and-a-half minutes to go. I'm going to be closing down here until we get well into the ascent. Gonna' have to close my visor. . . .

Liz, with the kids and escort Rich Clifford, was watching again from the launch control center roof. In the growing light of dawn, she saw the gout of orange exhaust flare beneath the orbiter and saw the steam billow from the flame trench as the engines spooled up to full power. Waiting for the liftoff, she heard the NASA announcer's excited voice: "We have a Go for main engine start . . . we have three main engines running! Three . . . two . . . one . . . and . . .

"We have main engine cutoff! RSLS safing is in progress! We have a cutoff of the main engines." The Redundant Set Launch Sequencer, an automatic software routine controlling the final count, had intervened to prevent booster ignition and liftoff.

As the rolling thunder of ignition rumbled past the LCC, Rich hurried to my wife's side: "You know what's going on, don't you?" he said. He wanted to make sure Liz understood the engine roar was not an explosion, just the delayed sound of the distant engine start. But Liz never heard the truncated roar. Her first raw emotion was anger at the prospect of going through all the launch preliminaries and another walk to the roof. But her anger quickly gave way to understandable concern.

In the spring of 1993 I had watched a similar pad abort from Banana Creek while escorting the STS-55 extended families. After that last-second shutdown, I had sprinted to the public address microphone to give some assurance that the astronauts were safe and coolly working through their checklists to complete the shutdown. It had been tough to outrun relatives' racing imaginations.

From my seat in the middeck, I noticed the gantry outside the hatch window seemed to be swaying back and forth; it was actually *Endeavour* in motion, still rocking from the "twang" induced by the main engines' initial shove of a million pounds of thrust. Jeff and I hurriedly unstrapped in case we had to abandon ship and head for the slide wires.

While we prepared to exit the orbiter, launch control confirmed a safe shutdown. There was no fire or danger of explosion, and Jeff and I relaxed a bit, commiserating over our bad luck.

Well, we just scrubbed at T-1 second to go! We had an Engine Number Three HPOT [high pressure oxygen turbopump] temp sensor "A" fail just prior to SRB ignition. I saw the flame outside the window—orange and red. I was waiting for the boosters to shove us off the pad but instead we got a master alarm.

Let's see what Jeff Wisoff, astronaut, thinks about that: "We should've been gone!" Yeah, all our families are out there going, "Holy smoke!"

...We were right down there, ready to go. I was ready to see the boosters light and I was looking out the hatch window waiting to see motion—ready to punch my button on the clock, and then we got the master alarm and the engines shut down. And we all know what that means—that's a three week turnaround, minimum.

Thirty minutes after the alarm, I was dangling my legs over the edge of my horizontal seat back while munching a peanut butter and jelly sandwich and waiting for the pad crew to open the hatch. We teased Dan Bursch mercilessly about his role in our misfortune: on his STS-51 mission the previous summer, he had endured four launch scrubs, including a last-second pad abort. (His crew joked that their favorite movie was *Groundhog Day*.) He lamented that no one would bother coming to another of his launch attempts.

After lunch with the families at crew quarters, we headed out of quarantine to Cocoa Beach. I was soon sipping margaritas in our family's beachfront hotel room with Liz, the kids, my mom and siblings, and some close friends who had come down for the launch. Yes, our liftoff was scrubbed, but I got to enjoy the 'postlaunch' party I would ordinarily never get the chance to attend. It eased the disappointment a little.

Ours was the fifth pad abort in shuttle history. Technicians isolated the cause to the Number 3 engine's high-pressure oxygen turbopump, which forces liquid oxygen into the main engine's combustion chamber at 4,300 psi. Just after I had seen the red glare of engine ignition out the side hatch window, two sensors on the Number 3 turbopump detected a dangerously high exhaust temperature. The turbine exhaust is a partially burned mixture of oxygen and hydrogen from an upstream preburner; these hot gases spin the turbine, then go on to the main combustion chamber. With the exhaust temperature exceeding the redline limits, the orbiter's computers decided just 1.9 seconds before booster ignition that it would be best not to go flying that day and ordered a shutdown.

Had the violation occurred a couple of heartbeats later, the bad turbopump might have caused an engine shutdown just after liftoff, forcing a dangerous return-to-launch-site (RTLS) abort back to Kennedy's three-mile-long runway. The RTLS is a daunting prospect for the crew: we would have to fly the orbiter and attached ET through half an outside loop, then ride backward through our exhaust plume at Mach 5. Ditching the empty tank, we would then try to make it back to the Kennedy runway. No shuttle crew had ever flown such an emergency approach; none wanted to be the first to try.

While we all teased Dan about his jinxed record, we recognized what had been a very close call. The turbopump was an amazing yet delicate piece of machinery. The size of a V-8 engine, it produced 310 times the horsepower. If the Number 3 HPOT, spinning at 28,000 rpm, had come apart in the aft engine bay, it would have been a very bad day for *Endeavour* . . . and her astronauts. John Young regularly worried aloud that a catastrophic engine failure would likely blow the back end off the orbiter.

Back in Houston, Trudy Davis, in charge of our families' Cape travel, told me she had her own suspicions about the failure: Father Don Grzymski, brother of my best friend, Tom, had been invited to our crew's Beach House barbecue a few days prior to launch, but the flight surgeons, noting he had a slightly inflamed throat, barred Father Don from attending. Trudy laid it on the line to the docs: "Next time you will *not* cause the priest to miss the barbecue!"

Returning to the Cape in late September for our next launch attempt, Dan climbed out of the plane wearing a Groucho Marx disguise, claiming it would help our chances if the shuttle didn't know he was around for the countdown. *Endeavour* had a new set of engines, and five of the six of us had colds. I came down with a head cold the night I went into quarantine, and soon all of us had symptoms except Jeff, who fortunately seemed immune to the bug.

While we fought off our colds, we took full advantage of the quiet at crew quarters and the Beach House. Walking the Cape sands with Liz, we came upon the remains of several battered rowboats and rafts that had carried the latest wave of Cuban refugees across the Gulf Stream. These primitive craft had almost nothing in common with the exotic technology of our own ship, now waiting just over the dunes. But the crews of both vessels shared the belief that their journey would lead to a better life for themselves and their families. I believed that a successful second voyage with SRL could help develop the tools we needed to monitor our global environment. Hard facts about our planet were essential to making the right policy choices to preserve it.

My cold was nearly gone as the Blue Shift waited out our last night on Earth. In the early morning hours of September 30, 1994, I was alone in the deserted O&C gymnasium, squeezing in one last workout before the Red Shift woke for launch. Stretching out on the rowing machine, I smiled when I saw CNN's Headline News flash an image of *Endeavour* on the screen, but the newscaster startled me with her pronouncement that the crew was already strapping into the cockpit for this morning's launch. *Holy Smoke! Had I missed my cue? My crewmates were already at the pad, and I hadn't even finished packing!* I laughed, finished my workout, and walked the empty halls back to crew quarters and our prelaunch breakfast.

From the launch pad's 195-foot level, I glanced back at the LCC and the lights edging the causeway and Banana Creek. My mom was there, God bless her, bravely manning the bleachers with the help of my siblings and a tranquilizer or two. Liz's sister Ginny was also there, along with their mother. The pair had been asked by Liz to stand ready to help her in case of a launch emergency—another lesson learned from *Challenger*.

Phoning Nora Ross from the launch pad again with a quick message for Liz, I was soon on the middeck strapping in next to Jeff. Yogi Berra said it best: "It's déjà vu all over again."

> Now it's 5:55 and they just closed the hatch; we're leaving the window [cover] open so we can get another look outside today while we go off the pad. It's nice and relaxed. Everybody is cracking jokes, having a good time up here. About an hour and 15 minutes to go. Everyone is feeling fine. So do I. Looks like we're really going to do it today. I hope saying that doesn't jinx us.

The remaining minutes of the countdown fled as I thought about Liz, Annie, and Bryce out on the roof, and the coming challenges of the mission. The August pad abort had swept away much of the tension in *Endeavour*'s cabin; we were rested, relaxed, and eager to get to orbit.

*Endeavour* trembled. The service structure outside once again turned from gray to deep red-orange as our three new engines roared to life, their horsepower matching the output of twenty-three Hoover Dams. Six seconds later, the gantry outside the hatch turned incandescent, and a torrent of white-hot flame blasted us free of Earth.

Two minutes later, I opened my helmet visor and glanced at Jeff, relief as plain on his face as it was on mine. We were off to a thrilling start, with six-and-a-half minutes still to go:

> OK, we got rid of the boosters. We're two minutes and fifty-eight seconds into the flight. And we're accelerating and ... it's a very smooth ride. ... A heck of a ride on the boosters. We saw the liftoff. Saw the boosters go off. We're running on three engines, accelerating uphill [at] 3:43. Just went over fifty miles.

...Four minutes into the flight. Mach 8. Negative return. The g's are starting to build. Pretty smooth, just a little bit of vibration. Slight rumble, and Steve says they got a nice view out the back window. Danno is saying that he can just see clouds—no land. Mach 10.... One and a half g's right now. Time is 5:12 [MET] and I'm beginning to feel like I'm starting to pitch back over my head now with my butt up in the air as we accelerate more. Everything is green [normal]....

About 5:45. Can feel the vehicle riding upside down here. Outside the hatch there's nothing but sunlight and dark blue sky.... Five minutes... six minutes now—two and a half to go. G's are building. 1.8 g's. Press to MECO [main engine cutoff]. Looks like we are going to make it to orbit.

6:23. G's are still building. About two minutes to go now. And it's starting to build. Mach 16. And we're really building up now—two and a half g's. It's seven minutes.... Now we're just holding on. Mach 18.

Engines, engines, engines! 7:24. About one minute to go. Three-g throttling, and I'm really heavy now. Got a little... a little pain in my side just like when you run a long way. Hard to breathe now—lots of pressure. Eight minutes now, thirty seconds to MECO. Hard to move your hands, but it's possible. Just hanging on.

8:11, twenty seconds to go. Very bright outside, lots of sunlight. Ten seconds to go, Terry's calling out the numbers. We're going to have MECO very soon. Here comes the bug—MECO!

We're in orbit! Congratulations, Dr. Wisoff! All right, we're here!

LETTER FROM HOME, LAUNCH DAY:
*Dear Tom:*

*The launch was great! It was very similar to -59's in that you could see the shuttle for a long time, and you could see the glint off the SRB's. But this time Scott [Horowitz] was standing right by me and told me when the roof started shaking, so I felt that this time. Scott kept up a running commentary for me. I tried to convince A & B to watch rather than take pictures, but they wanted to take pictures.*

*You were obviously busy with the messages last night. I got one from Nora and notes from Olan [chief test team engineer Olan Bertrand] and this time Mike L.A. delivered the hug! Thank you.*

*You must have been praying for me again because I was calm this morning. The roof was still a "treat," but luckily it's only for a short time. You know I was praying for you, too. Alece [crew secretary Alece McIntyre] says everyone at the family viewing area had a good time. She's not real sure about your mother, though. . . . I hope you're feeling well and fast asleep by now. (We're on the plane home.) We're all very tired of course, but happy. Now that the launch is over I'm looking forward to talking to you sometime during the mission, landing in Florida . . . and a more normal life.*

*I love you,*

*Liz*

I was glad to hear Liz had been surrounded by friends on the rooftop. In August I had asked astronaut Mike Lopez-Alegria to deliver a hug to her there, but in the rush of events he had forgotten. This time he had come through with a warm hug for my wife. I had also asked Scott Horowitz to stick close to Liz during liftoff, a mission he had executed admirably. Liz had found the sight of the astronauts' blue flight suits comforting on this, her fourth emotion-laden trip to the roof. It didn't really matter who the wearer was; she knew that there would be help if she needed it.

Upon opening *Endeavour*'s payload bay doors an hour after liftoff, we made a worrisome discovery. Looking aft out the cabin windows, we discovered that one of our thermal protection tiles, atop the starboard orbital maneuvering engine pod, had been partially shattered by debris impact during launch. The culprit was another small tile that broke away from an overhead window during ascent. Fortunately for us, the damaged tile was in an area that would see relatively low temperatures during entry. We surveyed the damage with binoculars, sent MCC some close-up video for analysis, and then got on with SRL-2's demanding timeline. (The parallels of our tile damage to *Columbia*'s demise in 2003 are chilling. As a crew, we were certainly aware of how lucky we had been in both our brushes with catastrophe.)

One encounter with space sickness had been enough for me: this time I had taken a Phenergan/Dexedrine combo capsule about ninety minutes before liftoff, and the drugs had worked perfectly. I actually had an appetite after MECO and experienced no motion sickness whatsoever. But our crew's medical officer, "Dr. Danno," still had plenty of customers for his needle.

Steve was three hours into the flight, feeling miserable, when he asked for a Phenergan shot. To aid Dan in locating the injection site low on his hip, Steve had had the foresight to get the flight surgeons to mark an X at the proper location (although Dan, a former A-6 Intruder bombardier, maintains he always had a good target fix). Steve might be forgiven for wincing at Dan's preparatory comment, "This is probably going to hurt," just prior to needle-to-fanny contact. But he really had second thoughts when Danno's first attempt merely bounced off Steve's tensed muscle, just pricking the skin. "Oops! Sorry, Steve." The patient was seriously considering plan B, a suppository, when Dan's second jab hit home. *Yikes!* I thought, observing from safely out of range. *I'm sure glad that PhenDex worked.*

My second adaptation to space in six months went well, but I certainly suffered from the residual effects of my quarantine cold. In free fall, gravity no longer helped congested sinuses drain; coupled with the usual orbit-induced fluid shift, I wound up, as did others, with a persistent stuffy nose and a daily headache. My recovery proceeded slowly, leaving me with that full-head feeling for nearly the entire mission.

Fortunately Jeff Wisoff was still healthy. Yawning and stretching, I checked in with him right after I woke from my abbreviated sleep period, anxious to learn how SIR-C/X-SAR activation had gone. Jeff was having a rough first day in orbit, and he looked a little harried. "It's going okay," said Jeff as he rifled one of the lockers for some needed photography gear. "But I've been so busy I haven't even had time to grab a drink from the galley." Bakes and Terry had apparently been fully occupied with the orbiter's cockpit demands, so the entire SRL activation had fallen squarely on Jeff's shoulders. With all six of us now turning to, we soon got the flight deck squared away for extended radar and MAPS operations.

Just after the shift handover we were treated to dramatic evidence of how quickly Earth's surface can remake itself. As our eighth orbit carried *Endeavour* across the Sea of Okhotsk and East Asia's Kamchatka Peninsula, we spotted something strange out the aft-facing windows. "Get up here now and take a look at this!" came the call from the Red Shift upstairs. A dark smudge of smoke cut the distant horizon along the Pacific coast. Perhaps a thunderstorm with its anvil streaming out far to the east? Or maybe dust lofted by high winds? But the terrain surrounding the smoke was clear; this didn't look like a weather phenomenon. We knew this was an active volcanic region, and soon we identified the source of the dark cloud. It was Kliuchevskoi, Asia's tallest volcano, in full eruption.

Kliuchevskoi, at 15,859 feet, is the most prominent of the string of active volcanoes forming the backbone of the Kamchatka Peninsula at the boundary between the Eurasian and Pacific tectonic plates. As the Pacific oceanic crust is carried under the overriding Eurasian plate to the west, molten rock rises to the surface and creates one of the most active volcanic fields in the world.

Kliuchevskoi is halfway down the east coast of Kamchatka; on SRL-1, we had photographed the region and observed a light dusting of ash on the twin peaks of the mountain, quiet beneath its blanket of April snow. Now the volcano had awakened.

By sheer serendipity, we had launched on the very day that Kliuchevskoi roared back to life. We could see little of its summit, entirely shrouded in dark clouds of ash boiling 50,000 feet into the stratosphere. The ash plume shot straight up from the mountain almost to the stratosphere, where it was caught by the jet stream and swept out over the Pacific. Satellites tracked the plume 350 miles to the southeast, where it later forced traffic controllers to divert airliners around the dust and ash.

We soon had every camera onboard zeroed in on the eruption as *Endeavour* gave us a dramatic, down-the-throat view of this impressive geology lesson. The northernmost peak was generating spectacular ash explosions, while Kliuchevskoi's southern summit added only a dirty plume of steam. The neighboring Bezymianny volcano threw its own wisp of steam into the massive plume. Dirty fallout from the eruption coated the downwind slopes of all three mountains.

While our photos of Kliuchevskoi in full eruption were spectacular, many of the details of what was happening near the summit, such as the extent of lava flows, location of the active vents, and changes in the mountain's snow and ice cap, were obscured by the spectacular ash and steam plume. But the radars easily penetrated the ash cloud. The science team reprogrammed their observations to scan the eruption three times a day for the first week of the mission. They successfully tracked the eruption's progress despite a snow-laden weather system that hid the mountain completely from our cameras. SRL's real-time monitoring capability may be a forerunner of future systems that can warn of eruptions at dozens of dangerous volcanoes threatening populous regions of the globe, from Naples to Seattle.

Kliuchevskoi was just the start of SRL-2's ten-day look at our dynamic planet. We were repeating most of the observations made on STS-59, looking carefully for natural and man-made change at our 572 science targets, including the nineteen supersites devoted to major Earth science investigations. Over Hawaii and the Galápagos we mapped lava flows to determine the history of past eruptions and track Kilauea's present outbursts. At Raco, Michigan, on the Upper Peninsula, the radar identified both the mix of trees and the carbon content of the northern forests. Across the Sahara, SIR-C's L-band radar penetrated the desert sands and revealed the ancient watercourses that still feed modern oases; above Libya's Wadi Kufra, the images created a subsurface map for archaeologists to locate artifacts and better interpret the history of early human populations and climatic cycles.

Just as on STS-59, we worked under the gun of the relentless mission clock, maneuvering the orbiter, changing data tapes (we carried 199 of the fifty-gigabyte cassettes), and carpeting the ground with our science photography. While our launch time and orbital lighting combined to treat Jeff and the pilots to passes over North America and Europe, the Blue Shift covered Asia and Africa during the mission's first few days. Although night passes were a bit more relaxed, we continued to call in fires in Australia and the East Indies to the MAPS investigators, even while gazing beyond Earth at the distant stars.

Okay, it's MET 2/22:45. We're south of New Zealand looking at the aurora with Dan and Steve. Off the nose . . . Orion's rising, Sirius is coming up. You can see the stars coming up through the airglow layer. It's fantastic. Over on the tail side we can see the big streamers of the aurora, green and white, curling back on each other and streaming up towards the orbiter; the tail is a black silhouette against those green and white streamers. Dan is snapping away with his camera. We've got the planets setting in the aurora, we've got the tail silhouetted against the aurora and I hope we are gonna' get some thruster firings in there, too. It just looks fantastic! . . . Now the aurora is going astern because we are flying out of it, but it was really something. . . . Gemini is coming up now. I can see Mars. . . . [W]e should be able to see Aldebaran. Thruster firing!

Dan and Steve were the perfect Blue Shift companions. Their generosity and teamwork made for a congenial atmosphere that just got better as the mission proceeded. There were no turf battles over who was king of the cockpit, and I had the same happy relationship with Dan and Steve that I had formed with Rich. During our time alone on the flight deck, I told both that it would be difficult to imagine a better set of space companions. Freed from those rookie pressures, this second dose of free fall had lifted a great weight from my shoulders.

For the Red Shift, too, mission events were unfolding as planned, with the exception of the dinner menu. One of Mike Baker's favorite entrees was smoked turkey, a generous pile of irradiated sliced breast meat warmed up right in its foil pouch. It was especially good folded into a tortilla sandwich. It seemed that in his preflight menu review, Bakes had shown an inordinate fondness for smoked turkey, including it in at least two of his three meals for each of the first four days! When Bakes asked what was for dinner, the answer nearly always was . . . smoked turkey! Noticing our escalating laughter, by the third day our commander was convinced he was the victim of a practical joke instigated either by us or some of his 1985 astronaut classmates. But we maintained our innocence, and postflight investigation showed the onboard turkey surplus was his own doing. Fortunately for Bakes, the pantry offered plenty of alternate meal selections, but I've always wondered what he ate for Thanksgiving dinner six weeks later.

The smoked turkey *was* pretty good, but I had my own favorites: appetizer—freeze-dried shrimp cocktail (the horseradish—whoo!— would open my sinuses for at least a half hour); entrée—MRE beef tips and mushrooms, or freeze-dried spaghetti and meat sauce; vegetable— freeze-dried buttered asparagus. Lemonade or grape drink with artificial sweetener rounded out the meal. For dessert there was chocolate pudding, freeze-dried strawberries, or those Tastykake chocolate cupcakes. Although the nutritionists may have winced, my little stash of junk food was just the thing for a midshift snack.

MET 5/02:57. We've just finished our 6th work shift of the flight. Getting ready to go to bed. I need about 7 hours of sleep and I'll get it if I go to bed in the next three minutes. There's not much to complain about except a stuffy nose that keeps me all stuffed up all night long. Last night I had a headache. . . . I think I can get through tonight if I'm not too chilly and cold in here. Bunk has a lot of cold air blowing into it. . . . [I] have to stop [it] up with a set of shorts stuffed into the air vent, so that I don't get a breeze blowing on my face all night.

The lingering effects of the head cold made me a frequent customer of our "doctor in a drawer" medical locker. I logged so many doses of decongestants and headache pills that my medication chart looked like a pharmaceutical catalog. But my minor discomforts disappeared whenever I gazed out at our magnificent Earth.

**MET 6/04:00.** Well, some of my sights today were spectacular passes over Asia where we did a lot of mapping of the river valleys and landscapes of east Asia. Still no snow in most of Asia yet except in some of the mountains. Some spectacular passes over Tibet's high plateau . . . then we had some looks at the volcanoes in Indonesia. Some good shots of Australia today. But I think the best pass of all was the very end of the day when we went down over Europe. We shot pictures of the Normandy invasion beaches. We shot a picture of Paris, and mapped the Orgeval and Montespertoli watersheds, two hydrology radar targets. Then we saw the entire coast of Italy laid out in the clear, so I was able to map the entire coast of Italy all the way down past Rome, past Naples, several shots of Vesuvius. Just an incredible pass.

So we've been up here almost a week. It was a pretty good day. When we woke up we found out that we'd lost one of the vernier jets, one of the small steering jets at the back of the shuttle. Because we lost those, we couldn't steer as precisely, and so we had to go to our bigger thrusters, which meant that we weren't able to do science for most of the day. . . . We had a lot fewer steering maneuvers to do. That gives us a little bit of free time, so we had some time to shoot some movie film out the window, shoot some video tape of going around the Earth. . . . But we also had a lot of small things that kept us busy. We had not only the failure of the steering jet but a lot of malfunction procedures and troubleshooting on that particular jet. It cost us a lot of time, making changes in the flight plan and science plan. We also had a failure of one of our primary thrusters, but we have at least two backups for each one of those so that was no impact. One of our TV cameras broke down—we fixed that—and then we also had an alarm due to the cabin nitrogen supply being replenished.... We have all been breathing oxygen in the atmosphere here in the cabin, which had reduced the total cabin pressure, so it has to be replenished by letting some nitrogen into the cabin, but that alarm when it went off startled all of us.

...Right now I'm in my bunk and I'm speaking into this tape recorder while it floats above me in space; it's floating...hovering, right above my mouth, as I lie back in my bunk. This is just marvelous to watch this little tape recorder just floating above my head as if it had a levitation device on it. Of course, it's just falling around the Earth with me. Here in this bunk it's hard to imagine that you're weightless because it's a small little space, but...something as simple as that can show how marvelous an environment I'm in right now. My head phones from my stereo are floating above my head.... It's just incredible.

If I had a nickel for every time I said "incredible" during a space flight, I'd have enough savings to pay my way back as a tourist.

**MET 9/20:45.** I have a tape change to do but first I just want to describe what we see of the cloudy East Coast coming up at night. It's the middle of Sunday night in Houston and I've got a clear view down, looking at Washington and Baltimore. I can see a lot of clouds around the eastern Chesapeake Bay and the Eastern Shore but there's Washington, Dulles airport area; you can see Baltimore harbor very clearly.... If I look west of Baltimore I can see Frederick and York, Pennsylvania.... Wow, there's Philadelphia, York and Lancaster and Harrisburg along the Susquehanna.... Montreal and Quebec are up there. New York I can see barely through the clouds. I can see Central Park and Manhattan and then it's getting awfully cloudy up towards the northeast. And if I see all the way to the northwest here, this must be Albany or something right along the river. Dan is saying "Holy Smoke! Great!" Now it's getting to the northern lights. Hey! The aurora is out here on the northern lights side! Look out the tail! It's gotta be the aurora.... I've seen the northern lights for the first time! I've never seen them except in a B-52 up there at 35,000 feet and they were always on the horizon.

Incredible! Incredible! And they stretch all the way around to the sunrise band coming up over the Atlantic. Whoa! This is worth turning off the lights for, I'll tell ya. [Dan: Come over here and you can actually see the circle.] Okay. [Steve: You can actually see almost all the way to the ground it seems like.] Yea, it's a whole ring. It ripples and curves right around. Wow! And here we've got the air glow and the blue of the horizon, that robin's egg blue, coming up. I need some lights for a tape change; I'm gonna' turn them on now. Incredible!

The radar echoes streaming in from the antenna filled tape after tape from our middeck lockers; we were swapping out the bulky cassettes every thirty to forty-five minutes. On October 9, one of the three digital payload recorders on the flight deck malfunctioned, leaving the remaining pair unable to cope over the long term with the flood of data streaming in from the Radar Lab. Jeff and Steve teamed up with Mission Control to replace the big recorder—the size of a two-drawer filing cabinet—with a spare from the middeck. Rerouting the data stream between the two remaining machines, the POCC and MCC targeted the repair for one of our passes over the Pacific, so broad that even at Mach 25 it took us thirty minutes to transit its empty reaches. With the radar inactive over the ocean, Steve and Jeff were well into the repair before we made landfall again. An hour later the swap was complete, and the pair had stowed their wrenches and screwdrivers with almost no loss in science data.

As our radar operations neared a close, the Blue Shift finally got a clear pass over the Himalayas on the tenth day of the flight. Neither Dan nor Steve had ever glimpsed Mount Everest, so I played tour guide as we readied our cameras and the radar scanned the glaciers and snowpack of Earth's highest mountain range. We were astonished at the vivid hues of the tan plateau, cerulean lakes, and dazzling glacier fields. Steve radioed back our excitement.

"Houston and SRL, *Endeavour*: Just to let you know that we had a beautiful pass over Mount Everest, wide open. We really plastered it with photos. . . ."

Story Musgrave answered: "*Endeavour*, Houston. You are making us real jealous down here."

Our final challenge of the mission was to test a technique for using radar to obtain three-dimensional maps of Earth's surface. In the early 1990s much of Earth's topography was imprecisely known, especially in remote regions of Asia, Africa, and South America. Using a technique called radar interferometry, which is roughly analogous to stereo photography, JPL could combine radar images taken from nearly identical orbits to construct three-dimensional topographic maps. With a vertical error of less than fifty feet, these maps are precise enough to steer a cruise missile, safeguard an airliner, or monitor the ominous swelling of a dangerous volcano.

The catch was that our repeat orbits could be separated in space by no more than 300 feet or so. MCC and our crew combined to perform the most precise orbital maneuvers ever seen in the shuttle program, putting *Endeavour* in an orbit for the first six days that nearly matched our SRL-1 flight path of the previous April. At times the two orbits differed by only thirty-three feet, well within the tolerances for creating successful interferometric images. For days 7 to 10, we lowered our orbit to a height of only 124 miles, an altitude that put *Endeavour* on a path matching its orbit of the previous day. With Bakes and Terry manually firing the orbiter's thrusters, this exacting navigation adjusted our orbital velocity to an accuracy of one part in two million, producing long swaths of closely overlapping images. The interferometry passes produced digital elevation maps that reveal not only the topography but also actual shifts in Earth's surface of just a few centimeters. Detection of such small-scale changes can help predict hazards in volcanic or earthquake-prone areas.

On the morning of our twelfth day in orbit, film gone and pantry nearly empty, we suited up to come home, but not without a reminder that Murphy's Law also applies in space. As part of a medical experiment, I had trained along with Dan to serve as an astronaut guinea pig during entry, wearing a device that would record blood pressure and cardiac activity as my body readapted to gravity. The two of us had spent several hours learning how to use the device and practiced with it in the simulator; we were more than willing to endure a little discomfort to help our colleagues returning from future space station missions.

Just before the Red Shift woke to get ready for entry, Dan and I broke out the equipment and floated up to the flight deck for donning. The EKG electrodes on our chests came first: Dan and I removed our clothes, put on our entry diapers, then applied the electrodes to each other's chests. We ran into trouble right away: the adhesive on the electrodes had dried out, so we reluctantly resorted to duct-taping them in place—anything to advance the cause of science. ("It won't hurt until later," we consoled ourselves.) Then Murphy struck again: the blood pressure monitor had been stowed before launch with the power switch on; its batteries were stone dead. With the clock ticking toward the deorbit burn, I scrambled to dig out a few spare AA batteries from our camera locker. Powering up successfully, I found that the menu on the display didn't correspond to our checklist steps. Rapidly losing patience, I pushed past this obstacle by trial and error, managing to restart the machine's internal clock. Next I had to test the cassette recorder that would capture my voice comments during entry. It was working perfectly, as evidenced by the stream of my invective that wound up on the tape. (I made a point after landing to apologize to the investigators in Houston before they played it back.)

Curious as to what the yelling was about, Steve Smith floated up to the flight deck. He coolly took in the scene: two astronauts cursing amid a tangle of tape and wires, wearing nothing but their superabsorbent diapers. He shook his head in mock disbelief:

"I never imagined that spaceflight would be like this: floating in the cockpit of the space shuttle *Endeavour* with two weightless, seminaked guys."

After that long last shift setting up the cabin and suiting up, the weather at KSC turned against us. Thick clouds led Mission Control to cancel our first two landing opportunities at the Cape that day. Deciding not to gamble on another day in space waiting for Florida's weather to improve, Flight Director Rich Jackson brought us home to Edwards Air Force Base. Liz and the spouses were sitting in the Cape crew quarters conference room when they got the news over NASA TV. Steve's wife, Peggy Brannigan, going through this for the first time, asked the room at large, "What's going to happen now?" Liz was resigned to the inevitable: "Well, pretty soon, Bob Cabana is going to come through that door and apologize. We'll watch them land on TV, then we'll head back to Houston." Sure enough, a few minutes later, Bob, now the chief astronaut, arrived shaking his head, telling the wives he was sorry that their husbands would be landing a continent away.

After the deorbit burn, Steve and I lingered on the darkened flight deck to watch for our Mach 25 encounter with the atmosphere. Visible to our west, the city lights of Japan were soon overwhelmed by our own spectacular glow.

> OK, it's MET 11/05:17. Positively pink outside the hatch window. We have a huge pink flame bathing us, creating a glow outside the hatch. Steve is looking out overhead. We're just about past EI [entry interface]. We are in a pink glow, inside a pink cotton ball outside. OK, what do you see?... Guys, it's like a yellow tube behind us. Yellow-purple, all pink and yellow-purple. Fantastic! OK, see you guys on the ground!

It was time to get downstairs to help Steve strap in and get seated myself. On the way past the side hatch I looked again at the glowing, cherry-red ball of cotton candy outside—the entry plasma enveloping the orbiter's nose. Deceleration made weight a meaningful concept again: I found I could lightly skip across the floor.

Half an hour later, my second mission to Planet Earth was over. I unstrapped, made a Herculean effort to stand up per the medical experiment procedures—and the blood pressure equipment failed. Well, what the heck! It was time for a cheeseburger and a shower.

# 9
# BACK IN THE POOL

*The perfect spacewalk is like a ballet.*

STORY MUSGRAVE, STS-61, *SPACE SHUTTLE:
THE FIRST TWENTY YEARS, 2002*

*ndeavour*'s hatch swung open just after 11 a.m. Pacific time on October 11, 1994, bringing our journey of 4,703,216 miles to an end. The suit techs and our flight surgeon stepped gingerly into the middeck, sniffing the air even as they shook hands with Steve Smith and me. We had been a reasonably shipshape crew, and the orbiter cockpit was clean and neatly stowed. Still, there was an unmistakable whiff of . . . something . . . in the air. Our vehicle integration and test team engineers had told us that every mission was different, but sometimes a distinct sour smell rolled out of the cabin at hatch opening. Put six people in a sealed volume the size of two minivans for eleven days, and no matter how good the activated charcoal filters and airtight trash bags, you'll get a certain amount of that lived-in aroma when you take the cork out of the bottle.

Our medical experiments and my role as a guinea pig continued for about three hours after landing. Dan and I had volunteered for an investigation that would test our bodies' adaptation to gravity immediately after landing. The good news was that I was immediately placed on a gurney upon leaving the orbiter, so the heavy feeling I experienced after

STS-59 was minimized. The bad news was that I was the focus of an entire medical team for three hours: they drew what seemed like quarts of blood, scanned my heart with an ultrasound imager, took a series of blood pressure measurements, treated my lungs to a whiff of carbon monoxide in order to measure blood volume, and finally swung me erect abruptly to see whether my heart could cope with the sudden load imposed by gravity.

I did fine until the team briskly swung my gurney to the vertical. My heart raced; I felt a slight tingling. The med-techs hovered around with serious expressions, measuring, monitoring, observing. Although I felt a strong desire to sit down, I noted no dizziness or tendency to pass out. I focused on wiggling my toes and flexing my leg muscles to keep the blood from pooling in my legs, a technique perfected through many hours on the Air Force Academy parade ground.

But the team wasn't done with me yet. Next up was a ride on the exercise bike, comparing my heart and lung capacity to preflight measurements. They gave me a pretty challenging ride, wired up and huffing and puffing into a breathing hose, but visions of that cheeseburger kept me going until the fifteen-minute spin was over.

The medical marathon ended at last, and my crew and I boarded the STA for a fast trip to Houston and a reunion with our families. My ride home was a physical challenge equal to anything I'd faced on the mission. Worn out by the long work shift before entry, the physical assault of gravity after touchdown, and the three hours of medical trials, I hoped to grab a few winks en route to Houston. But the exuberant Red Shift of Jeff, Bakes, and Terry joined Richard Jennings, our flight doc, in storytelling and raucous laughter all the way home, waking me each time I started to nod off in my seat.

The results from the flight were more than satisfying. In our 11 days, 5 hours, 46 minutes, and 8 seconds in space, we had recorded 910 data takes with the SIR-C and X-SAR radars (about 80 hours of radar imaging). If we transfered all that data to today's compact discs, the images would fill a stack of CDs more than sixty-five feet high.

Perhaps the most promising result was the successful demonstration of interferometry to create topographic maps of Earth's surface. During SRL-2, we imaged more than a million square kilometers using repeat-pass interferometry. Our images of creeping lava flows and the changing shape of Kilauea volcano, along with those of rapidly advancing glaciers in Chile's Patagonian ice fields, showed the great potential for satellite interferometry in mapping global change. The JPL radar team immediately began planning a third radar mission, this time with the goal of mapping most of Earth's topography in a single eleven-day flight. Our efforts on STS-68 led directly to the success of the Shuttle Radar Topography Mission in 2000; maps from that ambitious project were put to immediate use in civil aviation and by US troops in the war on terror.

To accomplish our radar and interferometric observations, we had maneuvered *Endeavour* roughly 470 times, including 44 nose-to-tail swaps, every one on time. Back in Houston, we spent days at the light tables in Building 8 with our Earth observations team poring over the 13,000 still-camera images of our planet from the flight.

All of us were still readapting to the unfamiliar pull of gravity. Sleeping was difficult for the first few days as I coped with a massive case of jet lag (the Blue Shift and the Johnson Space Center were on nearly opposite schedules). At the same time, my sense of balance was seriously, if rather humorously, impaired. For nearly three days after landing I couldn't stand erect without using my eyes to determine the up direction; without them I was in imminent danger of keeling over. If I accidentally dropped something, I could pick it up only by bracing myself against a wall or table; bending over for it would quickly cause me to topple. My inner ears didn't start working well again until about seventy-two hours after touchdown.

Fully recovered two weeks after landing, I was working on our crew report when Bob Cabana stuck his head into our crew office. "I've got the urge to go set my hair on fire in a T-38. Anybody want to go flying?" Strapped into Cabana Bob's backseat, headed out over the Gulf of Mexico, I discovered too late that he meant to set *my* hair on fire, too.

Cabana wanted to explore a particular corner of the T-38's performance envelope: How many aileron rolls could one do at maximum stick deflection before having to pull out of the ensuing dive? I honestly don't remember the answer: I was too busy trying to keep my stomach from being hurled out of the cockpit as Bob raised the nose slightly, checked the airspeed at 300 knots, and smoothly put the stick against his left knee. Together we counted the rolls as the nose fell through the horizon; the spinning continued until we were pointed straight down at the blue waters of the Gulf. The ensuing pullout took about 5 g's, then Bob set us up for another trial. I think we averaged about sixteen complete rolls during each run, but my interest in counting decreased rapidly in inverse relationship to the gurgling sensation in my stomach. "Bob, let's fly straight and level for a bit." I went to 100 percent oxygen, Bob gave me the controls, and I flew straight and level back to Ellington, my stomach protesting all the way home.

Over the ensuing weeks, the crew of Space Radar Lab 2 wound up its formal debriefings and published its mission report. JPL's radar processing lab released a stream of imagery augmented by our photography, which drew a complex portrait of our changing world. The two missions had proven that a space-based radar in permanent orbit could significantly improve our ability to monitor Earth's complex and dynamic surface. I was particularly interested in the archaeological results from SRL: the "radar rivers," still-watering Saharan oases; ancient settlements along western China's Silk Road; buried traces of that country's original Great Wall; caravan tracks pointing back to Arabia's lost city of Ubar; and the intricate thousand-year-old waterworks surrounding the Cambodian temples of Angkor Wat. But we found no trace of Cambyses and his Persian army lost in the Sahara 2,500 years ago.

After SRL-2, Bob Cabana gave me just the job I wanted: capcom. The capsule communicator job in Mission Control dated back to the Mercury program. The Original Seven astronauts handled radio transmissions from the control center to their colleagues in space, conveying advice and instructions to the crew in orbit, and providing the flight director and controllers with an astronaut's insight into the situation in the spacecraft. Next to actually being on the orbiter, the capcom job was the closest you could get to the actual experience of spaceflight. In fact, the best way to do the job was to figuratively put your brain in orbit, thinking right alongside the astronauts.

The Mission Control Center is located in Building 30 a couple of hundred yards across the campus from the Astronaut Office. In 1995 shuttle operations were run out of a single flight control room (FCR), first used in mission operations for the flight of *Gemini 4* in 1965. Both the second-floor FCR and an identical room upstairs were used during the Apollo era: second floor, Saturn IB Earth orbital missions; third floor, Saturn V moon flights. The third-floor control room witnessed *Apollo 11*'s historic Moon landing, *Apollo 13*'s dramatic rescue, and the subsequent voyages of lunar exploration. When the shuttle debuted, that FCR was converted to a controlled-access, secure control center for

Department of Defense missions. By the time I became a capcom, this upper room had been decommissioned, and the second-floor FCR handled all simulations and flight control.

The FCR had terraced ranks of consoles that stepped down toward three large display screens at the front of the room. The capcom sat adjacent to the flight director, one level down from the viewing room at the rear. The other dozen or so console positions culminated in the front row "trench," where controllers known as FDO (flight dynamics officer), Booster (main engines and SRBs), GNC (guidance, navigation, and control), and Prop (reaction control and orbital maneuvering engines) handled the most time-sensitive phases of flight, ascent, and entry.

When training for my first mission, I had little concept of what MCC operations were really like. My initial introduction to MCC came through dozens of hours spent in the shuttle simulator; MCC was a mysterious but omniscient entity, delivering crucial advice via the disembodied voice of the capcom. My visits to the FCR itself were infrequent, limited to short visits leading VIP tours or escorting spouses during a mission. I knew next to nothing about how MCC digested telemetry and crew inputs to make decisions. What I *did* know after my two SRL flights was that MCC was a reassuring partner in the cockpit. The control center was the key to fixing STS-59's galley problem and in developing a software fix that recovered the failed vernier thruster on STS-68. Mission Control always seemed to deliver. All I had to do was pick up the mike to tap into the minds of the most talented space operators on the planet.

Training as a capcom put me in the middle of that professional team. As a mission specialist I would train for the orbit phase of operations; pilot astronauts served as the ascent-and-entry capcoms. Flight directors and controllers for both simulations and actual spaceflights were assigned by the Mission Operations Directorate; the Astronaut Office supplied the capcoms.

Like all the other controllers, I worked for the flight director, whose word was law inside the control center. While I represented the interests of the crew, the flight director relied on me to communicate his decisions and advice promptly and accurately. It was vital while on console that I follow the operational chain of command. We could always discuss any differences of opinion later outside the control room.

Back from his mission to service the Hubble Space Telescope in December 1993, Story Musgrave was now chief capcom, and he soon had me training alongside more experienced astronauts on console. During my first six months as a capcom, I handled a string of generic simulations while preparing for actual orbit operations. In a typical day at the flight control room, the flight director would have the team go to "run" on the simulation at about 8:30 a.m. After eight hours of working with the crew and their instructors, we would freeze the sim, then regroup for a debriefing about fifteen minutes later. The flight director would usually ask the console positions to individually explain the major failures they had dealt with, analyzing any mistakes they had made or any new wrinkles that the simulation supervisor, or Sim Sup, had thrown at them. Tied in via intercom, the crew could break in at any time to ask a question or comment on how they had reacted to malfunctions.

> **JANUARY 11, 1995:** It's remarkable how quickly we get de-focused on shuttle systems. I've forgotten so much. The computers and guidance systems tax us the most, and they're the first to evaporate from memory. Capcom will be a great way to retain what little I picked up during flight. Haven't been in a "flight deck" sim for so long (March 1994?) that the words are almost unfamiliar. This deorbit prep sim today with Curt Brown is like a bucket of cold water—a shock how much I've gotten rusty. But it does come back....

The flight director and crew commander were cocaptains of a team that worked seamlessly together to execute the mission safely and successfully. Circling the planet at Mach 25, the crew commander was responsible for the orbiter and crew's operational safety and for precise execution of the flight plan. The flight director ensured the readiness of the entire mission operations team and assumed the larger responsibilities of mission success and flight safety.

My first mission as lead capcom, responsible for communications during the most important phases of a mission, was for STS-70, a week-long flight in late spring of 1995 that would deliver a Tracking and Data Relay Satellite (TDRS) to orbit. The TDRS would join several others already in orbit to relay communications and data from the shuttle and other satellites, such as the Hubble Telescope, to ground stations. Working with flight director Rob Kelso, I rehearsed with the flight control team for the satellite deployment. Ready to support the crew in late May, I was as amazed as Tom Henricks's crew when flicker woodpeckers at the Cape chipped away enough foam insulation from *Discovery*'s external tank that it had to be rolled back to the VAB for repairs. The crew finally launched on July 13, 1995.

A couple of hours after liftoff, Kelso's shift replaced the ascent team in the FCR, and I took my seat across the aisle from Rob. Because of its limited battery life in carrying the TDRS to its geosynchronous orbit, the inertial upper stage had to be launched from *Discovery*'s payload bay within a few hours after liftoff. Coping with a short circuit in the deployment control panel on *Discovery*'s flight deck, Hairball Don Thomas used backup activation systems to help the crew get the satellite on its way to the proper orbit.

Each day of a shuttle mission was covered by three overlapping shifts, each lasting about ten hours: eight hours for the work on console plus an hour on each end for handover. Two shifts in MCC handled the crew's daily activities in orbit, and a third, planning shift worked the consoles while the crew slept, preparing the revised flight plan and working up new procedures to deal with any orbiter or payload failures. Once assigned as a capcom for a specific mission, I could generally count on working the same hours for the entire duration of the flight.

The camaraderie of working with the controllers and the other astronaut capcoms made my stints in MCC a pleasure despite the occasional graveyard shift for simulations or spaceflights. The controllers were young, extremely professional, and so much more expert in orbiter systems and operation than me that at times I was embarrassed to sit in the same room with them. Even after two spaceflights, my knowledge of the technical aspects of the space shuttle rarely matched their in-depth expertise. But the flight controllers were always willing to give me a quick tutorial on the intricacies of an orbiter system. Whatever I learned could be useful to the crew, and if I asked enough good questions during sims, I would be much better prepared to deal with the unexpected once "my" crew was in orbit.

Story was my favorite mentor as I learned the basics of the capcom job: he was funny, irreverent, knowledgeable, and completely approachable. Once qualified, I enjoyed sharing the handovers with him when our shifts overlapped. With his twenty-eight years of experience going back to the Apollo program, Story could summon up a tale for any occasion. I was interested to know what he thought about George W. S. Abbey, who had just returned to Houston as deputy director of the Johnson Space Center under Dr. Huntoon. (We all guessed his sights were on the director's job.) Abbey had a good habit of paying a casual visit to the flight control room every afternoon just to see how the mission was going. I wanted to know more about what made George tick.

Story had watched George's rise in the JSC hierarchy over the years, through Apollo and Skylab to his tenure as chief of flight crew operations during the first half-decade of the shuttle era. From his staff position on the National Space Council, George had helped champion Dan Goldin as the successor to NASA administrator Dick Truly. Abbey had played a key role in redesigning the Space Station program in 1993 and had supported bringing in the Russians as major partners in the project. With his position as deputy center director, he was clearly in line to influence both the shuttle and Station programs. Watching him roam the control room every evening, both Story and I wondered how George would remake JSC and the human spaceflight effort. During George Abbey's long career at NASA, he both made and broke astronaut careers, but I have never had anything but straightforward dealings with him.

After a rewarding year working on the capcom console and "flying" a series of interesting shuttle missions with my colleagues, I got that marvelous call from Bob Cabana:

## Astronauts Selected for STS-80 Shuttle Mission

*From NASA Press Release 96-6, January 17, 1996*

Kenneth D. Cockrell will command the third flight of the Wake Shield Facility (WSF) aboard *Columbia* (STS-80) scheduled for November 1996. He will be joined on the flight by Pilot Kent V. Rominger (Commander, USN), and Mission Specialists Tamara E. Jernigan, Ph.D., Thomas David Jones, Ph.D., and Dr. Story Musgrave.

STS-80 will mark the third flight of the WSF that flew on STS-60 and STS-69 and the second flight of the Orbiting Retrievable Far and Extreme Ultraviolet Spectrometer (ORFEUS) satellite. Both satellites will be deployed and later retrieved during the mission.

The saucer-shaped WSF is designed to fly free of the
Shuttle, creating a supervacuum in its wake to grow thin film
wafers for use in semiconductors and other high-tech
electrical components. The ORFEUS instruments are
mounted on the reusable Shuttle Pallet Satellite and will
study the origin and makeup of stars.

Astronauts Jernigan and Jones will conduct a spacewalk
during the mission to continue the flight test and evaluation
of hardware for future spacewalks or extravehicular activity.

I was as surprised as anyone when Bob told me of my next flight assignment. Although I felt technically ready to move into shuttle training once again, I hadn't expected another flight so soon. The mission
was complex: for the first time the orbiter would fly formation with two
independent satellites, and the sixteen-day flight would stretch our fuel
and life-support margins to the limit. STS-80 would include a demonstration of just about every capability the shuttle possessed: formation
flying, rendezvous, robotic arm satellite launch and retrieval, onboard
scientific research, and EVA. In addition to launching the two free-
flying scientific satellites and bringing them home again, we would conduct a pair of EVAs in a shakedown of construction techniques for the
International Space Station (ISS). Tammy and I viewed the space walks
as the most difficult challenge of the mission. Fortunately we had the
right crew to get the job done.

Ken Cockrell, our commander and my Hairball classmate, was uni-
versally known as Taco (only his family and NASA Public Affairs called
him Ken; even the capcoms used "Taco" over the radio). He had already
completed two spaceflights, most recently with the Wake Shield on
STS-69 in September 1995. An Austin native and former Navy attack
and test pilot, this would be his first flight as a commander, and he was as
skilled in the cockpit as they come.

Kent Rominger was our pilot, making his second flight. A Navy veteran of the 1991 Gulf War, "Rommel" had flown F-14 Tomcats in the
fleet and at Patuxent River. He had joined NASA in 1992 and piloted
Columbia's fifteen-day US Microgravity Lab 2 mission in the fall of
1995.

Tammy Jernigan had been with the Astronaut Office since 1985. A
research astrophysicist by profession, she already had three shuttle mis-

sions under her belt. A stint as deputy chief of the Astronaut Office in the early 1990s had given her an insider's perspective on the ins-and-outs of our hallway politics. We were already friends through our common interests in space science and our work together in the Mission Development branch. In meetings she would quickly get to the crux of any issue on the table; that no-nonsense approach would be needed to accomplish the dozens of tasks crammed into the two six-hour space walks.

I had gotten to know Story well during our 1992 flight training in the Citation business jet and then on the capcom console. He had done everything and knew everybody, most recently leading the spectacular EVAs to repair the Hubble Telescope in late 1993. As my capcom boss for a year, he had taught me volumes about how to work with the controllers and the crews while in orbit. Although a generation older than the rest of us, Story was showing no signs of slowing down as he approached his record-tying sixth flight at age sixty-one.

STS-80 promised a packed training schedule for our five-person crew, with everyone taking on at least two jobs on this mission. In a new role for me, I was to be the flight engineer, or MS-2, backing up Taco and Rommel through every one of our demanding ascent-and-entry simulations. I would also be Tammy's assistant robot arm operator as she deployed the ORFEUS telescope, then we would swap roles for the launch of the Wake Shield. I was Story's backup for our onboard control of that satellite during its three-day formation flight with *Columbia*. Finally, Story would be our in-cabin space walk coordinator while Tammy and I were working outside. We began to plow at full speed through a relentless training schedule. By the early autumn of 1996, our launch date rapidly approaching, the pace had accelerated to the point where I barely knew what would be on tomorrow's schedule.

As I began the final training run for my first space walk just three weeks before liftoff, I was headed not up but down. This last underwater rehearsal, a demonstration of tools and construction techniques developed for the new International Space Station, would finalize our EVA plan and test procedures. The highlight of our EVAs was the test of a telescoping Space Station construction crane. Setting it up on the edge of *Columbia*'s payload bay, Tammy and I would load it with a dishwasher-sized ISS solar power battery and put the crane through its paces. Each of us would take turns swinging the battery across the payload bay and retracting and extending the crane's boom—moves the crane would later undertake at the Station. The first space walk would take us more than six hours to execute, even if all went smoothly.

From the first day I slipped beneath its startlingly clear waters, the WETF always gave me a secret thrill. An oversized swimming pool 78 feet long, 33 feet wide, and 25 feet deep, its nearly half-million gallons of water produced a realistic simulation of the working conditions in free fall. The water tank was large enough to hold a submerged mock-up of the shuttle payload bay, giving shuttle crews the chance to practice all the necessary repair tasks that might confront them. It was the primary training facility for teaching EVA procedures.

Our morning at the pool began around quarter past seven with a briefing from Glenda Laws, our tireless instructor. She had designed the overall plan for these EVAs, she had trained us here at the WETF, and during the flight she would assist us from Mission Control. After meeting with Glenda, we underwent a quick physical exam that checked heart, lungs, and ears, which got a real workout as we changed our depth, and thus suit pressure, in the pool. By eight o'clock or so we began the laborious process of donning our 250-pound space suits.

In dressing for an orbital space walk, the first item of our ensemble was also the most humbling: an adult-sized disposable diaper. There were no rest-room breaks during a seven-hour EVA. Synthetic long underwear and cotton socks covered us from neck to toe, protecting us from the cold of orbital night. Over the long johns came a frothy creation of white nylon taffeta and spandex called the Liquid Cooling and Ventilation Garment, or LCVG. This close-fitting body suit is laced with a network of spaghetti-like plastic tubes, bringing chilled water from the suit backpack close to the skin. The water removes body heat that would overwhelm an astronaut inside the well-insulated suit in just a few minutes. Ducts stitched outside the LCVG also drew warm, humid air from the forearms and ankles, pulling it out of the suit and through the air conditioner and carbon dioxide scrubber in the backpack. The suit fan pumped fresh oxygen back into the suit from an outlet just behind my neck.

Striding in our LCVGs past our trainers, divers, and technicians on the WETF floor, Tammy and I arrived at the donning stand at pool's edge. I sat down on a plastic mat, and with the assistance of John

Williams, my suit technician, wriggled into my space-suit trousers, tugging on the metal waist ring to get my toes and heels seated firmly in the close-fitting boots. Hobbled by the weight of the stiff, multilayered pants, I ducked under the suit's torso mounted with its backpack on the donning stand. Now came the hard part, in both senses of the word: the torso, sculpted from metal and tough fiberglass, fit as snugly as a knight's armored breastplate, and the process of squeezing into it always brought me uncomfortably close to claustrophobia.

The hard upper torso, or HUT, was barely wide enough to let my shoulders pass upward through the waist ring. From an awkward crouch beneath the open oval, I raised my hands high and thrust them upward into the inside openings of the suit arms. Rising slowly, I pushed my arms up and out through the suit while my head rose slowly toward the helmet ring above. Rocking my shoulders from side to side to inch my arms out through the sleeves, I got a close-up view of the HUT's fiberglass interior and the pouch of drinking water passing an inch from the tip of my nose. For about ten seconds of this slow squeeze up through the HUT, I felt like a chimney sweep stuck in the flue above a fireplace, unsure if I'd ever scramble out the top. But I knew from experience that the worst was over once my armpits came even with the HUT shoulder bearings. From there, I could scissor my arms down and, red-faced and short of breath, pop my head up through the helmet ring to daylight and open air.

Standing upright at last on the donning stand, I relaxed and turned things over to John. He hooked my LCVG into the backpack air and water connectors. Hoisting my heavy suit trousers up against the HUT's waist seal, he threw the locking lever home, joining the upper and lower portions of the suit. Now for the gloves.

Space-suit gloves come in several standard sizes and can be adjusted to accommodate the wide size range of astronauts hands. True comfort in the gloves is elusive, though, and it took me many underwater training runs and repeated adjustments to finger length and width before I was satisfied. By 1996 I had logged five years of working in my gloves, and by now they were old friends. By taping knuckles with moleskin and wearing a pair of thin silk glove liners, I emerged from most WETF runs free of blisters, pinches, and numb fingers—all occupational hazards of EVA. As I pushed hard into the gloves to get my fingers all the way into their thin rubber tips, John carefully locked each one onto the metal wrist rings on the suit forearms. I wiggled my fingers; encased in tough white fabric, the stubby digits moved only grudgingly.

The helmet came last. Before donning the gloves, I had pulled on the communications skullcap, referred to as the Snoopy cap because of its resemblance to the cartoon beagle's World War I aviator's flying helmet. It supported earphones and a pair of boom microphones perched at the corners of my mouth. First checking for good airflow from the fan outlet, John crowned me with the double-layered polycarbonate helmet, slipping it past my nose and locking it down on the neck ring.

On a real space walk, the suit's backpack provided all power and life-support functions, but here in the WETF, electricity, air, and cooling water were fed to the suit through a thick hose, called the umbilical, from pumps and chillers topside. A life-support technician monitored airflow and water temperature from a console alongside the pool. She made sure the internal suit pressure kept up with outside water pressure, and she could vary the cooling water flow at my request.

Suspended back to back on the donning stand, Tammy and I watched with interest as Story saddled our chests with heavy racks of tools. With a grin and a muffled, "Have fun, Jones!" through the intercom, he hopped clear and joined Glenda at the test conductor console. In late 1996 probably no one on the planet was more expert at spacewalking than Story. He had a no-nonsense approach to spaceflight, but he also brought an impish grin and a mischievous sense of humor to the job. With his shaved head and wizened brow, Story was old enough to be my father, and his experience gave me additional confidence about the flight.

I gave the hoist operator a thumbs-up as he swung the donning stand off the pool deck and carefully lowered us into the water. Our safety divers unlatched our space suits from the stand and herded us, blimp-like, to the thick buoy lines that guided us in our hand-over-hand descent to the bottom.

The WETF's warm blue waters, lit from above by ranks of brilliant mercury vapor lamps, closed gently over my helmet, submerging me in rippling shadows and drifting bubbles. The space suit gurgled as trapped air escaped from seams and creases in the fabric. Fresh air streamed steadily into my helmet, swirling around my cap and tickling my nose. I shivered momentarily on the way to the bottom, twenty-five feet down, as chilled water began coursing through the LCVG. For the next seven hours, I was totally dependent on the safety divers and life-support techs keeping watch topside.

Tammy drifted downward a few feet away, shepherded by her own pair of scuba divers. Suspended between bottom and surface in the 90°F water, we worked in session after session to acquire the skill, strength, and dexterity required by a space walk. In the WETF, just as in space, a careless push with a finger could set me adrift in the pool, out of reach of shuttle handrails just inches away. With a twist of my arms I could vault slowly into a handstand to reach work sites in the submerged mock-up of the shuttle's cargo bay. Unfortunately that heads-down attitude also guaranteed me sore shoulders and a headache if maintained for more than a few minutes. To my dismay, such topsy-turvy work was nearly always on our STS-80 training agenda.

The constant presence of our safety divers reminded me of the ever-present hazards involved in EVA training. The water in the WETF was only twenty-five feet deep, but it was still dangerous. Astronaut Mark Lee had a close call in the WETF during a 1995 training session, preparing to service the Hubble Telescope. The circular bearing linking the right arm to the suit's fiberglass upper torso gave way without warning, and the internal suit pressure instantaneously blew out the failed shoulder joint. As air rushed out, water flooded in, cascading down into his boots and trousers. As the heavy suit sank, the water rose to his armpits, but fortunately his umbilical continued to feed air into his helmet above the breach, keeping it from flooding as the safety divers dragged him to the surface. Had Mark been working upside down, his helmet would have filled immediately, and it would have been a race against time to get him upright, feed him air, and get his heavy, flooded suit to the surface. After the incident, NASA beefed up space-suit inspections, searching for hidden signs of wear and tear; a similar structural failure in orbit would have been instantly fatal.

All went well that morning as Tammy and I reached the bottom, our divers swarming over us like pilot fish around a couple of fat groupers. Popping small weights in and out of pockets on the front and back of the suits, they fine-tuned our buoyancy, arresting any tendency to drift up or down. I wriggled my shoulders and relaxed while they worked, watching the divers' bubbles stream lazily up to the quicksilver surface. The weigh-out complete, our divers towed Tammy and me over to the airlock mock-up in the submerged cargo bay. Through our headphones crackled the voices of Glenda and Story, who would take us methodically through this final dress rehearsal. We negotiated the cramped exit from the airlock, got additional tools from the orbiter's payload bay storage box, and swung into our familiar EVA-1 timeline. With Story running the checklist from his perch next to Glenda in the test conductor's booth, we were off and running. After setting up the ISS crane, we used some new cargo-handling tools to wrestle the 400-pound solar array battery out of its launch cradle. As bulky as a large-screen TV, the battery demanded brute strength in maneuvering it through the water, then finesse in mounting it on the crane. Operating the experimental crane was a similar, daunting blend of thread-the-needle delicacy and plain old muscle power. Although packed with Styrofoam for buoyancy, the crane and battery were not perfectly balanced underwater, and lugging them into position had us both breathing hard with exertion.

For nearly seven hours we wrestled with the litany of tasks in a variety of body positions: sometimes sideways, sometimes upright on a portable footrest, sometimes upside down. Inside my helmet, I gulped water from a drinking straw positioned just below my chin; the drink bag helped me keep up with a constant thirst aggravated by the steady flow of dry air whistling in through the umbilical. The LCVGs did keep us reasonably cool, but the varying workload had me alternately chilled or perspiring. Although protected by pads sewn into the LCVG, my shoulders soon became sore from being forced repeatedly against the hard metal of the HUT's arm bearings and neck ring. Throw in tired muscles and hunger, and the experience was complete. When we were finally hoisted from the WETF at the end of the run, damp with perspiration and drained of energy, Tammy and I slid out of the confining suits with undisguised relief.

Working against the stiff, pressurized suit gloves for more than six hours, our hands suffered the most. The intensive flexing and gripping guaranteed tender fingers, sore palms, and forearms that ached with fatigue. It took time to recover. When we first started our STS-80 training, my hands and arms were so wrung out after a WETF run that I couldn't use a computer keyboard until the next day.

The bruised bodies and stiff hands were worth enduring, though, because we learned in the WETF how our suits and tools would behave in free fall. We were expert at handling the crane and herding a bevy of "weightless" tools between our work sites. I couldn't wait to take the space suit for a real spin. We were ready, and with Story backing us, we were confident we could meet any challenge the EVA could throw at us.

Our crew completed a successful countdown rehearsal on October 23, 1996. Both the Wake Shield and ORFEUS were snuggled into *Columbia*'s payload bay. Back at JSC, I was eating my way through a case of Hostess Snoballs, my junk food choice for this flight. The space food folks had purchased the orange marshmallow variety, anticipating our original launch date near Halloween. But management reviews of a problem with our solid rocket motors had slipped the launch date a week. Snoballs or no, *Columbia* was scheduled to lift us to orbit on November 8.

# 10

# Go for EVA!

*How come I'm stuck here looking at the Earth and sky
through glass? Let me see what they look like out there!*
Yuri Romanenko, *Soyuz 26*, as quoted by crewmate
Georgi Grechko in *The Home Planet*, 1988

Taco, Rommel, Story, Tammy, and I entered quarantine on Friday, November 1, 1996, a week before liftoff. At last I could get off the training treadmill. My head was packed with details of the STS-80 mission, and my crew notebook, the only personal reference material I could carry into orbit, bulged with facts, diagrams, and reminders. On flight day 1, I would help the pilots get us through ascent and post insertion. Later that day, Tammy would use the robot arm to launch OR-FEUS into orbit. On flight day 3, we would inspect and test our space suits for the EVAs a week later. Late on flight day 4, after a demanding series of maneuvers, the crew would deploy the Wake Shield satellite. It would be my job to release it from the robot arm high over the payload bay. Three days later Taco would guide *Columbia* into position so that I could snare Wake Shield and haul it back aboard. Days 10 and 12 were allotted to our two space walks, followed by a second rendezvous on flight day 15 to grapple and berth ORFEUS. If all went well, we would complete sixteen days in orbit and be back with a month still left to prepare for Christmas.

A weekend in the JSC crew quarters had helped us all unwind a bit, but on Monday there was bad news from the Cape. The Flight Readiness Review on November 4 had assessed some worrisome evidence from STS-79's recovered solid rocket boosters. During postflight inspection at the Cape, technicians found the throat insulation in one of the SRB nozzles had experienced greater than normal erosion from the passage of the 5,000°F exhaust. A breach of the insulation might destroy the cone-shaped nozzle and lead to a catastrophic failure. Even as we prepared to fly to the Cape the next morning, word came in from the Florida meeting postponing our November 8 launch. We broke quarantine and returned to our families.

Engineers would take a few days to mull over the evidence from the SRBs. We were told that the insulation on our boosters was probably okay, but the experts needed time to examine the matter more closely. Further research revealed that the erosion process presented no additional risk to the flight's safety. It reassured me that the shuttle program manager, Tommy Holloway, had given the booster team an extra week to finish their inspections and analysis of the eroded insulation. *Take all the time you need.*

The Flight Readiness Review reconvened on November 6 and concluded that there was sufficient confidence in the boosters to reset the launch date for November 15. On November 8, Friday, the five of us were back in quarantine.

Father J. J. McCarthy, the pastor of St. Bernadette's Catholic Church, generously agreed to join me one cool afternoon at crew quarters after he passed the required physical. We met under the live oaks in back of the building, where he heard my confession, then laid out an altar cloth on a picnic table to celebrate Mass. Both Story and Taco joined us under the trees, and I was gratified that they would take time to pray with us for the safety of our crew, the comfort of our families, and the success of our mission.

Quarantine cemented the close relationships among the astronauts and spouses. Liz knew Taco's wife, Joanie, from all those years with the Hairballs, and Mary Sue Rominger grew from just a Clear Lake neighbor to become our good friend. Story was seeing Sharon Daley, one of the WETF divers, and Tammy, to my delight, invited my former crewmates Linda Godwin and Jeff Wisoff to our dinners. I wished we could take all of them along on *Columbia*—that would be the ultimate trip.

One of the memorable pleasures of the STS-80 quarantine was an invention from the fertile mind of Dom Del Rosso, one of the engineers behind the space station crane slated for testing on the first space walk. His "Dom-inator" was an oversized blender designed around an under-sink garbage disposal. Howling like a main engine turbopump, the device would instantly transform ice, whole limes, and tequila into a high-volume stream of space-qualified margaritas.

We recovered from the Dom-inator's output in time to fly to the Cape on Tuesday morning, November 12, followed closely by the spouses. But a major storm off the coast with high winds, heavy surf, and low cloud cover made the outlook for Friday's launch so pessimistic that shuttle managers suspended the countdown. Negotiations with the Air Force's Eastern Test Range gave us a date on the following Tuesday, the nineteenth, giving the storm a wide berth. It made no sense to fly back to Houston for just a few days, so the quarantined astronauts, our spouses, and kids settled in for a quiet week in Florida.

The delay enabled Tammy and me to visit the launch pad, where *Columbia*'s cargo bay was still open for inspection within the enclosed pay-load change-out room. Clad in bunny suits, we could scale the various work platforms and reach most of our EVA tools mounted in the anti-septic cleanliness of the cargo bay. We also examined both the ORFEUS and Wake Shield satellites, knowing that the next time we would see them, we would be lifting them out of the cargo bay in the high vacuum of space. Exploring *Columbia*'s interior on the pad heightened my antic-ipation of the launch, leaving me eager to clamber about the payload bay in a space suit a few days hence.

On Friday, our original launch day, the weather was terrible, with high winds, rain, and heavy surf pounding the Cape beaches. We waited out the storm with Beach House visits, movies at crew quarters, and daily workouts at the gym.

Fair skies and calm winds returned in time for launch day. We woke at about 8:00 a.m. for our 2:52 p.m. launch time; these were bankers' hours compared with my night-shift duty on the Radar Lab flights. After the traditional photo op around the crew quarters breakfast table, I headed for the suit room to dress for the ride to the pad.

In my crew quarters room, a small but comfortable space kept metic-ulously clean by the staff (the old Apollo-era rooms had been renovated and enlarged in 1995), I packed my Earth clothes in a small suitcase to await my return. I readied a duffle bag with a flight suit and sneakers to be sent ahead to Edwards Air Force Base in case of a landing diversion—in my experience an all-too-frequent occurrence. Finally I donned my pull-up diaper and a set of thermal underwear. Over them went a set of expedition-weight long underwear laced with water cooling tubes to be plugged into the seat-mounted thermoelectric chillers in the cabin.

Padding down the hallway in thick wool socks, I joined the rest of the crew in the brightly lit suit room. I tracked down Olan Bertrand, chief of the vehicle integration and test team, and turned my wallet over to him. As head of the VITT, he was the traditional keeper of the astronauts' personal effects during flight; we always teased him that with this financial incentive, he would be a prime suspect if anything ever went terribly wrong on a mission.

I sank into a large brown recliner to shimmy into my orange launch and entry suit. Completing a ritual dating back to Apollo days, I endured the pressure checks and filled my pockets with survival radio, spare wristwatch, pen, class ring, handkerchief, Swiss Army knife, barf bag, and thermal mittens. The suit-up is a quiet time for the astronaut to focus one more time on the equipment, make small talk with the suit tech, and reflect on what lies ahead. About three-and-a-half hours before launch, it was time to head for the Astrovan.

But not before observing one more tradition. Taco, Bob Cabana, Dave Leestma, and Olan Bertrand crowded around an equipment table and dealt a hand of five-card stud. Custom dictates that the crew can't leave for the pad until the commander loses a hand. Fortunately Taco managed an early and graceful exit. Inoculated now against further bad luck, we were off.

In the blaze of camera flashes and TV lights outside, we just had time to wave to a few familiar faces in the crowd opposite the Astrovan: payload engineers, orbiter technicians, members of the VITT team, old friends from around the Cape. Then the door closed behind us. Replacing that last burst of human contact was the isolation of the launch pad and orbit beyond.

Passing the Vehicle Assembly Building and launch control center, where our families would be arriving shortly, Cabana halted the bus and rose to leave. He would be flying the STA this afternoon, assessing weather and visibility conditions in case we had to fly an emergency return to the Cape. Bob wished us good luck, then led us in the Astronaut's Prayer: "God, please don't let me screw up." Thinking about all the chances I would have to do just that, I never uttered the words more sincerely.

Waving to the TV camera in the White Room, I held up a sign greeting Liz, Annie, and Bryce before turning to crawl through the hatch. Squirming to the right onto the flight deck, I looked around at this storied spaceship, the first of the shuttles. Twenty other crews had flown *Columbia* before us; what spaceflight tales these walls could tell! For the next two weeks I would be living in this cabin, and it occurred to me that one day I would be visiting this historic spacecraft, flown by John Young and Bob Crippen on STS-1, at the National Air and Space Museum. I noticed some retouched paint and the scuff marks of fifteen years of space service as I stood up on the back wall of the flight deck. Already in their seats, Taco and Rommel craned their necks around to say hello. Story looked over from his MS-1 seat. "Welcome, Jones," he said, eyes crinkling in greeting.

I hopped onto the seat back to strap in. *This was something new.* As Mission Specialist 2, the flight engineer, I would be perched on the flight deck just behind Taco and Rommel, my seat centered between them, to the rear of the center console. Soon the four of us were comfortably situated, and Tammy, alone on the middeck, checked in on intercom.

Inside the final five minutes, *Columbia* began to shake as the APUs limbered up the elevons, body flap, and rudder. Next the main engine nozzles swiveled in preparation for ignition. The vibration reminded me how fragile our craft was, balanced precariously on the booster skirts far below.

> TACO: This is just lightweight shakin', Story. The real shaking is going to come in a minute or two.
> STORY (*with a resigned chuckle*): Yeah, I know.
> TOM: I think it's those rats on the treadmill downstairs. [Fourteen lab rats in a middeck locker colony were catching a ride to orbit with us.]
> TACO: Those poor buggers. They don't know what's coming, do they?
> ROMMEL: OTC, PLT. Caution and warning memory cleared. No unexpected errors. [We had a clear fault summary display.]

ORBITER TEST CONDUCTOR: *Columbia,* OTC. Close and lock your visors, initiate O$_2$ flow, and enjoy a weightless Thanksgiving.

TACO: We sure appreciate all your work on the flight and this morning. We'll see you in a couple of weeks.

OTC: One minute thirty seconds.

TOM [*to the front-seaters*]: Go get 'em, guys.

TACO: Be ready for anything, everybody . . . and nothing's going to happen. [*Funny, I thought.* That's a good way to sum up my STS-68 pad abort.]

ROMMEL: One minute 'til we're outta Dodge.

TACO (*after a moment's thought*): I'm glad they don't do live heart monitors anymore.

STORY: *I'm* glad they don't, man. They'd abort this launch. [This set off a ripple of laughter in the cockpit. I suddenly remembered Jim Voss's story—about Story—waiting for the STS-44 launch. Tom Henricks, the pilot, who was about to launch for the first time, noticed that three-time-flier Story wasn't participating in the lighthearted banter on the flight deck. Tom said, "Story, how come you're so quiet over there?" His voice uncharacteristically serious, Story replied: "Because I'm . . . scared . . . to . . . death."]

OTC: Hold the clock at :31!

We listened intently as the launch control team engaged in a rapid-fire exchange of critical information about an apparent hydrogen gas leak in *Columbia*'s engine compartment. *Holy smoke! A hydrogen fire . . . how do we rate that?* Excess hydrogen trapped in the aft could touch off an explosion at engine ignition and do God-knows-what damage to the back end. The high readings came across the radio in a tension-filled litany: "Steady at 425 [parts per million] . . . 450 . . . 370 . . ."

"Come on down!" Taco pleaded.

". . . 590 . . . 570 . . . 570 . . ." Amid the readings, controllers discussed their diagnosis:

"This looks like low pressure leakage. So this is an acceptable amount of leakage for launch, sir."

". . . 500 right now . . . 575 . . . 610 . . . 600."

The propulsion engineers hurriedly conferred and the lead controller called the NASA test director (NTD): "It appears to me that we're on the edge but that this is an acceptable condition. My recommendation is that we continue." If Story had been scared on STS-44, what was he thinking now?

"Launch Director, NTD. Based on the trending, it's maintaining around 600. The launch team recommends continuing with the countdown."

The reply was instantaneous: "Copy that, NTD, you're clear."

As we tried to absorb the meaning of that launch control exchange, the NTD removed all doubt, moving smartly to proceed: "Copy, resume on my mark: three . . . two . . . one . . . Mark!" The digits began counting down again on the cockpit display. *Jeez! We're going ahead!*

"Go for autosequence start."

":25 . . . :20"

Taco said, "Okay, motors . . ."

"Come on, Rommel." I urged our pilot to give us three good engines.

Last words from the OTC: "Go for main engine start."

I braced for the rumble now shivering up the stack. "Three at a hundred!" Rommel cried, and then the boosters crashed into life. The shock of liftoff allowed me just a strangled "102, 102!"—my confirmation to Taco that the correct software was now guiding us in powered flight. To our left, the brilliantly lit gantry dropped from view like an imploding skyscraper. There was no doubt—we were gone.

"Auto, auto. There goes the tower!" Taco exulted. "That looks like a roll."

A few seconds later, "Through the clouds!" as we flashed through the scattered cumulus deck littering the blue November sky. The thin howl of the engines rose from below as *Columbia* arrowed east out over the Atlantic.

"Here comes the throttling," said Taco.

"Three at sixty-seven," confirmed Rommel.

"Coming up on Mach 1 . . . at forty-three seconds! We're hauling!" Taco verified what we all could feel. "Excellent," I agreed, as the engines climbed back out of the throttle bucket and eagerly squeezed us into our seat backs.

"Feel that throttle!" Taco called, and Rommel agreed. "Comin' on!"

"We're serious now, man!" I managed to add.

Houston called: "*Columbia,* go at throttle up!"

The oldest orbiter soared through 50,000 feet at Mach 2, accelerating upward even as the last hundred tons of propellant blasted through the booster nozzles. Taco watched the g-meter. "Tail off . . . $P_c$ less than fifty."

With a simultaneous bang and brilliant orange fireball outside the front windows, the empty boosters announced their departure for the Atlantic. "Bye-bye, boosters," said Tammy, coming on the loop from downstairs for the first time. The first two minutes were behind us. Our visors were up, we were off oxygen, and *Columbia* was riding on silk.

The flight deck ride was living up to its advance billing, I thought: "That was something, guys . . . Mach 5.8, going for 6!"

A few seconds later, I piped up again: "How quiet! That was fantastic so far!"

"What a ride!" said our first-time commander.

"That was something," I agreed, and Story chimed in, too: "That's the easiest one I remember."

"What a view out that back window! That was fantastic!" I had my mirror in my left hand, aimed loosely out the overhead windows. At liftoff I had sneaked a peek back into the flame trench. I was almost sorry I did: a gray jet of steam had rocketed out from the pad, laced with yellow tendrils of main engine exhaust until the boosters overwhelmed their efforts with a burst of pure white incandescence.

Story said, "Yeah, the beach, the different-colored water, even the clouds. All the way up to SRB sep[aration] you could see clouds in the back window."

I was still stealing glances out the back. "Look at that. Look at that!"

Story added, "That's nice clouds and water out the back window."

"Mach 9. What a beautiful ship!" I replied. *Columbia* was living up to her reputation.

Taco and Rommel compared velocity indications at Mach 10. Guidance was right on track. Our commander glanced at the altitude tape: "We're above fifty miles." We were all veterans, but he smiled and added, "Anybody that wasn't, you're an astronaut now."

*Columbia*'s three engines kept churning, faster and faster.

"Getting a little g now . . . a bit of debris, high off the tank," noted Taco.

"Yeah, I saw that, too," answered Rommel.

"Got the horizon out the front window, goin' for speed," Taco observed. "Boy, it's pretty over the Atlantic. I can see blue water and white clouds. Looking great."

Rommel reported: "Mach 15, 2 g's."

I checked my mirror again. "You can see the flame out the back end!" Expanding in the vacuum, our main engine plume was licking up the orbiter's tail and fuselage. "There are orange flames coming up from the back window. Great!"

"Look at the glint and glitter on the water," Story added, amazed.

"About two-and-a-half g's," noted Taco.

"This is a higher g than I remember from last time," Rommel commented.

Taco looked over at him, amused: "2.8 g's is higher than you remember last time?" Every shuttle ascent profile was nearly identical; Rommel meant that this g-load just *felt* higher.

"Mach 18," I called as the velocity tape slid past perceptibly faster on the instrument panel.

"Okay, Rom. Shutdown plan?" Taco was teasing. In every simulation he had always asked Rommel how our many malfunctions would affect our engine shutdown plan. This was unlike any sim we had ever had: *Columbia* was clean as a whistle.

Behind a big smile, Rommel said, "It's nominal, Taco."

"Roger, thank you for that report," and we all had a good laugh. The g's were heavy, but the mood was light.

"Three g's. Expect throttling," said Taco.

"Through Mach 20, about another minute to go," my voice now reflected the effort it took to breathe against the g's. "Twenty-one Mach . . . twenty-two!"

Taco spoke for all of us: "There's a gorilla right on my chest!"

I was still sneaking looks out the back. "Can still see the fire out there, that's cool. . . . Through 24 Mach, looking for 25.9."

"Twenty-five," called Taco. "We're at idle," Rommel confirmed. A few seconds flashed by, and suddenly all five of us were in free fall.

"Whooh! We're in orbit!" I gushed. I felt the familiar lightness under my harness. With a bang and the cannon-like thump of the primary thrusters, *Columbia* spat flame and edged down and away from the spent ET. The longest shuttle mission in history was under way.

Tammy did a remarkable job on launch day, quickly checking out the remote manipulator system (RMS) and using it to deploy the Orbiting and Retrievable Far and Extreme Ultraviolet Spectrometer Shuttle Pallet Satellite; everyone gratefully called it ORFEUS/SPAS. The entire crew worked flawlessly together that first day, with Story setting up the middeck, the pilots managing flight deck activities and maneuvering *Columbia*, and Tammy and I working out the back windows to ready the satellite for launch. She grappled and held ORFEUS aloft, about twenty feet above the payload bay, while the science team at the Cape used the orbiter's communication links to check out the spacecraft and telescope systems.

Riding on the SPAS carrier spacecraft, ORFEUS was an astrophysical observatory designed to explore the very hottest stars and the coldest interstellar gas in our universe. ORFEUS's one-meter-wide ultraviolet telescope had to operate above the atmosphere to capture light from these exotic sources and relied on the shuttle to set it adrift for its fourteen-day observing program. The satellite, which also carried a high-resolution spectrometer called the Interstellar Medium Absorption Profile Spectrograph (IMAPS), would trail *Columbia* by about thirty miles as it independently studied the stars, galaxies, and cold interstellar gas. Although not as powerful a light-gatherer as the Hubble Telescope, ORFEUS could see farther into the ultraviolet region of the spectrum and at greater spectral resolution than its more famous orbiting cousin.

After an hour delay for additional checks of the telescope and satellite, Tammy gently released ORFEUS into orbit. Flying from the aft station on the flight deck, Taco nudged *Columbia* down and away from the satellite to set up our thirty-mile separation. Our first day in space, a very long one, concluded more than eighteen hours after our wakeup at the Cape that morning. We were all bushed.

We had a very beautiful and delicate deploy.... As *Columbia* very slowly crept away, we had a very long view of the SPAS out our overhead windows. And that was probably around 8 hours, 15 minutes (MET) on the deploy. Since then we've been doing a lot of housekeeping, putting the ship to bed for the night, and getting ourselves ready for bed. I'll get maybe about eight hours altogether from bedtime to wakeup. Gorgeous trip out of Florida today. Hope my family really enjoyed it; it was sure spectacular from this end. The sensations of launch were really tremendous: big shaking on the pad when the main engines lit; big lurch at liftoff; then a real acceleration when we came out of the throttle bucket and headed past Max Q and up towards SRB sep[aration]. A real surge from *Columbia*. There was some vibration in second stage, some steady throbbing from the main engines. Also some spectacular views out the back windows: saw flickers of flame from the main engine plume creeping up the back of the vehicle.... And as we head for bed, just looking at the delicate blue—the eggshell blue—of the Earth, the thin blue atmosphere from 190 nm nautical miles [218 statute miles]. The atmosphere looks a little thinner this time versus the last, as I recall [due to my higher orbital altitude on this flight]. Lots of clouds over the Pacific now, not much land in sight at all. I'm going to head on to bed. It's been a long day and I'm sort of tired. Still floating with some ice crystals in formation with us, too.... Amazing they can still be here lingering around the vehicle, 12 hours after launch.

On flight day 2, Tammy and I set up and laboriously tested the Space Vision System (SVS), a TV-computer combination designed to provide precise robot arm guidance for Space Station assembly crews. In this early experimental version, SVS used two laptop computers and what seemed like a mile of power and video wiring.

**MET 1/10:44.** It's early morning over [north] central Africa, the Sahel, and we're looking over a lot of clouds over central Africa today. We're in a belly-forward, left-wing-to-the-ground attitude. SPAS is behind us. Not a very good Earth-viewing attitude, because we only have one window that's facing the ground, and that's the commander's window. It's been a long, long day: got up over 15 hours ago. We spent almost a good eight hours of that working on SVS today. Very slow, methodical work with computer and camera, nothing very exciting there. And a huge amount of wiring to do. What an incredible amount of overhead for a system. Did have some fun flying the arm today, but because of a headache, and backache, and stuffy nose, I didn't really enjoy most of the day that much because of the way I was feeling.

**MET 1/12:05.** It's about an hour before sleep on flight day 2. I just opened up the special flight data file locker, and I got the pictures of Liz and the kids out, and the drawings by Bryce and Annie. Those were really appreciated. Nice to see those things from home, along with a bunch of jokes from our instructors. I'm sitting sideways, looking right out the window...the side hatch window...at the Earth below, and clouds and early morning sun coming in across the Earth. Pretty bunch of clouds there this morning. This is probably the eastern Atlantic.

I didn't have much of an appetite today, thanks to the stuffy head, backache, headache and queasiness. Some of the odors of the food seemed to provoke queasiness...an upset stomach this evening and this afternoon.

Later I found that my upset stomach was due to a reaction to the decongestant that I'd taken. My appetite returned when I switched to a different medication. I treated anything that bothered me with a visit to the medical locker, my motto for space ailments being "Better life through chemistry." The flight surgeons dispensed advice freely during our daily radio chats, and I soon felt fine.

**MET 2/12:00.** We've been up two and a half days in orbit.... Finishing up my third workday in space. Most of today...was eaten up by checking out the space suits and preparing the airlock for EVA. All three suits work fine. It was sure a mess in the middeck while we checked 'em all out, but we finally did get everything put back together in the airlock. The suits are ready for their trip outside, in about a week.

[Flight Surgeon] Rainier Effenhauser...just told me to watch out for the Hostess Snoballs. Guess I'll have to eat some of those tomorrow. It's actually becoming enjoyable in orbit today, as my body adapts to the experience.

I'm enjoying floating a lot today, enjoying the weightlessness and the extra dimension it gives you for sharing space, and working together. I was able to maneuver these big spacesuits that weigh several hundred pounds out of the airlock with some finesse.... Listened to some stereo today: Natalie Merchant's *Tiger Lily* album, *The Rocketeer* soundtrack, and...the sounds of rain and a thunderstorm.

After nearly a year of training, the five of us were at ease together. Taco had been a friend for more than six years, and I respected his flying and leadership skills. Story was tremendously experienced in all things

EVA, so I knew our space-walk preparations were in good hands. Tammy and I had trained together under water for so long that we could practically read each other's minds. Rommel was not only an extraordinarily competent pilot but was also armed with an inexhaustible supply of jokes. He and Taco once worked through an entire afternoon telling us all about the funny call signs, or Navy nicknames, they'd known (like their pilot colleague, last name of Wise, whose call sign was, of course, "Notso"). They soon had my stomach muscles aching from near-constant laughter. STS-80 *Columbia* was a happy ship.

Flight day 4 was my first big test on the mission: deploying the Wake Shield satellite. Wake Shield was designed by NASA and the University of Houston to be a free-flying factory for thin-film semiconductor wafers, the basic raw material of computer chips. Using its saucer shape to plow through the wispy remnants of the atmosphere 200 miles up, the satellite would create an ultrahigh vacuum in its wake, a thousand times emptier than those in Earthbound laboratories. Wake Shield's contamination-free semiconductor films, built up layer by atomic layer, might outperform terrestrial chips and eventually lead to a thriving manufacturing operation in space.

That was the theory. To test it, we would have to clean Wake Shield's surfaces and launch it free of any contamination from our exhaust-spewing thrusters. Kicking off a delicate five-hour ballet in orbit, I grappled the satellite in the aft end of the cargo bay just before Taco shut down *Columbia*'s steering jets. We were in free drift, the orbiter's attitude controlled only by atmospheric drag and its own angular momentum. I had a tense few minutes as I tried to lift Wake Shield from its berth. The satellite seemed to be stuck on the seals that protected its manufacturing surfaces, and it took me a little extra maneuvering with the robot arm to break Wake Shield free and hoist it over the orbiter's port side. We hung the satellite "out in the breeze" for several hours, letting the upper atmosphere's highly reactive atomic oxygen scour its surfaces and clean them for the three-day manufacturing run.

With checkout complete, I delicately pulled the arm away, turning the satellite loose about twenty-five feet above the cargo bay. Under the control of its engineers, Wake Shield would now fire its own cold nitrogen thruster to move off into its trailing position some thirty miles astern. We would keep our thrusters off until it was several miles away.

But the Wake Shield engineers delayed the thruster firing, waiting to be absolutely sure the satellite's attitude control system was functioning. The delay threw quite a scare into us. Wake Shield and *Columbia* moved inexorably along their two distinct paths, and orbital mechanics would bring them together in space about half an orbit—forty-five minutes—later. We hadn't planned for that, and as the command went up to trigger the nitrogen thruster, we noticed Wake Shield ever-so-slowly descending toward our overhead windows. Taco couldn't fire the orbiter thrusters to keep clear without contaminating the now-clean Wake Shield, so we gritted our teeth and watched the four-ton spacecraft draw ever closer. All eyes on *Columbia* were riveted on the edge of Wake Shield's saucer-shaped disk, and watching me snap photo after photo of the looming satellite, Taco suggested I might want to start planning how to grab the thing instead. But it was already too late for the arm. The only way to grapple Wake Shield was to fire our thrusters and move back under it, delaying its mission by a day and expending its limited battery power. Taco decided to wait it out.

Closer . . . closer. . . . Now the satellite's rim was a mere ten feet from our cabin roof. Its forward motion might carry it clear over *Columbia*'s sloping nose, or a sudden dip might force Taco to move away under thruster power. "This thing needs curb feelers," Story said as sweat broke out on our collective brows. When Wake Shield closed to within five feet of our windows, Taco nearly threw in the towel, but now its motion was almost parallel to the orbiter's fuselage. We might just skate free. The satellite drifted by so close we could see the rivets on its EVA handrails; it filled the viewfinders on all our cameras. Wake Shield finally cleared the nose and sailed off on its three-day mission, unperturbed at the tension it had created in the cockpit.

**MET 6/2:00.** Just a gorgeous view of the Wake Shield, much more brilliant now than the SPAS, moving lower in the sky against the Earth as we get closer to our rendezvous, heading towards the $T_i$ [a key rendezvous] burn. The shuttle is floating through space surrounded by very bright, sparkling crystals of thruster residue. Against that foreground constellation, flying with us against the dark cloudy Earth below, we saw the Wake Shield brighter than Venus, and the SPAS trailing it behind. . . . Really something.

Taco, Rommel, and Story pulled us up twenty feet below Wake Shield on flight day 7. Taco lined up the grapple fixture in the TV monitor, fired a thruster burst to kill the drift rate, and said, "It's all yours, Tom." The grapple was the moment of truth for an arm operator. Taco had carefully lined up the end of my robot arm with the grapple pin, and now it was up to me. The dozens of hours of practice in Building 9's arm trainer and the shuttle simulator kicked in. I squeezed the hand controllers and moved in for the pickup.

The arm's initial motion surprised me, though. The initial kick of the arm's joint motors toward the satellite set up a vibration, a circular wobble that I could see on the TV monitor as I closed in. The fifty-foot arm, weightless in free fall, continued to oscillate as I moved steadily in over the grapple pin. Waiting for the motion to damp wasn't an option, as Wake Shield was drifting slightly, too. I just had to work through it. Heart pounding with adrenaline, I moved the end-effector gingerly over the end of the grapple pin. I cringed at the prospect of bumping Wake Shield and forcing Taco to come around again. *Please, Lord, don't let me blow this.* A foot to go, then six inches. I finally halted just three inches from the satellite, that pesky wobble still noticeable but almost gone. Good enough. I squeezed the trigger on the right-hand controller, and the snares snapped shut over the grapple pin.

With Wake Shield back aboard, I thought about those last few feet of the grapple, the most anxious moments I'd yet had in space. If I had bumped the satellite into a tumble, we probably would have recovered, but it would have cost us precious time and fuel. Not for the first time, I was conscious of how many people, here on *Columbia* and back on the ground, were counting on me to deliver the goods. The pressure was real. Very few situations in spaceflight gave one the luxury of saying, "I'll do better next time."

LETTER HOME, THURSDAY MORNING, NOVEMBER 28, 1996:
*This has been a fun day setting up space suit gear and checking out all our tools. Everything went well. Did you see us on TV setting up all the stuff in the middeck? Since this morning we've been studying and taking pictures and cleaning up the middeck after making a mess this morning. I feel fine and have exercised every day since day 3, though I'm taking today off to rest up for tomorrow. Stuffy head seems to be the worst of my*

*problems. . . . We let some air out of the cabin last night to get*
*our bodies acclimated to the lower suit pressure and reduce our*
*body levels of nitrogen, which might cause decompression*
*sickness. No worries. What a kick to go out the airlock*
*tomorrow!*

*Our reunion is just a week away, and amazingly the mission*
*is over half over. I am amazed at how we adapt to living in such*
*a little space as the crew cabin for over two weeks. Wish you*
*were all here to play. . . .*

*Liz, you are in my dreams. I am soooooooo lucky!*

*Love*
*Tom*

I turned in for sleep looking forward to a memorable Thanksgiving, marked by that long-awaited space walk.

OUR STRUGGLE IN the airlock and the cancellation of the EVA on flight day 10 made the traditional American day of thanks an empty holiday for our crew. I tried to strike a light tone in the note I hurriedly sent down to Liz that evening.

LETTER HOME, THANKSGIVING EVENING
(FRIDAY MORNING, 2 A.M. IN HOUSTON):
*Well, Lizzie and Kids, it's been an interesting day. On the*
*plus side, I've successfully gotten into an EVA suit and managed*
*to survive in it while I let all the air out of the airlock. So you*
*could say I've been on an EVA. And I successfully repressurized*
*the airlock (well, Tammy did most of the switch throws) and got*
*back out of the EVA suit. Now, on the down side, we didn't*
*manage to get out into the payload bay. The outer hatch is*
*jammed, and we couldn't budge the handle open enough to even*
*crack it. It moves about 30 degrees, and then stops. After about*
*2 hours at vacuum (I'm sure you were watching), we gave up*
*and repressurized. The experts are looking at things and maybe*
*they'll figure out a way to unjam it. I have my doubts. If the*
*mechanism is jammed inside the hub, we can't get to it, and of*
*course we can't get to anything outside to unjam it. Tammy*

*pushed as hard as she could on the handle, and so did I. I even pushed my foot against it and it wouldn't move. So it's not a strength problem. There's something wrong with the little bugger. Locking it back up if we get it open is no problem . . . the air inside the airlock pushes it shut with tons of force, even if we can't latch the handle again.*

*Well, I don't know where our spacewalks will end up, but Mission Control will do their best to crack that hatch. I'll talk to you again tomorrow, I hope after the problem is solved and we manage to get our EVA in. I'm in good spirits and I'm thinking of all of you. I hope you saw us eating our turkey on TV, as well as part of suiting up. Thank you for your daily letters and keep sending that love out this way. I miss you all and look forward to seeing you very soon.*

*It occurs to me that this is actually the first disappointment I've had on a mission. Now I know what the Tethered Satellite guys felt like. [The tether holding the research satellite had either jammed (STS-46) or snapped completely (STS-75), and both scientific missions had ended in disappointment.] Hope it doesn't play out that way. I can imagine what the Friday morning and evening news shows are going to do with this story.*

*All my love,*
*Tom*

The morning after the canceled space walk, I woke only reluctantly from an exhausted and anxious sleep, peering out groggily from the bottom sleep station at my crewmates on *Columbia*'s middeck. The wake-up call from Houston offered no hint that the ground had come up with a solution to unlock our jammed hatch, but I wasn't alert enough yet to have much appetite for a technical discussion with Mission Control. Half out of my sleeping bag, I rubbed my eyes and tried to stay optimistic that an answer to the hatch jam would soon materialize.

While we conferred with MCC, Tammy Jernigan and I could only stare out the rear flight deck windows at our EVA hardware in frustration. We wondered how we would get out the door and what we would be able to accomplish now that 50 percent of our space-walking time had just evaporated. It was just twenty feet from the crew cabin to our work site outside, and I found it inconceivable that we wouldn't somehow find a way to traverse that distance.

This Thanksgiving had taught me a new definition of the word *memorable*. Back on Earth, there was a mad scramble to discover the cause of the jammed hatch. The only flight-like orbiter airlock hatch at JSC was in the Building 7 vacuum chamber I'd trained in before liftoff.

Unfortunately, the mechanical linkages that may have failed were all on the inaccessible far side of the hatch, in a small vestibule connected to a larger vacuum chamber by fifty yards or more of narrow ducting. With his uneaten turkey dinner grown cold on his desk, Dom Del Rosso joined the troubleshooting effort. At about 1:00 a.m. he tied a rope around his waist and crawled into a manhole heading to the back side of the test airlock, dragging a TV camera and cable through the cramped passage to show flight controllers how a fouled mechanism might be causing our orbital jam.

Dom spent hours crammed in the tiny vestibule next to the hatch, operating the handle mechanism and linkages over and over to identify the possible failure modes. He remembered that there was no panic or hysterics, just "people buckling down and getting the work done." It became clear to him and the flight control team that the crank mechanism itself was the problem, but it was accessible only from the payload bay side of the hatch. There was no obvious way the astronauts could repair it from inside the shuttle's airlock.

Caught up in the supercharged atmosphere permeating both MCC and Building 7, he was ready to stay in the pipe indefinitely, or to suit up for repair work in a vacuum chamber run. "It was the reason I joined NASA," he said, but he later confessed he harbored one secret fear: "I hoped my blender had nothing to do with this problem."

Augmenting the work of Dom's team, astronauts Leroy Chiao and Jerry Ross had flown to the Cape to examine an actual orbiter hatch, and a team of structural engineers systematically reviewed possible causes for the failure. Next morning aboard *Columbia*, we expected to wake to news that the team had found the key to unlock the jam. Tammy and I had high hopes that we'd suit up later in the day to tackle that stubborn handle.

Instead we spent several hours taking measurements around the hatch's interior surface and the airlock bulkhead, looking for a misalignment. Tammy and I fitted makeshift tethers to a crowbar and mallet from *Columbia*'s toolbox, then found the spots on the hatch rim that would give them the best purchase. If we got the go, we were confident that we would be able to "nudge" open the hatch on our second attempt. We ended the day after Thanksgiving readying our equipment for the big breakout.

Flight day 12 was originally scheduled for our final EVA, but we still had enough orbit time to pull off both space walks if we got the door open. As I listened to Houston's wake-up music that morning, my spirits rose. Dom Gorie, our capcom, was playing the Doors' "Break on Through to the Other Side." But my hopes of following that musical advice were shattered by his first words.

"The Mission Management Team [meeting] just finished up. . . . The sense right now is that EVAs are not going to be a player for this flight."

I glanced around the cabin. Tammy floated motionless, holding the microphone. Her shoulders seemed to sag even in free fall. Drifting from my bottom bunk, I shook my head at Tammy and cursed privately. From his sleeping bag, Rommel said he couldn't believe it. Taco and Story were already adrift in the middeck—their eyes searched our faces for a reaction. I'm sure they saw quite an interesting display of emotions—my year of underwater training had just gone down the drain.

The ground "Tiger Team" attacking our problem had weighed the evidence and concluded there was an obstruction inside the handle mechanism itself. Once convinced of the point of the failure, NASA managers made the only call possible. The airtight cover over the handle's internal gearing was not designed for removal in space—breaching it would depressurize the airlock. Even in our space suits we had no way to remove it without prying or sawing it open, since the fasteners were on the hatch's outer surface. If we had forced open the cover to fix the jam and couldn't reseal it on our way back in, Tammy and I would be unable to repressurize the airlock. We would be marooned at vacuum on the wrong side of a pressure bulkhead. At that point, we would have only a few hours left before our suit carbon dioxide scrubbers became ineffective. To save us, our crewmates would have to execute an emergency reentry with us in the airlock, abandoning in orbit the multi-million-dollar ORFEUS/SPAS telescope we had deployed on launch day. Breaking through to the other side just wasn't a smart option.

The frustration onboard was palpable, and it led to some anxiety over an upcoming press interview that followed a day later. It wasn't too hard to guess the focus of most of the reporter's questions. We went on camera the next day, flight day 13. The CNN inquisitor was friendly, but he didn't pull any punches, either. After a few questions about the impact of the lost EVAs on NASA's plans for the Space Station, he got to the point:

"You seem pretty much subdued as we talk right now. Am I right about this, or is my question leading you down a very serious and somber path today?"

Tammy acknowledged the question with a forced smile and a nod, her brunette ponytail bobbing high behind her head. "No, I think certainly that we are feeling some combination of disappointment at the failure of the hatch, yet pleasure at being part of this mission that has been, in every other way, very successful."

I was asked if I had any final comments.

"Well, disappointment naturally comes into our feelings, and I hope for better times on future missions . . . when I'm assigned to a space walk. And I'll be an optimist about that too!" (I threw a not-so-subtle hint to my boss back on Earth.)

"But you can't take away the fabulous nature of this experience, and last night, after our EVAs were canceled, I spent an entire night orbit of the Earth, looking down at thunderstorms and out at the Southern constellations . . . and watching the lightning flicker off the surfaces of the orbiter. And that's such an experience—unreachable on the ground— that I'm very privileged to be here. I have no basis for complaints."

MET 13/12:20:16. . . . When they canceled the EVA yesterday, that was the last chance we had to go out into the payload bay. I'm fairly philosophical about this, but after having put in almost a year of training to get EVA-qualified for this mission, and not have it happen, I think I'm justified in feeling rather frustrated. I don't look forward to another year of EVA training in order to get another flight with this sort of WETF training. . . . I don't know what's in the cards. . . .

Our disappointment over the lost EVAs was aggravated by the nagging worry that some vital component in the payload bay might fail. We had no way to get out and fix it. Our hatch jam was what engineers call a "smart" failure—one not easily overcome by simply switching to a backup system: there was only one outer hatch, it had only one operating mechanism, and its workings were unassailable from inside. The payload bay doors were our biggest concern: it was critical that they close for entry, yet without access through the airlock hatch we had no way to clear jammed hinges or attach backup door clamps. But after a couple of days of brainstorming, Mission Control had faxed us up a preliminary strategy for enabling a last-ditch EVA.

Glenda Laws, our lead EVA instructor, had worked with the rest of the flight control team to find another way to get outside. First they considered putting the rest of the crew in their launch and entry suits while Tammy and I depressurized the cabin and went out the side hatch. Very dicey. Many critical orbiter electronic systems needed air cooling to survive; we might very well cripple *Columbia* while trying to repair her. Worse, a pressure suit failure in the cockpit would have been fatal. Plan B made more sense but still left us wondering if we could pull it off: Tammy and I would depressurize the airlock, then use hammer and chisel to punch through the thin aluminum hatch in two places. Inserting a hacksaw blade through the holes and cutting the hatch handle linkages on the outer surface, we would thus bypass the jammed mechanism and yank the hatch off its seals. Once our orbiter repairs were complete, we would replace the hatch, patch the two holes using putty and our metal wrist mirrors, and repressurize. In a worst-case scenario, unable to close and seal the hatch, we would have ridden to the ground in the airlock as the crew executed an emergency deorbit. That would have been one heck of a reentry experience—one to tell the grandchildren.

We had a lot of time in those final four days in orbit to mull over these failure scenarios. In truth, my most contemplative hours in space came during those last few days aboard *Columbia*. Before turning in each night, we would turn down the cockpit lights and drift with our faces just inside the flight deck windows, taking in the view of Earth and sky for a whole ninety-minute orbit or more. One by one, my friends would head below for sleep, until finally just two of us were left to drink in the unparalleled view. I tried to capture a sense of the wonder of it all in my diary.

I just spent a marvelous night pass upstairs with Story, sweeping across the Indian Ocean, out across the Philippines. If I can describe it: I saw more stars than I've ever seen from the ground. The stars were brilliant in the Southern Hemisphere—the Southern Cross was skimming the southern horizon the entire time, right down through the airglow. Saw the Large and Small Magellanic Clouds, and while I was doing this I was upside down in the commander's window, so I was floating with my belly towards the Earth. The shuttle was upside down, so there I was lying on my belly, upside down in the shuttle looking at the horizon. Thunderstorms over southeast Asia, the spangled cloth of the Milky Way stretching from my lower left to my upper right in my field of view towards the Magellanic Clouds. The Southern Cross on the horizon. An occasional meteor below. Saw the city of Bangkok all lit up. Just a marvelous sky. I [have] told school audiences that if you go out on a dark night you'll see as many stars as you can from space, but it's not true. There are more stars in space than you can see from the ground. And they're all bright—the brighter stars don't stand out as much. Part of a marvelous night scene from the orbiter *Columbia*...."

Thanksgiving had finally come to *Columbia*'s crew.

# 11

# THROUGH THE FIRE

*That was a real fireball, boy.*

JOHN GLENN, *MERCURY-ATLAS 6*, FEBRUARY 20, 1962

Weightless for a few final minutes, the five astronauts on *Columbia*'s flight deck toasted the end of the longest shuttle mission in history. But the champagne of choice that last morning in orbit on December 7, 1996, wasn't vintage bubbly. Instead it was a hard-to-stomach combo of water and salt tablets. After nearly eighteen days our bodies had adjusted to the lower blood volume required to function in free fall. In space, fluid no longer pools in the legs and gut, so the kidneys dump the excess in the first few days in orbit. Under gravity's returning onslaught, a fluid deficit could result in nausea, dizziness, or even fainting. My goal was to down at least forty ounces of water in the hour before "entry interface"—that arbitrary boundary at 400,000 feet where a reentering space shuttle first begins to sense the atmosphere. So, "Cheers!"

Drinking this stuff was a nasty business. Every fifteen minutes I sucked eight or ten ounces of chilled water out of a foil pouch, chasing each drink bag with a pair of salt tablets to aid fluid retention. I kept the silvery pouches floating close at hand in a mesh bag tied off to the back of the commander's seat; after each salty cocktail, I could feel my stomach swell with its growing fluid load. Capacity was a real problem, since my g-suit was especially snug around the waist. The orange launch and entry suit and cinched-down parachute harness straps added nothing to my comfort level. Bloated, I drifted gently under my lap belt and shoulder harness, perhaps half an inch between my LES-clad bottom and the seat cushion.

Our fluid loading routine seemed justified this morning, because after two previous days of bad-weather wave-offs, conditions at the Cape were finally a go. With spare gear packed away behind panels and locker doors, the cabin was neat and tidy—so different from the cluttered decor of orbit. Entry-and-landing checklists drifted lazily at hand on their tethers, ready for the reappearance of gravity. Crisp, business-like chatter among the flight deck crew replaced the easy banter of the last few days aboard Columbia. Downstairs on the middeck, Story Musgrave ran through the last few items on the checklist to prepare the lower cabin for entry. As the countdown for the deorbit burn progressed, he powered down the waste management compartment for good. Internally awash in salt water, I longed for one last chance to visit, but I was already strapped in for landing.

Columbia soared at five miles per second toward a sunset over Australia, the last we would see from our 220-mile-high orbit. Houston had given us final approval to initiate the deorbit burn on schedule. In preparation we commanded our four general-purpose computers to fire thrusters and gently maneuver the orbiter to a tail-forward attitude. Pointed into the direction of flight, the twin engine nozzles of the orbital maneuvering system (OMS) were ready to brake us out of orbit. We five faced backward along our flight path, our heads pointed down toward the Indian Ocean, but we felt as comfortable in this orientation as any we had experienced over the past eighteen days. Up, down, backward, sideways: it was all the same to us. Out the front windows, Earth floated quite naturally above our heads. In our final moments apart from it, all was right with the world.

In the flight engineer's center seat just behind the pilots, I mentally reviewed our preparations for deorbit. To the tip of my "swizzle stick" (the aluminum baton that enabled me to reach cockpit switches while strapped in), I taped a lipstick-sized video camera, its thin cable trailing back to a recorder stowed in a nearby locker. By aiming the camera with the end of my wand, we hoped to shoot video footage inside the cabin without having to heft an increasingly heavy camcorder. (On my first reentry three years earlier, a video camera felt like a fifty-pound sandbag to my weakened muscles.) To supplement the camera footage, Tammy Jernigan, strapped in a seat to my right, readied an intercom recorder to tape our cockpit conversation.

Our impending departure from orbit had more than a hint of the bittersweet to it. From a scientific standpoint, our mission had gone very well. We had retrieved both of our satellites (ORFEUS-SPAS and Wake Shield) after their successful free-flying missions. Our nearly eighteen days in orbit had been that rarest of gifts: we had watched the world go by at an unhurried pace envied by our colleagues in Houston. After two bonus days in space, we were now bound at last for a happy reunion with our families at the Cape. But we would return from orbit still stung by the disappointment of two lost space walks, and we harbored a nagging worry that the jammed airlock hatch might work perfectly once back in 1 g. In that event, Tammy and I would have "a lot of 'splainin'" to do.

As the GPCs ticked off the final seconds to the burn, Taco offered me the honor of starting us on our way home, hitting the EXEC button, which would tell the computers to proceed. I was delighted at the gesture, but strapped into the MS-2 seat, I was about a foot shy of the necessary reach, and the swizzle stick was brought up short by the TV cable.

Taco accepted my regrets with good grace, and his finger was already on the keypad when, with fifteen seconds to go, Columbia's computer display flashed a blinking green EXEC signal. With the push of a button, Taco turned the burn over to the computers. Now it was up to them and those two stubby engines beneath Columbia's tail.

We watched on our systems displays as the engine valves opened and helium surged into the propellant tanks, forcing fuel and oxidizer into the feed lines. On command and on time, propellant isolation valves slapped open; rocket fuel sprayed into twin combustion chambers and exploded on contact. Ignition!

We heard and felt a firm thump behind us as the OMS engines blasted their exhaust against our forward motion. "Clock is running," I called. The chamber pressure gauges showed good thrust. In the brightly lit cabin, we could feel the shuttle slowing, and our bodies settled gently against our seat backs under about a tenth of a g of deceleration. Five pairs of eyes zeroed in on the display to catch any sign that the engines were malfunctioning. All perfect so far.

Our deorbit burn would last just a bit more than two minutes, the

12,000 pounds of thrust from the two OMS engines slowing us by a mere 200 miles per hour. But it was enough. Our minimum orbital altitude, our perigee, had already dropped well below the 220-mile starting point, heading toward a low just 47 miles above the surface.

"Through one-twenty for eighty," I called, noting our rapidly shrinking perigee altitude ($H_p$, or height of perigee). In twenty minutes or so our path would dip low enough to encounter the upper reaches of Earth's atmosphere, and *Columbia* would once again be a captive of the home planet. We watched engine instruments, our perigee altitude, and the clock steadily ticking down toward the end of our burn. "We're below Engine Fail $H_p$," Taco called. If an engine quit now, we would continue burning with the other OMS engine to hit our deorbit target.

The perigee kept dropping, now below eighty miles, too low for a safe orbit. Atmospheric drag would pluck us from space on this next revolution. If we lost both motors now, we would use our small thrusters to reach our target perigee.

Through seventy nautical miles now, heading for forty-seven. "Can you believe it, Jones?" Taco glanced back at me. "You're heading to KSC. That's where your wife and kids are." He was tweaking me about my two previous landings in California while my family waited in vain at the Cape.

"Incredible," I murmured. After the turmoil of the wave-offs the last couple of days, finally heading for KSC was almost too good to be true. No one was happier than me to be headed for Florida (except Liz, perhaps), but I had deliberately curbed my hopes until now. Evidently the canceled EVAs had fulfilled my quota of bad luck for this mission.

The burn continued. "Twenty miles to go," Rommel called. The remaining altitude melted away, vaporized in the silent blast from the OMS engines. Fifty, forty-nine . . . target $H_p$ . . . *Now!* Cutoff!

The twin chamber pressure needles slid back to zero on the gauges. We sensed weightlessness again. Taco threw the two toggle switches out of the ARM-PRESS position. The old OMS had performed its last job flawlessly. We were coming home.

Curt Brown's voice crackled over the radio. "We see a good deorbit burn, good residuals," he called laconically from Houston. Our free fall toward the top of the atmosphere had begun.

"There's our attitude. Let's maneuver to it." Taco punched in an "Item 27" and commanded a nose-up rotation to get our belly pointed forward into the anticipated Mach 25 slipstream. I ticked off checklist items with Rommel, shutting down the OMS system for good. Next I activated our in-cabin camera. Sunset was coming on, and the crew was quiet for several minutes. Light faded from the sky outside as we watched, weightless, enjoying these last few minutes of orbital free fall.

Story drifted near us on the flight deck, having monitored his last deorbit burn. He was suited for his final trip home, his seat and parachute waiting below on the squared-away middeck. Its neatness testified to Story's Marine training some forty years before. Soon he would be strapping in downstairs, but the adventurous Dr. Musgrave remained true to form: he wanted to stay on the flight deck long enough to see the first traces of the reentry plasma marking our hypersonic plunge into the atmosphere.

*Columbia* fell toward Earth, her nose raised forty degrees above the horizon, her tiled belly ready for the atmosphere. We had no view of Earth, in darkness out the side windows, but on the computer screens we watched the seconds march down to entry interface. While Taco donned his helmet, my eyes were drawn to a weird sight outside Rommel's right-hand window: luminous crystals drifted forward past the cockpit, glittering against the surrounding blackness. We were engulfed in a cloud of John Glenn's famous "fireflies." I had the momentary impression that we were flying *backward* through a soup of glowing snowflakes. Although I knew we were still racing forward and down at 17,500 miles per hour, the optical illusion was a powerful one. It reminded me instantly of how an adjacent car drifting slowly forward at a stoplight could cause me to jump on my brakes to arrest my backward "motion." My guess was that the fireflies were frozen flakes of OMS engine exhaust, perhaps knocked loose by our thruster firings.

"Thunderstorms," Taco murmured as the dark Earth below flickered out his left side window. Rommel smoothly set up a final firing of our nose thrusters to consume the last of their now-surplus fuel. For twenty-four seconds, *Columbia's* forward jets lit up the blackness outside the cabin. "I see a bunch of fire out this side," said Taco as the burn sputtered out. While he positioned checklist cue cards on the dashboard Velcro, Rommel accepted a salt tablet from Tammy and downed another bag of water.

The next checklist item was "Exercise rudder pedals." It had been one of Taco's favorite moments in our entry sims. He relished this chance to crack us up with a goofy impression of his old commander, Dave Walker, and we loved to be entertained. Stretching to his full height of five feet, eight inches, he tromped on the rudder pedals in exaggerated fashion while flailing his orange-suited arms. His uncoordinated, jerky motions were the exact opposite of the *Right Stuff*, steely-eyed test pilot image.

Still laughing, I activated my g-suit by rotating a small knob on the left knee of my space suit. My legs and lower torso felt a familiar squeeze, forcing more blood to my head and brain. I was determined to keep ahead of the coming onset of gravity, and the leg pressure felt good. Tammy locked her helmet in place and checked in on the intercom. "Welcome aboard, Tammy," Taco responded. She answered with mock formality: "I'm excited to be here with you fine gentlemen." She expressed an emotion felt by us all: we were privileged to have the honor of bringing *Columbia* safely home.

Taco got back to business: "Okay, at five minutes out, Ops 304 PRO, and the machine comes alive. I love it!" He looked forward to engaging the reentry software. "I think of it as a living beast already, but it really comes alive at 304."

Five minutes before *Columbia* felt the tickle of the sensible atmosphere, he keyed in the crucial command:

"Okay, Ops 304 . . ." We had just commanded the orbiter to forget about spaceflight and start behaving like a flying machine. I confirmed it, and Taco punched the Proceed key. The five of us watched the transformation with rapt attention. "Three-oh-four, here come the tapes," I observed. Flight instruments whirred to life: the Mach, angle of attack, and vertical velocity tapes slid smartly to the correct readings. Computer screens flashed and brought up displays of dynamic pressure, velocity, and drag. Attitude indicators spun into airplane mode, and the compass needles in front of each pilot swung toward Kennedy Space Center, 4,400 miles ahead of us. The airspeed tape settled out at a reading of Mach 24.5.

The g-meter was still pegged at zero, but that would soon change. I yanked my mirror out from beneath the small clipboard on my right knee and glanced out the two windows above and behind me. Against the black sky, an occasional flicker of faint white light began to shimmer against the window frame. "There's a little bit of plasma being generated 'cause I can see the jets reflecting [from the slipstream]," Taco noted. It was the first glimmer of a silent, eerie light show that would grow to envelop us in the next twenty minutes.

Beneath our feet, an ever-growing avalanche of air molecules slammed into *Columbia*'s heat-shielded belly. At first it was no more substantial than the near-vacuum a hundred miles higher, but here at 400,000 feet and descending, Earth's atmosphere was real enough. Unable to get out of the way of the hypersonic orbiter, the molecules smacked into the wall of tiles at 25,000 feet per second and instantly converted a fraction of the shuttle's kinetic energy into heat. The stalled molecules crammed themselves into a shock wave just off the surface of the tiles, where the accumulated energy from billions of these collisions heated the air to over 2,500°F. The shock wave radiated this heat both into the tiles and into the surrounding air, stripping away electrons and causing the atoms to glow with ever-greater intensity.

Outside *Columbia*, the shocked and heated air wrapped our cockpit windows in an eerie glow. Brilliant bursts of light materialized in our overhead windows, intermittent at first but soon resembling the barrage of flashbulbs outside a Hollywood premier. The front cockpit windows began to take on a pastel orange hue, a soft, almost fuzzy glow, reminding me of neon tubes wrapped in a fog bank. The startling brightness of the plasma was a marvel. Whenever *Columbia* fired a thruster to maintain attitude, the enveloping glow rippled and writhed outside the nearest window. Occasionally sparks would flit past the front panes and zip down the lengthening plume streaking out behind us. I tried to capture the unfolding scene on our miniature camera; Story held it up to the window edge and recorded our artificial meteor trail. The light show was hitting its stride.

"Good-bye, zero g," Rommel observed.

"'Gravitationally Challenged Are Us,' coming to a theater near Kennedy Space Center," said Taco.

"Mach twenty-four and a half," I noted. We had barely begun to slow down.

"A minute to air. I don't believe it, though: we're still seventy miles up!" Taco said in amazement.

"Yep, we're gonna hit that "wall" at 400K," Tammy answered. You could sense the anticipation in her voice. Already, the plasma pulses out back were coming faster—bright flashes of green and white against the black sky. Story was braced against the window frame, intent on capturing the mesmerizing scene behind us.

"You're gettin' some good flashes there," I said, checking the monitor.

"Barely starting to show up out front now," Taco responded.

"Zipping through the Earth's atmosphere in a spaceship . . . what a deal!" Tammy said out loud what all of us were thinking.

Breasting the tenuous gale outside, *Columbia* was finally responding to the effects of the thickening atmosphere. Even this high, the slipstream exerted enough drag to allow the four GPCs to begin navigating. They cranked the drag numbers into their calculations and actively began correcting our glide toward Kennedy Space Center. The computers showed us right on the money for touchdown at the Shuttle Landing Facility (SLF) 2,000 miles away.

"Our descent rate is slowing as we kiss the top of the atmosphere." Taco was watching the gauges. Between now and Mach 12, he and Rommel had little flying to do; *Columbia*'s computers and autopilot were handling that duty, and our crew's primary role was to monitor for proper guidance, navigation, and control. Although he could take over manually at any time, Taco's real flying job would begin in about fifteen minutes.

"What a phantasm of light," Tammy said, awed by the view outside. As we streaked northeast toward Central America and the Caribbean, the front windows were bathed in a pink-orange glow stoked occasionally by the white blast of thruster firings. Overhead, streaks of yellow-and-white incandescence streamed behind to form an intense, radiant plume. Flickers of green, purple, and orange whipped down the length of the wavering tube, while the flashes increased in frequency until they created a strobe effect. I was impressed not only at the light show but by our flying machine, bringing us safely through temperatures that could melt and even vaporize the orbiter's aluminum structure. Thanks to its tiled heat shield, *Columbia* was not only surviving but actually *flying* through this inferno, weighing drag and lift and dipping delta wings to steer a precise hypersonic course. It had been fifteen years since this spaceship first flew; no other nation on Earth had a vehicle that could match this performance.

"Rommel, we're on fire!" Taco laughingly said to his copilot.

"What do we do now?" Rom replied with mock worry.

"Hey, the g-meter is *quivering!*" Taco noted. While we had watched the light show outside, the little needle on the accelerometer had nudged up from zero and now indicated a fraction of one-tenth of a g. Objects in the cabin gave us additional indications that our time in free fall had ended. My entry pocket checklist—the emergency procedures for this phase of flight—had been drifting lazily on its nylon tether near my left elbow. Now it began a slow descent toward the cockpit floor, hanging straight down for the first time in over two weeks. To Tammy's right, our remaining water bags sagged to the bottom of their brown mesh sack. I noticed that I was no longer bobbing loosely under my lap belt but feeling gentle pressure from the seat cushion. *Sitting* was a word that actually had meaning again.

"Closed loop guidance," Taco called. On the display, the glowing little orbiter symbol chased a target box that had just appeared, confirming that the GPCs were now computing the path to the landing site.

"How does it *do* that?" Taco joked about the guidance algorithm.

"How do it know?" Rommel pondered in mock amazement.

My feet now settled to the floor. "Sort of an unusual feeling."

"First weight on my butt!" Taco laughed. "Start turning your head around, Rommel. Vestibular adaptation." Some pilots contended that their sense of balance returned more quickly if they made repeated, deliberate head movements during entry, recalibrating their inner ears to accurately sense accelerations. With *Columbia* planing down through the atmosphere, using its barn-door bottom to dissipate our once-vital speed, we would soon be at 1.5 g's. Once weightless, our suits now seemed to encase us in lead. Arms sagged into our laps, while helmets sank heavily onto our shoulders. Weight blanketed us even as the plasma show outside reached a climax. The orange gauze surrounding the front windows brightened, and reflections from the trailing plasma flashed off the white domes of our space helmets.

"It's just a blowtorch back there. Great video, Musgrave!" Story was getting it all on tape, and I couldn't pull my eyes from the dazzling plasma visible in my mirror. "Can you see that little doughnut at the top of the plasma there?"

"I sure can," Tammy replied, similarly fascinated. Sparks raced down the white-hot plume trailing us, a scorching whirlwind blazing at over 2,500°F.

"About one-third of a g, Story," I called. "Twenty-three-and-a-half Mach."

The plasma tube behind us now featured a brilliant white core surrounded by a thinner gas layer of yellow-orange, punctuated occasionally by sparks racing off into the distance.

"There's a little white hot spot in that plume," I called out. "Yeah, it's from that missing tile," Taco joked, whistling past the graveyard.

Tammy, like the rest of the crew, wasn't worried. "It's . . . awesome."

"What color does aluminum make when it burns?" Taco teased her, Tammy laughing unconcernedly at his joke.

Something on the front windows caught my attention. "See that bright spot on window three, Taco?"

"Yeah, it'll burn through. It's going to be a pinhole plasma coming through at Mach twenty-two and a half!"

"I think not," said Tammy. In 1996, on the eightieth mission of the space shuttle, reentry held few fears for us. Not with this ship; not with this crew.

Six-tenths of a g. "Feelin' good," Rommel said. "Sparks!" Story exclaimed in amazement. Over my shoulder a shower of sparks zoomed lazily back down the plasma tube, receding into the distance. "Just collisional excitation," Tammy, our astrophysicist, replied with more laughter. The flame out back angled off to one side as *Columbia* went into her first roll, banking to increase our descent rate.

Tammy nudged me from the right. "Almost 1 g."

"Down to Mach 20," I nodded. "Twisting my head is a doozy."

"Story, you're going to get an Emmy for this," I said in genuine admiration. I had seen the plasma before, but I'd *never* seen the sight of Story Musgrave *standing* next to me on the flight deck during entry. Yes, the flight rules called for Story to be strapped in downstairs long before Mach 20 or so, but since he had no official duties during entry, he had decided to remain on the flight deck for the light show. We were at 1 g now, yet Story stood braced against Taco's seat in front and my seat in the middle rear. His weight with suit and harness was probably something like 250 pounds and increasing steadily with our deceleration. I shook my head at his sheer stamina. While we focused on the cockpit tasks, Story braced the camera against the overhead window frame to capture the full view of the plasma display outside. I kept expecting him to call it quits, thump me on the shoulder, and clamber downstairs to his seat (or just collapse to the floor). But he made no move to leave. This was his sixth reentry, his last, and he looked determined to get the most out of it. So he stood, sweating and breathing hard, but still braced solidly and holding onto that camera.

"See some city going by here," Taco commented. "Mach 19. It's whizzing by, that's for sure. One g."

Tammy was exploring this new concept of weight, too. "Okay, bod, this is where you were born. You can do it."

"Just went through Mach 18," I said, reluctantly pulling my gaze away from the windows. There was little movement around the cockpit: we were all feeling too heavy. Someone's g-suit pressure relief valve cracked open, groaning like a dying bagpiper. I'd already dialed in three full "clicks" on my g suit, getting the maximum constriction around my legs and abdomen from its oxygen-filled bladders. The squeeze was firm, almost painful. I felt good, energized.

"Only 960 miles to go." Rommel was grinning. "Incredible!" The plume behind us was now a hot, pure white, no longer gauzy but a solid sheet of flame. At Mach 17 the plume acquired a faint orange halo, perhaps a sign that we had passed through peak heating. The pace of events in the cockpit began to pick up.

"Man, this time goes quick," Taco said.

"One point two g's," I managed to blurt against the g suit.

"Should be going by Houston soon." Rommel noted that our range to the runway was about the same as the distance between Ellington and the Cape. What took us an hour and a half in a T-38 would now take *Columbia* just fifteen minutes.

"Look at that plume on the front windows!" I pointed out a shifting ribbon of orange playing across the forward windows like a stream from a garden hose.

"Wow . . . that is wild!" Rom replied. "Pulling about 6 g's now?" We all laughed, or tried to, our legs and diaphragms squeezed by the g suits. We were at just 1.4 g's, but Rommel's estimate felt close to the mark.

Taco was philosophical: "This is as bad as it gets, and it's not bad."

"Cool as a cucumber," Tammy added.

We were at Mach 15 and still trailing that white hot plume. At 200,000 feet, our speed was 10,000 miles per hour, just 700 miles from the Kennedy runway. I remembered Sid Gutierrez's comment during our

STS-59 entry: "How are we *ever* going to slow this thing down in time?" Sid had half worried that we would overshoot Edwards and land somewhere in Arkansas. But now *Columbia* was dropping fast into the lower reaches of the atmosphere, where thicker air would brake our spacecraft even more efficiently. The orange glow was fading from the front cockpit windows, while the plasma tube behind us began to pale and glow orange, like the dying embers of a once-hot campfire. *Columbia*'s nose was dropping toward the horizon, our angle of attack gradually assuming a more airplane-like stance.

"Think there's some sunlight coming through the front now," Taco observed. We couldn't see the horizon yet, but Venus, the brilliant morning star, was just visible in the eastern sky.

To my left, Story had his legs braced wide against unfamiliar gravity. We were at 1.5 g's, and I was amazed he was standing under that seemingly crushing load. Although he wasn't plugged into the intercom, he was close enough to hear me. "Story, you all right?" I asked. A quick nod, then he went back to watching the plasma.

Mach 12 slid past on Taco's airspeed tape. We were slowing rapidly now, our velocity decreasing forty-eight feet per second during every second of our descent. Inside fifteen minutes to landing, Taco was already flipping the radiator bypass switches on his side panel to cut cold Freon into our cooling system. Held in reserve inside the payload bay radiators, the slug of chilled Freon would take over from the flash evaporators to keep cabin and electronics cool.

"There goes the Moon in this roll reversal," Rommel broke in. "You ought to see the sunrise! Oh!" The beauty outside had arrested even the attention of a test pilot.

Everyone's mood heightened further, perhaps juiced a bit by g-load-induced adrenaline. I had a sudden urge to talk about this marvelous flying machine and how well it was performing on this dive toward Florida. On the intercom tape I'm heard calling out every Mach number, even if no checklist action was required. Nervous energy fed my urge to chatter.

"Mach 10.2," I blurted excitedly.

"Okay, we made it down to Mach 10 without burning up," Rommel replied, laughing.

"We got a good chance," Taco admitted. Consciously or not, we relaxed at having passed safely through the peak heating zone.

I leaned to my left and double-checked the cooling unit that circulated cold water through Taco's suit and mine. Just sitting up under 1 g was hard work, but with the cabin temperature holding steady below 80°F, I was still reasonably comfortable. The water-cooled underwear under our launch and entry suits was a huge improvement over the old cooling fans used on my first two flights. Those suit fans merely circulated cabin air around the torso, and if the cabin got hot, you were out of luck. By contrast, the new cooling units used thermoelectric chillers, each supplying two crew members. Pumping cold water through tubes in our long johns, they were very effective at keeping us comfortable all the way to landing.

Out ahead I could see the brightening dawn reflecting off a solid deck of clouds over Florida. Behind us, the plasma plume had vanished. We were approaching the most critical part of the descent, and *Columbia* was feeling the air around her, testing her wings.

"There they are . . . TACANs," called Rommel. The displays confirmed all three TACAN receivers getting good range and bearing data from the tactical air navigation beacon near the runway.

"*Columbia*, Houston. Take TACANs." Mission Control liked what they were seeing, and Rommel keyed in the command to accept the external sensor data. The computers would now use this ground truth to correct any errors in our calculated inertial position. We had a good lock on the field.

"Through Mach eight and a half," I noted. "How you doin', Dr. J?" I double-checked her cooling as Tammy came back with a quick "Okay."

Capcom Curt Brown radioed again: "*Columbia*, Houston. Energy, ground track, and nav are go."

"As far as you know." We laughed at Rommel's use, one last time, of our favorite wisecrack. His irreverent comeback never went out over the radio, though; the joke was "for internal use only."

Three hundred miles out now, at 150,000 feet and Mach 8. *Columbia*'s nose was still so high that the horizon was invisible forward, and we were too busy scanning the instruments to look out back anymore. Up on the ship's tail, the rudder was opening into a wedge-shaped speed brake, deflecting the slipstream to control drag and energy. *Columbia* arrowed toward the Cape, dipping the left wing earthward in another roll reversal to control our descent rate into the thicker air below. Taco's side window, now facing Earth, gave us a glimpse of darkened cloud tops sliding by in our hypersonic pass over the Gulf of Mexico. Cooking along at nearly 6,000 miles per hour, the sensation of speed was startling. Not only did it look as though we would overshoot Florida, but we might be too fast to make Puerto Rico! From outside the cabin walls, a faint scream began to seep into my consciousness. Dense air was rushing past *Columbia*'s hull for the first time in eighteen days.

Mach 7. Florida was still invisible in the murk below, but as we continued in our bank I could see the cloud tops drawing ever closer. I caught a glance from a grinning Story as he hunched behind Taco's seat, craning to see forward. At my knee, I could take in Rommel's view on the tiny LCD monitor being fed by the dashboard-mounted camera. When we came within range of the Cape's FM receiver, our TV signal would go out live, giving family and friends a glimpse of what we were seeing as *Columbia* charged toward them.

"TACANs look great in a roll to the right. One tiny little star out there." Rommel could see Venus above the brightening horizon.

My right leg twitched involuntarily with excitement as my big moment as flight engineer approached. "Getting ready for the probes, guys."

"That's a good idea," Taco answered calmly. At Mach 5 I would have the job of throwing the switches to deploy the two air data probes on either side of *Columbia*'s nose. These small pitot tubes, similar to those on any aircraft, would rotate out from behind protective tiles to feed altitude, airspeed, and descent rate to our computers. "Your big moment in life is coming up, Tom."

We could detect more wind noise now, hissing past the cabin walls. *Columbia* rolled farther right, the sky out front still black, a faint band of blue atmosphere lining the horizon.

"At 5K, guys!" I practically shouted out. Taco was hunched over the control stick, his eyes intent on the guidance needles and flight instruments.

"Okay, probes," he answered quietly, a smile in his voice.

Leaning forward under the heavy pull of my helmet, I lifted each probe switch, just in front of my left knee, to the deploy position. "Comin' out. . . . One . . . and . . . two."

The faint scream from outside began to build into a roar. Although slowing constantly, our space shuttle was still traveling half again as fast as an SR-71 Blackbird, the world's speediest jet. Twenty-four miles high now and dropping like a rock.

"I can hear some wind noise now," said Taco, eyeing the guidance needles.

"Yup, we're smokin'. Mach 4," Rommel confirmed.

Having survived probe deployment with reputation intact, I wanted to play, too. "Mach 4! Columbia is doin' the job, man!"

"Oh, yeah!" our pilot chimed back.

Columbia's wings rocked back to the left for our last roll reversal. The ride was still smooth, but the roar of supersonic air past our cabin was steady. We were aiming for the west coast of Florida now, just south of Sarasota. A hundred miles ahead, at the approach end of our three-mile-long runway, our families were waiting. I imagined Liz, Annie, and Bryce looking up into the growing dawn, waiting for a first glimpse of our returning spaceship.

On our navigation display a bright green circle began to expand ahead of the small orbiter symbol: the outline of the heading alignment cone (HAC), leading to the approach end of the runway. In a few minutes, Taco would follow that spiral down to glide onto Runway 33. For now, our cockpit team was focused on hitting the HAC precisely and following it right down to our rollout on final.

Columbia's angle of attack had now dropped to 20 degrees, and we could sense our descent steepening even as the airspeed continued to fall. Taco and Rommel cycled through the four air data sensors, comparing the altitude, airspeed, angle of attack, and descent rate information from the fully deployed probes. Out the front windows, the pink prelude to sunrise was spreading on the horizon.

"Altitudes, airspeeds—everything's identical," Rommel observed, satisfied that the numbers from the probe sensors matched the GPCs' internal estimates. As the airspeed tape slid below Mach 3.4, we were ready to feed the "real world" data into the waiting computers. "Houston, we're ready for air data," Taco radioed.

The reply came back instantly from Curt: "Columbia, Houston, take air data." Rommel complied, and our supersonic glider now took the pulse of the Mach 3 slipstream outside.

"One point four g's, marginally down," I said.

"Under 100,000 feet, over the west coast of Florida," Taco noted.

We were just sixty miles out from the shuttle landing facility. At 80,000 feet, now below Mach 3, we were once again within the realm of the world's fastest airplanes. The cloud tops came up at us as if on an elevator.

"Good morning, sun!" Rommel squinted out at the brightening horizon.

I scanned down the checklist. Time to make sure all our hydraulic systems were in order. "Okay, Mach 2.7, guys. Check the APUs."

"APUs are running like champs." The call was Rommel's old standby, confirming that his three auxiliary power units were in the green. In the simulator his call usually produced an immediate APU failure fed in by an alert instructor with a devilish sense of humor. Today, though, we were having a rare "nominal" run, just the way I liked it. Things were so quiet that I worried I might be missing something.

There! Alarm bells went off inside my head. The roar outside was getting even louder, a danger signal to any pilot. More wind noise meant we were accelerating, picking up speed in the dive. My seat-of-the-pants sense was that our plunge was too steep, on the verge of losing control. I felt the urge to get the nose up, to reduce our dive angle, but I forced myself instead to scan the instruments. It was only the thickening air outside, finally giving voice to our passage through the atmosphere. All was well, still right on the profile.

"Through Mach two and a half," I said as much to calm myself as anything else.

As if in answer to my unspoken worries, Curt came up on the radio: "*Columbia*, Houston, we show you on energy at TAEM [Terminal Area Energy Management] interface. No change to weather. Winds are two-two-zero at four, peak six."

The horizon out front was sharply defined now, glowing orange against the clean indigo above. "Mach 2, slowin' down. Kinda cloudy across Florida this morning," Rommel observed matter-of-factly.

*Columbia* at last sliced down through a thin cloud layer, slowing below Mach 2. As the ship eased toward the sound barrier, the g's backed off to a mere 1.1 times the force of gravity. The subtle tug of our lap belts restrained us from sliding forward in our seats.

"Sixty-five thousand feet, Mach 1.6," Rom called.

Taco picked up the familiar patter: "Descending into the Kennedy Space Center . . . will everyone please check their seatbelts fastened?" He glanced back at Tammy, Story, and me with an easy grin.

*Columbia*'s computers lowered her nose smoothly below the horizon at last, aiming at the HAC intercept point up ahead.

"One g, feels like fifteen!" Rommel grunted.

"Mach 1.2," Taco added.

"Hey, you gotta *do* something, Taco," Rommel teased, knowing Taco had been itching to take the controls for the last half hour.

"Man, we're gonna be having some fun," our commander answered, his right hand closing around the rotational hand controller (RHC), his joystick. "One point one [Mach]. Here comes the rough road ride."

The roar outside was joined by a rumble, growing slowly in intensity to a rapid-fire, turbulent drumbeat. This was the Mach buffet, created as shock waves thrummed across the slowing orbiter's fuselage and control surfaces. Our ship vibrated like a dump truck on a washboard-rutted dirt road. My heavy limbs shook, the rumble intensifying as we approached the sound barrier.

We rattled through Mach 1 as the eastern horizon brightened to the coming dawn. Bam! Bam! Two rapid-fire sonic booms smacked across the Cape and the SLF as the veteran orbiter announced her arrival back at the spaceport. There, Liz and the kids jerked their gaze upward, searching for some sign of *Columbia's* presence in the lightening sky.

"Mach point nine seven," I called out as we whipped into a cloud bank, the horizon disappearing in the predawn gloom.

"I'm IFR, hoping to see something soon," said Taco. Blind out the window, he was flying strictly on instruments.

"Okay, Taco. HAC coming. You are looking mah-velous," Rommel said assuringly.

"Amen," added Tammy.

Ahead, the clouds gave way to clear skies over the Cape, but we could see nothing but darkness below. No signs of ocean or coastline. Somewhere down there was our landing strip.

Once below Mach 1, Taco had taken over from the autopilot. He was now flying *Columbia* through the RHC; the orbiter's fly-by-wire control system electronically translated his stick inputs into surges of hydraulic power that moved the elevons and rudder. The approach called for a descending right turn around the HAC, taking us over the beach and out to sea before circling back northwest for landing. Rommel slipped into his professional coach's voice just as he always did in the simulator.

"Five seconds, four, three . . . one . . . HAC!"

Taco smoothly rolled *Columbia* into a right turn as we fell through 50,000 feet. The bank angle in our downward spiral was about 30 degrees, a tight turn for this big glider. I strained with the effort of sitting upright as the g's built to 1.7. That gorilla I had imagined sitting on my chest during launch eighteen days ago was back, this time hunched on my shoulders. A quick check to my left: Story was still there, leaning on Taco's seat back. I could hear the buffet, still with us even subsonic, a reminder of *Columbia*'s poor aerodynamics: slab-sided fuselage, bumpy tiles, knobby OMS pods. Yet Taco, the old attack pilot, was handling her like a fighter.

"Okay, Skid Strip in sight," called Rommel, spotting the old Cape runway just inland from John Glenn's Mercury pad. My view directly to the front was blocked by the windshield's wide central pillar, but to either side I could see the darkened scrubland of Florida slipping past the nose. Only a few lights were visible in the vacant landscape to our west as we swept around the HAC.

"Point seven Mach now." I called. "Altitude 24,000 feet."

"What a fine flying machine," Taco said in admiration. "A beautiful morning."

Rommel was right with him. "It definitely is daylight."

Taco looked past his heads-up display (HUD) at the dark ground rushing up. He was having none of this "day" talk: "Shoot, I'm logging a night one!"

We were all laughing now. "Twenty-three thousand feet," I ticked off the milestones on the unwinding altimeter. "Twenty thousand, 285 knots."

"*Columbia*, Houston, on at the 90," called Curt from MCC. Just 90 degrees to go in our turn to the runway, Taco was on course and on glide path.

"Houston, we have the runway." Taco radioed. The SLF was in sight to our right.

"Sixteen thousand. MLS is in." I confirmed that our computers had picked up the data from the field's precision microwave landing system. "There's the VAB!" The huge Vehicle Assembly Building loomed reassuringly just off the approach end of Runway 33. Rolling out on final now. The flashing strobe at the runway aim point slid into view on the monitor and stabilized as Taco picked up the guidance cues on his heads-up display. Our descent rate was more than 7,000 feet per minute.

Nine thousand, eight thousand; the altimeter was unwinding with renewed zeal as we hurtled down the 19-degree glide slope. The families at the runway caught sight of *Columbia*, nose down and diving for Earth with a will.

"You're looking on and on, Taco." Rom verbally patted his commander on the back: on airspeed, on glide path.

"Seven thousand," I chimed in.

"Six thousand feet, 300 knots," called Rommel. The radar altimeter tapes came alive on both sides of the cockpit as the sensors locked onto the ground echo.

"Both radars are in," Rommel remarked, his voice cool and reassuring. The two readings compared well with the adjacent barometric altimeter.

Out front, xenon searchlights shot fingers of light onto the three-mile-long runway, the gray-white pavement standing out from the dark scrub on either side. The single aim point strobe flashed insistently. The guidance was taking us right down the pipe. Pilot Rommel was doing all the talking now.

"Four thousand feet. Three thousand, speed brake goin' to 28 percent. Two thousand, preflare next."

"We're in the preflare. Arm the gear," Taco directed. Rommel punched the button. The gear warning light in the "arm" button glowed in confirmation.

The slipstream roared past the cabin as Taco rocked the stick gently back to slow our dive toward the runway. *Columbia's* nose came up to assume a shallow 1.5-degree glide slope, slowing the descent and bleeding off airspeed. Now only the pilots could see the runway.

"Good pull, Taco," said Rommel of the smooth flare. Taco was flying visually now, bringing the lighted "ball-bar" visual glide path indicator on the left of the runway into his scan. The only instruments that mattered now were the radar altimeter and airspeed tapes, both steadily decreasing.

Through Taco's window I glimpsed the glare of the searchlights to the west of the runway. Only in the monitor could I see the concrete strip itself, swelling rapidly to fill the screen.

"1,000, 800, 700, 600, 500."

The commander saw 400 feet slip by on the radar. "Gear down," Taco interjected.

"Gear's coming," Rommel acknowledged, simultaneously hitting the gear button.

"Three hundred, fly the ball." Rommel kept Taco focused on the ball-bar, tracking the inner glide slope.

"Two hundred, two-eighty-five." Altitude and airspeed.

Taco wanted to bring *Columbia* over the end of the runway about twenty feet off the ground, with 220 knots of airspeed. If he flew the ball-bar precisely, *Columbia*'s airspeed would decrease steadily from nearly 300 knots down to the desired 200 at touchdown. All I could do was hold my breath and watch the monitor. Rommel's calls were nearly continuous now.

"Forty. Thirty. Twenty. Fifteen, two-thirty. Ten, two-twenty." We were skimming the runway.

"Hold the nose up, hold it up, keep holdin' it up."

"Ten, two-twenty."

The hundreds of spectators gathered at the SLF saw *Columbia* flare gracefully over the runway, her boxy form silhouetted in the brilliance of the xenon searchlights. As the main gear tires sought the concrete, twin tornadoes of mist swirled from each wingtip, tracing the orbiter's sweeping path toward touchdown. Just off the approach end taxiway, the little knot of our families and their escorts watched *Columbia* sweep past, the roar of her passage announcing another safe return to Earth.

"Greased!" Rommel exulted.

Taco rolled the wheels onto the runway just a few hundred yards past our families; smoke spurted from rubber hissing onto concrete, glowed in the powerful xenons, then swirled away in the wake of our passage. Landing came with an almost imperceptible bump; it was as sweet a touchdown as any shuttle commander had ever flown—a "greased" landing. In the cockpit the landing was so subtle that only the change in orbiter software modes convinced us we were actually on the ground. Tension whooshed out of my lungs in a rush.

"Chute!" Taco called.

"Chute's comin'," said the pilot. Rommel thumbed the deploy button atop the glare shield, and 100 feet to the rear, a mortar fired under *Columbia*'s tail to deploy the small drogue chute. The drogue streamed out behind and quickly yanked the main parachute canopy into the slipstream. As we rolled down the runway, chute billowing behind, the fire trucks and convoy of emergency vehicles roared onto the strip.

"One-eighty-five [knots], comin' down." We felt the tug of the inflating main chute as the nose gear fell inexorably toward the concrete, touching down with a solid thump. A few seconds later, *Columbia* rocketed past the bleachers halfway down the field.

"Okay. Midfield, cleared to brake," called Rommel. "Prettiest landing I've ever seen." His compliment triggered a flood of praise from the rest of the crew; our words tumbled over each other in congratulations.

"What a job!" I added, laughing with released tension.

"Nice job!" Tammy chimed in. Story clapped Taco on the shoulder.

Rommel called out the ground speed. "Ninety, eighty, . . ."

"Chute," Taco commanded, and his pilot hit the jettison button.

"It's gone," said Rommel.

*Columbia* now rolled sedately toward a final stop on the centerline stripe. Taco was justifiably delighted and understandably drained.

"My . . . feet . . . are . . . *quivering!*" he said.

"That was a beautiful landing, Taco," said Rommel.

"My knees are *jumping!*" was all Taco could say, smiling in relief.

"Beautiful," Tammy said. All of us were wowed by the near-perfect approach.

Taco toed the rudder pedals with one final squeeze and brought the orbiter to a halt. "Houston, *Columbia*. Wheels stop," he radioed.

The reply came back promptly from Curt in Mission Control: "Roger, *Columbia*, wheels stop. Welcome home after your record-setting mission." At seventeen days, fifteen hours, and fifty-three minutes, we had set a shuttle endurance record that still stands.

"Thanks, Curt. It was a fantastic mission. We feel privileged to have been a part of it."

Through the forward and overhead windows, the soft gray light of dawn filtered into the cockpit, our eyes drinking in the muted colors after eighteen days in the stark sunlit glare of orbit. In the distance, flashing lights atop the convoy vehicles signaled a welcome as they approached. I turned to Tammy; we had been through the mill together on this one, and never had we shared wider grins of exhilaration and relief. With Story still standing to my left, I flipped my checklist to our postlanding procedures, weary but exultant at the spectacular homecoming we had just experienced. For the first time in nearly three weeks I could think about earthly affairs, about an upcoming reunion at crew quarters. I imagined meeting my family again and the prospect of hugs, kisses, and even a shower. I was back from space, back with my people, back on the good Earth. I had come through the fire, wrung out but intact, cradled by an amazing spaceship. Hail, *Columbia*.

I HAD FINALLY MANAGED to land in the right place. The elevator doors opened on the third floor of the O&C building, just outside the entrance to crew quarters. A shriek of delight erupted from the families gathered in the hall as the five of us, now clad in blue flight suits and still trying to find our Earth legs, stepped into the arms of our loved ones. Liz was right there, with Annie and Bryce, huge smiles on all their faces. There was a chorus of happy voices and smiles all around as we excitedly shared impressions of the landing for a minute or two, got in some serious family hugs, then parted: me to a physical exam, the family to breakfast.

The checkup didn't last long. The worst of it was my own fault: I was handed a big, cold tumbler of purple grape juice by one of the nurses . . . which immediately slipped through my uncoordinated fingers. The clinic carpeting was brand-new, and I'd just christened it with a very prominent, perhaps indelible, purple stain.

I was still a little wobbly when I made it back to crew quarters a half hour later. The heavy feeling I had always experienced immediately after landing had passed, but that familiar teetering sensation was back. Liz lingered behind as I headed down the hall to my room. Trudy Davis, noticing her hesitation, encouraged her: "Go ahead! You're married, after all; you can go with him!"

It was a wonderful shower. I leaned against the tiled wall for support as I let the hot water cascade over me, all the while exchanging news with Liz. She kept a careful eye on my wobbly stance as I got dressed, and we headed back to the dining room for some food: I had already been up for more than eight hours, and I was hungry for a real Earth meal.

It was vegetable lasagna instead of a cheeseburger, but I wasn't complaining. The fresh fruits and salad didn't come out of a plastic or foil pouch, and they didn't float off my plate. Tammy and I recounted to Liz the story of our almost space walk and our worries about what engineers would find when they finally examined the hatch.

What terrified us both was that the techs would find the hatch operating normally after landing, leading to the inescapable conclusion of human error. But all five of us had operated the hatch handle in the airlock after the space walks were canceled. The handle had jammed hard at that 30-degree point in its rotation, not giving an inch.

At crew quarters, Tammy and I waited for words on the hatch from the orbiter folks. But we also needed some sleep. Given a clean bill of health by the flight surgeons, all of us headed for our families' Cocoa Beach hotel rooms, where I collapsed for a long nap. That night we all met in one of the rooms to relax and tell space stories, the laughter still spilling out of us after nearly three weeks in space. That was the one consumable the STS-80 crew still had in ample supply.

Still exhausted from the long day and the spent adrenaline of landing, I turned in early that night. It was a very strange sleep period. I was desperate for some rest, but I kept dreaming I was weightless. My legs felt particularly light: I was certain that if I let go of my pillow, I would instantly float to the ceiling and be stuck there for the rest of the night. I found myself holding desperately onto my anchor with both arms, and it worked—I managed to stay firmly atop the mattress all night.

By morning I was feeling much more Earthbound, although the soles of my feet, largely unused for eighteen days, were surprisingly tender. The crew had decided on an early departure for Houston, right around sunrise, and we enjoyed a full day at home to recuperate after landing at Ellington.

Monday brought good news, at least to Tammy and me. Kelvin Manning, *Columbia*'s vehicle manager at the Cape, called to report the hatch inspection results. The handle still wouldn't budge. *Thanks be to God!* As troubleshooting continued, X-rays of the mechanism showed a machine screw missing from the handle gear train. When technicians finally removed the handle and gear assembly, they found the half-inch-long screw lodged in the teeth of one of the planetary gears. With the gears jammed, the only way Tammy and I could have freed the handle was to have forced it until we either dislodged or destroyed the screw. KSC determined that the screw had come loose due to a human error in assembling the gear mechanism, and NASA instituted new tests and inspections to preclude a recurrence. This particular failure would never happen again.

Liz drove me around for the next few days after my initial attempt to drive to work on Monday for our returned film review; I had promptly run my right tires over the curb executing a simple turn. My internal gy-ros were still getting recalibrated. The other crew members and I spent the rest of the week writing our report and examining the 7,000 or so photos we had taken on the mission.

Just a few days after our return, I was already wondering what my next job assignment might be. Bob Cabana had pulled Tammy and me aside at the Cape just after landing: "I want you to know I'm going to do everything I can to get you two another shot at an EVA." He thought there might be a chance to add us to the crew of STS-87, a mission that included a scheduled space walk. I was grateful, but I wasn't holding my breath: I wasn't anxious to undercut someone else's shot at a space walk.

In orbit, while the crew filmed a public service announcement en-couraging organ donation, Rommel had teased us about what body part we might trade for the chance at an EVA. I laughed and joked that I'd have given my left nut to get that hatch open. Tammy quickly added: "And I'd have given his right one."

Well, I was still whole, but it wasn't at all clear whether I'd be as-signed another EVA. In January 1997 Cabana called us both in and told us he had some thoughts on another space-walk opportunity, but in the meantime he wanted us both to represent the Astronaut Office in the International Space Station program. Tammy would become the chief of the Space Station Operations branch, and I would be her deputy. The two of us would replace Ellen Ochoa and Dan Bursch. Dan was, in fact, headed off to train for a Space Station expedition scheduled perhaps two years down the road.

I had mixed feelings about working on the Station program. I had stopped paying attention to it when I went off to fly the 1994 Radar Labs, and my knowledge of the program was slim. Now it was time to catch up with three years of Station developments. I was in for a shock.

# PART THREE
# STATION

# 12

# THE INTERNATIONAL
# SPACE STATION

*We believe that when men reach beyond this planet, they should leave their national differences behind them.*
JOHN F. KENNEDY, NEWS CONFERENCE, FEBRUARY 21, 1962

The transformation of my attitude toward the International Space Station came about slowly and grudgingly. As late as the STS-80 mission, I'd viewed the Station with indifference, a distraction from my focus on the space shuttle. My goals were to train to be a good team member, perform well in orbit, and help any way I could to ensure mission success. If those STS-80 space walks advanced the Space Station's prospects, so much the better. But I still had little emotional investment in the progress of the Station.

More than once in the early 1990s, I thought NASA might be better off if it cut its losses and canceled the ISS. That would give the agency and the country time to upgrade the shuttle and think seriously about what our next goal in space should be. Was a station required to get us out of low Earth orbit and on the way to the Moon, asteroids, or Mars? Or was it a dead-end distraction that would doom NASA to a cash-strapped future endlessly circling the Earth?

My real interest as a planetary scientist was seeing humans and their machines explore the solar system together, and I wasn't convinced we needed a space station to do that. At least, not *this* station, the ISS. It carried so much political baggage that I wasn't sure it could ever get off the ground.

When President Clinton's administration took office in 1993, his Office of Management and Budget had proposed canceling both the Space Station and the Superconducting Supercollider projects. NASA had told the new administration that the Freedom station would need an additional $750 million over the next three years to cover budget overruns. But Clinton's staff wasn't interested in funding a notoriously troubled space station program. NASA was told to come up with fresh ideas if it hoped to preserve a future destination for its shuttle.

Still, in March 1993 the Astronaut Office's Dave Finney told us not to write off the Freedom program—that "no one knows the whole story." It seemed that NASA administrator Goldin, a holdover from President George H. W. Bush's administration, was determined to persuade the new administration to keep the project alive. Clinton may not have been interested in NASA's proposed Freedom, but perhaps a revamped and less-costly version might survive.

A central thrust of the new efforts was to build on the joint space agreement made with the Russians in 1992. The planned joint flights on the shuttle and *Mir* might be used to interest the Russians in participating in Space Station Freedom. Because the new administration seemed most interested in space as an adjunct to foreign policy, NASA hoped the Clinton administration would see joint space programs as a way to keep Russia's technical workforce from selling its missile know-how to North Korea, Iraq, or Iran.

In mid-March, the Russians proposed a *Mir 2* as the new international station, essentially offering to provide all the hardware and expertise, if only the Americans would pay for it. This was too much for Administrator Goldin, but he pursued the idea of making some use of Russian hardware to lower the cost and accelerate the schedule of the Freedom program. Perhaps the best of both countries' ideas could produce a viable station design.

Astronauts were specifically excluded from the redesign effort, ostensibly because we would "try to design a Cadillac, instead of a Chevy." As the closely held talks on the station's future proceeded, confusion reigned at NASA, among the international partners, and among the

Freedom contractors. By June 1993 NASA had come up with three candidate redesigns. Option A was a bare-bones Freedom with a sticker price of $16.5 billion. Option B was essentially the existing Freedom with a few design changes, the most expensive of the three at $19.3 billion. Option C was the cheapest, at $15.1 billion, and the most radical concept: a wingless orbiter core supplemented by additional modules and placed permanently in orbit to serve as a small laboratory. This alternative would strip *Columbia* of its wings and transform it into a *Skylab*-style workshop. Launched to a Russian-accessible orbit, it could be outfitted with seven decks, living quarters, a shower and toilet, observation windows, docking ports for the partner modules, and a lifeboat. The drawbacks would be considerable, however: it would be much smaller than the original design, offer reduced research capacity, and perhaps most damning, it was about five years too late.

Clinton selected Option A, beefed up with some Freedom design elements. NASA was ordered to ensure that the new design had significant Russian participation. But that was no easy task. The *Freedom* station couldn't be launched into a 51.6-degree-inclination orbit, which was compatible with Russia's Baikonur launch site, without major design changes: its modules would be too heavy for high-inclination launch on the shuttle. An intensive three-day redesign effort at former astronaut Tom Stafford's office in Alexandria, Virginia, hammered out a proposal for a feasible compromise station. A scaled-back Freedom design emerged that used Russian *Mir 2* hardware, unmanned Progress cargo ships, and Soyuz crew transports to save money and speed up construction. The new design would fly in the Russian-compatible 51.6-degree orbit, trading reduced shuttle performance for inexpensive Russian launch services. Russia would contribute a propellant and storage module, a habitation module, and a steady flow of Progress supply transports. The United States would build a docking node, lab, and expansive solar arrays, using the shuttle to assemble them in orbit.

Clinton's staff approved the new design, and by September 1993 the Russian prime minister and the American vice president signed agreements formalizing the new station program. *Freedom* would never fly, but construction on a new International Space Station would begin in 1997. In the meantime, the United States would pay the Russians $400 million to host four American astronauts on separate long-duration flights aboard *Mir.*

The deal was done. The Russians had joined the International Space Station partnership. In fact, their participation was now on the critical path: the new *Alpha* station could not fly until the Russians launched the first element, the "functional cargo block," or FGB (the Russian acronym). Afterward, *Alpha* would continue to be dependent on Russian Progress cargo ships and the Soyuz crew transport for emergency escape.

The new station plan fundamentally altered the career prospects of the US astronaut corps. Instead of flying on the shuttle and eventually *Freedom*, many of my colleagues would learn the details of Russian hardware and training practices, first for *Mir* flights and then on the joint *Alpha* station. It was a task I tried to defer for as long as possible.

What I heard of the shuttle-*Mir* program over the next two years didn't improve my confidence in the new station project. Norm Thagard, one of the original shuttle astronauts, impressed us all by volunteering for the first astronaut mission to *Mir*. He soon vanished into the maw of the Moscow training establishment, where NASA's support for him and backup Bonnie Dunbar was, to put it charitably, minimal. Norm struggled on. He learned Russian largely on his own and translated his Soyuz technical manuals during long nights in a run-down apartment in Star City, the Russian cosmonaut training center outside Moscow. He became the first American to ride a Soyuz, making it to *Mir* in March 1995. With limited access to NASA and his family through Moscow's "TsUP" mission control, Norm spent more than ninety days in cultural isolation aboard *Mir*, able to communicate only sparingly about technical issues with his two *Mir 18* crewmates. Limited menu choices and difficulties tracking food consumption for a medical experiment caused him to lose seventeen pounds before his weight stabilized. In summing up what he had learned about Russian long-duration spaceflight operations, he repeated to us the philosophy of his crewmate Gennady Strekalov when dealing with the TsUP: "Tell them what they want to hear; do what you want to do."

Shannon Lucid prepared to follow in Norm's footsteps. Kevin Chilton's STS-76 crew (including Rich Clifford and Linda Godwin) would take her to *Mir* in early spring 1996. Chief Astronaut Bob Cabana was looking for more *Mir* volunteers, grappling with a problem that had faced the Astronaut Office for more than five years: How should we conduct simultaneous shuttle and station operations? Although most of us had been lukewarm toward the prospects of this redesigned space station, the reality was that the ISS would fly and we astronauts were going to have to crew it. Focused on flying three shuttle missions in rapid succession, I had paid little attention to station developments, as the debut of ISS always seemed lost in the indeterminate future. But with the Russians now involved and shuttle-Mir (Phase I of the ISS) a reality, I had no choice but to take notice.

Bob made no secret of the fact that he needed volunteers to train in Russia and fly on Mir. As an incentive, in October 1995 he pointed out at our Office meeting that those who flew on the later Mir missions would fly a Soyuz spacecraft to the new ISS and help with early station assembly. However, because only 57 percent of the Astronaut Corps had the physical dimensions to fit into the cramped Soyuz, he had only a small pool of potential volunteers. If too few stepped up to this challenge, Bob said, he'd have to assign people to shuttle-Mir. The business of this Office is spaceflight, he continued, and if you turn down such an opportunity "you ought to go look for a job elsewhere."

I was still working as a capcom in 1995 when Bob called me into his office. He came right to the point. What would I think about volunteering for the shuttle-Mir program and perhaps a Space Station flight later? I had anticipated the conversation, because at five feet, eight inches, I was smack dab in the middle of the Soyuz physical requirements, one of the lucky astronauts who could fit in the Russian machine. I was sympathetic to Bob's position, but I had to answer him honestly: "When I came here in 1990, I was ready to fly on the shuttle and the Space Station. I love flying on the shuttle, Bob, and I'd very much like to help build the Station. But I have no desire to leave my family and work in Russia." I told him I understood that he might have no choice but to let me go. That was fair enough. In the meantime, I would look forward to an assignment to help assemble the ISS on a shuttle flight.

In the end, Bob did manage to get his volunteers. I was prepared to walk away from NASA if necessary, but some of my Hairball colleagues didn't have the same option. Active-duty military astronauts, for example, served in the astronaut corps at NASA's pleasure. If a military astronaut was asked to return to his parent service, he or she would probably face a series of transfers, ending the stable domestic life enjoyed in Houston. And if to avoid such a move a military astronaut resigned from the service short of the twenty years needed for retirement, this colleague would lose his or her military pension. These practical considerations, along with their genuine commitment to see the space program succeed, persuaded a number of the military astronauts to volunteer for Mir. By 1997, Hairballs Dan Bursch, Susan Helms, and Carl Walz were all headed for Russia, along with Wendy Lawrence, Jerry Linenger, and Jim Voss. Mike Foale, Scott Parazynski, Andy Thomas, and Dave Wolf were among the few civilian astronauts who volunteered. Knowing the personal sacrifices they were making, I respected and saluted them for their dedication.

Instead of letting me go, Bob had assigned me to STS-80. With my return from that flight, my career in Houston had now carried me through three missions over six years, and I had plainly outrun my promise to Liz: "Two flights, four years, and we're outta here." But the Radar Lab flights were so similar that I had wanted a third mission. Bob's offer of the STS-80 space-walk assignment had come with a caveat: I would have to commit to two flights so that the Office could take advantage of my EVA experience for the Space Station. Liz and I discussed the offer; if I ever needed proof of Liz's love and support, it came when she agreed to stick it out for those *two* additional flights, along with the attendant separations and stresses.

With the STS-80 disappointment still fresh, the opportunity for a space walk would have to wait for another mission. In the meantime I would pay my dues by working on the ISS program with Tammy. We plunged into a crash course on its intricacies. ISS had its own language, its own acronyms, and its own management structure, all new to me. Even as we struggled to come up to speed on the new ISS program in Houston, our astronaut colleagues continued to learn lessons—some quite painful—on Mir.

The Phase I program was well under way when I returned from STS-80. NASA had delivered its first two annual payments of $100 million, and two Russian cosmonauts, Sergei Krikalev and Vladimir Titov, had flown on the space shuttle. Three US astronauts had been to Mir and back. The two most recent visitors to Mir had returned with widely different experiences. Shannon Lucid's buoyant personality withstood a six-month sojourn on the Russian outpost, and she returned eager to fly on the ISS. But John Blaha, an easygoing fighter pilot and highly regarded shuttle commander, struggled with the language barrier and depression during his stint in orbit. Jerry Linenger was now aloft, carrying on Blaha's science program and hoping his four months in orbit would put these joint operations on a clear trajectory toward the ISS.

Instead, Jerry and five *Mir* cosmonauts nearly perished when an oxygen-generating chemical "candle" caught fire aboard the station on February 23, 1997. Burning like a blowtorch at 1,650°F, the jet of flames from the lithium perchlorate canister barred the way to one of the Soyuz lifeboats and threatened to breach *Mir*'s aluminum hull. The fire persisted for fourteen minutes, immune to water and foam fire extinguishers, until it burned itself out. Smoke from the fire forced the crew to don respirators for hours until condensing water vapor and *Mir*'s ventilation filters began to clear the air. The Russians didn't tell NASA about the fire until the next day and in their discussions minimized the potential danger to the crew.

As if Jerry hadn't endured enough of a scare, two other incidents almost forced the crew to abandon *Mir*. During a manual redocking experiment on March 4, 1997, a Progress cargo ship nearly collided with the station. Only desperate thruster firings by the commander averted a catastrophe. Later, a cascading series of failures—leaking ethylene glycol coolant, failure of the oxygen generator, high heat and humidity, the breakdown of the primary $CO_2$ scrubber, and an ineffective backup system—caused high carbon dioxide levels that again nearly forced an emergency departure. Jerry returned to Houston in late May 1997 from one of the most harrowing spaceflights an American has ever undertaken. His frank report on the deteriorating conditions aboard *Mir* and the difficulties of training in Russia sent shock waves through the Astronaut Office. Did we dare send another astronaut into that environment?

NASA decided to do just that. Dan Goldin and George Abbey approved Mike Foale's launch to *Mir*, deciding to let *Atlantis* commander Charlie Precourt make the final decision about whether to transfer Mike aboard or return him with Linenger. We knew that the Russians had a financial incentive to keep the astronauts coming to *Mir*; a safety rejection by NASA would be a severe setback for them. So, as Jerry later reported, the *Mir* crew spent weeks getting the station, which was more than five years beyond its design life, to look its best. Linenger called it the Potemkin Village strategem and noted that the Russians would never abandon *Mir* "unless their clothes were on fire." Mike Foale moved in to begin his event-filled stay as the fifth American aboard *Mir*.

On June 25 Mike witnessed a near-replay of the Progress docking experiment that almost torpedoed *Mir* back in March. This time, though, luck ran out for the commander, Vasily Tsibliyev, and he couldn't prevent the out-of-control Progress from colliding with the Spektr module, puncturing the hull. Foale felt his ears pop as the pressure began to drop, and he and crewmate Aleksandr Lazutkin prepared the Soyuz for an emergency departure. But the crew managed to slam the Spektr hatch closed, sealing off the leak (but losing all of Mike's personal gear and much of his scientific equipment). Foale helped perform an EVA inspection of the Spektr module, but its hull was damaged beyond repair. Mike returned safely to Earth aboard *Atlantis* in October 1997 after 145 days in orbit.

I doubted the value of sending my classmate Dave Wolf to replace Foale. We knew the real conditions on *Mir*: Spektr and its scientific gear were lost; glycol coolant, a known chemical hazard, was oozing from the core module's plumbing; the station's moldy, dank interior smelled like a locker room; tons of discarded or obsolete gear cluttered the station, with no way to dispose of it; computer breakdowns regularly tumbled *Mir* out of control in orbit; and the crew spent most of their time on maintenance, leaving almost no time for research. NASA and the Phase I managers, however, felt they had little choice early in October but to send Wolf aloft to *Mir*. Pulling out on the Russians would threaten the budding ISS partnership and the difficult lessons we were learning about coping with failure in space. It was experience we would find invaluable for ISS operations. As Charlie Precourt put it to us in a briefing on September 15, if we abandoned *Mir*, we would cut ourselves off from our most valuable source of information on space station operations. Like it or not, we had to stay engaged and learn to work with our principal ISS partner. The Russians were learning from us, to be sure, and staying engaged would enable us to share with them our own hard-won experience. And they were teaching us volumes about how to handle crises in orbit. A space station was like a ship at sea. You don't abandon it lightly; instead you stay aboard and struggle to fix it until it sinks from under you.

Such was the atmosphere Tammy and I faced as we struggled to understand the complexities of the International Space Station. Bill Shepherd, who had participated in the 1993 Space Station redesign meetings and was once ISS deputy program manager, had been named the commander of Expedition One, the first crew slated to live aboard the ISS.

He was not a happy astronaut. Supposedly only eighteen months from launch, he reported regularly to the Astronaut Office on the shortcomings in both the Russian and American ISS elements. In May 1997 he listed a host of difficulties confronting him: Russians and Americans were developing separate operations plans; the crew's computer displays were a mess; US training for Space Station systems was unsatisfactory; and the Russians were treating him like a *Mir* "guest cosmonaut," giving him only dry *Mir*-era systems lectures and denying him any study documents. Shepherd said we were headed for a situation where "operations are being run from the ground only. We are riding a vehicle someone else is driving . . . and we're going backwards." The situation was "Unsat," totally unacceptable. And Shep wanted the Astronaut Office, meaning Tammy and me, to fix it.

Shep's troubles stemmed directly from the decision to bring the Russians in. The move had imposed a tremendous amount of drag on the US space program, with little in the way of actual benefit. The five-year-old US–Russian partnership, I felt, had been driven purely by NASA's need to salvage the Space Station (and NASA's bureaucratic prospects) under an indifferent administration. That had led the agency to capitalize on the Clinton team's unfocused idea of engaging the Russians in some post–Cold War joint technology effort, a foreign policy exercise. NASA's redesigned Space Station had filled the bill, but it was a Faustian bargain: the realities of working with the creaky Russian space establishment were dragging the entire program down. The training and hardware difficulties confronting Shep were just the outcome of NASA's unwillingness to deal firmly with its stubborn partner, I thought.

My cynicism slowly yielded to pragmatism as I recognized some hard facts. The administration would never give NASA the authority to jettison the Russians, no matter how difficult the partnership became. That would be a minor foreign policy disaster for the Clinton–Gore team, particularly as the vice president considered the cooperative space agreements his personal achievement. For NASA, the road to the ISS had to go through Moscow. My responsibility to NASA lay with moving the project forward if I could. If we could make it fly, perhaps a well-executed ISS program would give policymakers a renewed confidence in the agency and open up the path to the solar system. There was no choice but to make this Station work.

The International Space Station was undeniably ambitious. Even truncated, descoped, and downsized, the ISS was still the largest international technical undertaking in history. The United States led fifteen other nations (Russia, Japan, Canada, the eleven nations of the European Space Agency, and Brazil) in planning a permanent research outpost in orbit. The ISS would take six years to build, circling the Earth at an altitude of 250 miles in an orbit inclined 51.6 degrees to the equator. When finished, it would have a mass of about a million pounds, by far the largest object ever assembled in space. The solar array truss would span 360 feet from end to end, and the habitable modules would stretch 290 feet from front to back. Almost an acre of solar arrays would power six state-of-the-art research laboratories. The crew of six would inhabit an interior volume equal to that of a 747 jetliner cabin. About 160 space walks would be required to assemble and maintain the Station in the first five years of its life, and the space shuttle and Russian boosters would need to fly a total of forty-six assembly missions to bring together its components.

The completed ISS would have two missions. First, provide a world-class laboratory for research, using the space environment (especially free fall, or microgravity) as a resource. Investigations into the fundamentals of physics, chemistry, and materials science would be joined by experiments in the life sciences, pharmacology, and human physiology. Researchers hoped that the new insights and processes discovered or proven in space would lead to technological innovations and economic benefits on the ground. The second ISS mission would be overcoming the obstacles facing the human quest to explore and live permanently off the planet. Using the astronauts and laboratory animals as subjects, investigators would tackle the problems of calcium loss from the bones, immune system depression, muscle atrophy, and radiation exposure. Engineers, physicians, and psychologists all wanted to know the best way to keep astronauts healthy, productive, and motivated on long interplanetary voyages.

The United States and Russia had combined the best elements of their space programs, promising their partners and themselves a host of technical and financial benefits. American and Russian advocates promised that the joint enterprise would speed up the Station's assembly, invigorate the Russians' cash-strapped space establishment, and save American taxpayers some $2 billion. But the savings had yet to materialize: NASA was already paying the Russians $100 million annually just for the privilege of visiting *Mir*. The hope was that cheap, reliable Russian-built hardware would jump-start ISS assembly and save money over the long run. But forging a workable space partnership involved much more than handshakes and signed agreements. The former space rivals had to learn to work together—and trust each other—on humanity's harshest frontier.

# 13

# A Clash of Cultures

*I hope the guys who thought this up knew what they were doing.*

Jim Lovell, Apollo 13, April 15, 1970

Depressurizing the crew lock now, crew member." The tinny voice of the test conductor confirmed that the hiss of air I was hearing through my helmet was indeed air rushing from the airlock headed for vacuum. A year after the canceled EVAs of STS-80, I was beginning a mock space walk in Building 32's Chamber B. The objective was to test International Space Station tools and mechanisms under space temperature and pressure conditions. For the next five hours, as far as my body and the space suit were concerned, I would be in orbit.

Working at vacuum was always a hazardous undertaking, even in the terrestrial confines of the vacuum chambers at the Johnson Space Center. During the Skylab program, a test engineer, James LeBlanc, was working at a pressure equivalent of about 250,000 feet when his umbilical popped loose from his modified Apollo EVA suit, instantly exposing him to near-vacuum. Taking the full brunt of an explosive decompression, Jim said his last memory before blacking out was feeling the saliva boiling off his tongue. As he went down, rescue personnel raced to his aid. After flooding the chamber with air and swinging open the heavy chamber door, they reached Jim just as he was beginning to come around. Although hospitalized briefly, he made a full recovery. As the technicians suited me up for my own run that December day, I paid close attention to every step in the donning checklist, determined to avoid becoming an OSHA statistic.

Supported by a suspension harness running on a small trolley overhead, I stepped through the now-open crew lock hatch into the work area. As on a real space walk, my suit was running on its own, independent of any oxygen or power umbilicals. Only a thin communications cable trailed me into the work chamber: radio links were unreliable within the metal walls.

Powerful vacuum pumps held the chamber at a simulated altitude of more than seventy miles, and the circular chamber walls, which enclosed a work area about twenty-five feet in diameter, were chilled to −298°F by liquid nitrogen, simulating nighttime orbital conditions. Over the next five hours, shuffling to and fro across the floor like Godzilla in an old science-fiction movie, I tested a variety of candidate ISS hardware, everything from docking mechanism components to latches on Station oxygen tanks.

Holding the chilled tools in my gloved hands, I remembered Story Musgrave's experience on a similar 1993 test in this same vacuum chamber. Checking the low-temperature operation of newly designed Hubble Telescope EVA tools, the usual tight fit of Story's gloves soon caused his fingertips to go numb. But the lack of sensation also masked the cold seeping in. Although aware of the cold, Story stayed focused on putting some balky Hubble tools through their paces. Dismissing his numb fingers as symptoms of fatigue and the tight glove fit, he handled the frigid metal continuously in a session that eventually lasted twelve grueling hours.

When he finally emerged from the chamber, flight surgeons discovered that the sustained cold had frozen the skin and tissue of his fingertips. Story insisted that had he known his fingers were frostbitten, he would have terminated the test, but the damage had been done. It took weeks for his injured fingertips to heal, even as he prepared for the Hubble mission. Three years later on STS-80, his scars were still visible, and Story's fingers remained sensitive to cold and pressure. His experience for me was both an example of the dedication needed to succeed in space and a warning against even unconsciously pushing the limits of safety. On this run, my gloves' battery-powered heaters, developed as a result of his accident, kept my hands comfortably warm.

Altogether, my suit, backpack, and body weighed 375 pounds, which made it difficult to stand erect even with the help of the suspension system. Coupled with the stiffness of the inflated suit's arms and gloves, that load gave me an inkling of the physical demands of an actual EVA. The most challenging physical task of the test came at the end when I faced the job of extricating myself from the chamber. Already tired, I had to lift my heavy legs and boots over the raised hatch threshold, step out into the crew lock, then use a metal shepherd's crook to snare my trailing communications umbilical and drag it out after me. My heart was pounding with effort by the time I pulled myself clear of the hatch and locked myself into the donning stand. Thoroughly wrung out, I sagged down into the suit and rested while the test team repressurized the chamber and helped me out of my rig. The bad news was that a real EVA would be every bit as challenging. The good news was that I proved equal to the workload, and I had renewed confidence in my suit, knowing it would protect me when and if I actually made it through an open airlock hatch.

Working on bits and pieces of the Station in the chamber, I had reason to wonder if I would ever see these components reach orbit on a real ISS. The first ISS module was slated to launch in the summer of 1998, just eighteen months after Tammy and I joined the Station effort. Yet major obstacles, both technical and managerial, remained before us. Almost every difficulty in getting the core of the ISS into orbit centered on the partnership between the United States and Russia.

The ISS would initially take shape using a module from each of the two countries, although NASA was paying for both. The first element headed for orbit was the so-called functional cargo block, or FGB. This 42,000-pound spacecraft was to serve as the initial control and propulsion module for the ISS. Forty-one feet long and thirteen-and-a-half feet wide with solar arrays, fuel tanks, and rocket thrusters, it was designed to anchor the Station until Russia's Service Module (SM), the living quarters, arrived in orbit. The FGB, named *Zarya* (Sunrise), contained a central passageway and a spherical docking hub at its forward end. Paid for with NASA funds, the FGB's on-time and on-budget completion proved the Russians could, given proper resources, deliver the goods.

The initial ISS assembly plan called for the FGB to be launched unmanned from Baikonur, Kazakhstan, on a Proton rocket. The shuttle would follow with the first US element of the Station, a relatively simple docking module called Node 1, christened *Unity*. Built by the Boeing

Company in Huntsville, Alabama, the Node sported a central vestibule and six berthing ports for linking later ISS modules. Fifteen feet wide and eighteen feet long, the Node bore a pair of docking tunnels at its fore and aft ports, bringing its overall length to thirty-four feet. These tunnels, called pressurized mating adapters (PMAs), were designed to mate with the FGB and the shuttle's docking mechanism.

With the FGB and Node joined in orbit, Russia would launch another Proton a few months later, carrying its major contribution to the ISS, the Service Module, named *Zvezda* (Star). The SM was essentially a modernized version of *Mir*'s core module. It housed life-support systems, two sleeping compartments, rocket thrusters and solar panels, and a multiport docking sphere at its front end. The SM would be the control station and living quarters for the early crews, handling ISS attitude control and reboost. Once the SM had linked up with the FGB and Node, the Expedition One crew would dock in a Soyuz to begin permanent occupation.

Bill "Shep" Shepherd, the Expedition One commander, was already spending half his time at Star City. As he grappled with his heavy training schedule, he dropped by my office in Houston early in 1997 to talk about his concerns with the FGB. Although NASA and Russia had both agreed on English as the official ISS operations language, the FGB's Russian builders hadn't thought to include a single English label or marking in its interior. No one was tackling this problem, Shep said, or the larger one of getting the Russians to produce English versions of their engineering diagrams, training manuals, or, most important, ISS systems classes. Somehow, Shep continued, we had to produce flight hardware and training materials using a common ISS standard. Without such an effort, the crews' valuable training time would be wasted, and they would arrive in orbit to fly the ISS with hopelessly confusing switches and computer displays. Since ISS program managers seemed unable to cope with these problems, Shep wanted the Astronaut Office to tackle the job.

In so many words, Expedition One's commander told me that he doubted I was up to the task. I politely thanked him for his frank opinion and promised him that as deputy of the Office's Space Station operations branch I would do my best. But I, too, had strong doubts that our Office could pull the integration effort off without top-level engagement from NASA's Station managers. They had been ineffective at forcing the Russians to live up to their ISS agreements, a chronic problem over the next five years. NASA simply lacked the leverage needed to force Russian compliance: we couldn't walk away without bringing down the whole ISS project.

The Astronaut Office realized that the Russian hardware builders and Star City training establishments had no institutional incentive to meet NASA halfway. Many in the Russian aerospace establishment felt that NASA's real intent was to hijack Russian space station expertise. Star City's instructors, veterans of the old Salyut and Mir programs, worried that teaching in English and translating their technical manuals would hasten the day when their jobs were lost to Houston. In Moscow the attitude seemed to be that if the Americans wanted to learn about the FGB and the SM, they could learn Russian and come to the Star City experts.

The shuttle-Mir program had taught us that nothing got done in Moscow unless a personal relationship was established with a Russian counterpart. So the Astronaut Office would have to set up shop in Moscow. We sent VITT engineers and our astronaut "Russian Crusaders" there to support Shep at Krunichev and Energia, the major Russian ISS contractors, and at Star City, an hour's drive northeast of the city. They joined a small cadre of ISS engineers and flight controllers to build the relationships that would eventually help the crews. It was a start: the beginning of a long-term effort to forge lasting technical partnerships with the Russians who built the hardware, trained the crews, and controlled the mission.

Shep had now been joined by other ISS crews, including some of my close friends. Susan Helms had joined Jim Voss on Expedition Two. Ken Bowersox was backing up Shep and would be a natural choice for Expedition Three. Carl Walz and Dan Bursch were aiming for Expedition Four, at least two and probably three years down the road. Like the shuttle-Mir astronauts, all spent about half their time in Star City learning Russian space station systems and training on the Soyuz. In like fashion, their Russian crewmates visited the Johnson Space Center to learn about NASA's station and shuttle systems.

These early ISS crews endured an inefficient training process that was high on hardship and low on useful technical content. Their syllabus would eventually stretch to four years in length, with nearly half that time spent at Star City. The logic of bringing the Russians into the ISS program was inescapable, but the pain of preserving the partnership was felt most keenly by the first half-dozen ISS crews. As delays pushed back their own expeditions, the ISS astronauts were trapped in a professional limbo, hostage to ailing Russian finances. The effects on their careers and families would be considerable.

THE ISS PROGRAM WAS struggling with delays in both American and Russian hardware and the predictable difficulties of trying to blend two very different human spaceflight cultures. Boeing was producing not only the Node but solar arrays, truss sections, and the central control station and laboratory for the Station, the *Destiny* module. Hardware was being built everywhere from California to Florida, but the pacing item was the critical Node and Laboratory software. It was behind schedule and well over budget.

The Space Station program manager, Randy Brinkley, based in Houston, had responsibility for the ISS program, but he faced real constraints on his management authority and ability to maneuver. On the NASA end, both George Abbey and Dan Goldin had veto authority over his decisions. The Japanese, European, and Canadian partners also could elevate disagreements with Brinkley to the "head of agency" level. And then there were the Russians, whose outlook on the ISS was both proud and defensive. They considered themselves the world's experts on space stations, and they were proud their proven hardware was at the core of the ISS. Yet they were defensive about their country's fiscal crisis, which forced them to solicit NASA funds, and fearful of a US move to shift the program lock, stock, and barrel to Houston. These feelings made a nightmare of Brinkley's efforts to keep the Russians and the ISS program moving forward.

In contrast to the NASA-funded, Russian-built FGB, the Service Module was a basket case. The SM had been slated for launch in mid-1998, but with delays in payments to Energia from the Russian government, it remained an empty aluminum shell gathering dust on its Moscow factory floor. The subsystems, the Russians said, were ready for installation, but in fact they were all still at the suppliers' factories awaiting the cold cash needed to purchase and ship them. Russian aerospace companies had learned the lessons of capitalism quickly: no bucks, no Buck Rogers.

The SM delays threatened to stop the Station's assembly in its tracks. NASA started funding a stopgap Naval Research Laboratory spacecraft called the interim control module (ICM). The ICM's purpose would be to keep the FGB and Node under control even without the SM, enabling ISS construction to press ahead. It also provided a hedge in case the SM was destroyed in a launch accident. The ICM's existence was an obvious prod to the Russian Space Agency: if they didn't fund SM construction, the United States was able and willing to proceed without them.

I had more than a little sympathy for Randy Brinkley's trials as program manager. He had George Abbey in Houston pressing him for progress, Dan Goldin pushing him from Washington, and the Russians pushing back just as hard from Moscow. Congress wanted explanations, too, for the Russian delays and the evident lack of ISS progress despite our $400 million investment in the Russian space establishment. After one particularly painful appearance before Congress, Brinkley got a call from his mother, who asked, "Are you really that bad a manager?"

Adapting a technique he had learned during Apollo, George Abbey instituted a new management tool to keep the ISS program moving forward. He convened a meeting each Saturday morning in his Building 1 conference room. Summoning the NASA and Boeing ISS managers and representatives of all major JSC organizations, George used this meeting, often lasting three hours or more, to shine his unique spotlight on problems and people meriting his personal attention. Seeing someone called on the carpet by George Abbey in this very public setting was not a pretty thing to watch, but there was no denying he got results. My worry was that it forced ISS managers, especially at Boeing, to spend valuable time each week making briefing charts instead of progress.

Cabana, Tammy, and I rotated the Astronaut Office Saturday duty. It was an education for me to see the director of the space center chastising a subcontractor for his company's late delivery of a handful of Node electrical connectors. Often the only thing to say in such a situation was something I had been taught at the Air Force Academy: "No excuse, sir"—not that it would get you off the hook. You didn't want to come back the next week with the problem unresolved. Abbey was making a point, and it wasn't subtle: stay on schedule or you would be publicly brought into line.

For a solid year a chronic issue at these Saturday morning sessions was the ISS program's failure to get quality training for Shep and the other Expedition crews. He was spending half his time at Star City, but he was learning little of value. Star City's managers wanted NASA to pay them to develop training materials in English; worse yet, Krunichev and Energia wanted Star City to pay *them* for the technical data needed to develop manuals, classroom materials, and simulators. Tied in by phone, Shep used Abbey's weekend sessions to air his complaints. If the Russians wouldn't provide it, he wanted both Mission Operations and the Astronaut Office to get their people to Moscow and ferret out the necessary information. If things didn't change soon, Shep said, he would stop training and come home.

Sadly, NASA and the ISS program were in a poor position to force Russian action. Training problems were supposed to be hammered out through negotiations between the chief of Flight Crew Operations and Star City's director. But Dave Leestma and his successor, Jim Wetherbee, were unable to get their opposite number, General Yuri Glazkov, to improve things for Shep. Glazkov was determined to prevent the transfer of his team's collective experience to a training library in Houston. Glazkov stonewalled. In the end, NASA was forced to pay Star City to get Shep and other crews their training manuals and technical materials.

My understanding of the clash of cultures besetting the ISS program took a quantum jump during two trips to Russia in 1998, part of high-level joint meetings between ISS managers and their Russian counterparts. Randy Brinkley wanted to address Russian hardware delays and funding issues. Kevin Chilton, back from STS-76 and his successful *Mir* docking, was now deputy ISS program manager for operations; he would try to get the two sides to accelerate and improve crew training.

In a Star City Training Readiness Review, Kevin tried to close the gap between what Shep needed to learn about the Russian segment systems and what Star City was willing to give him. The results were mixed. Colonel Yuri Kargapolov briefed Chilton on the crew's training status, using view graphs prepared entirely in Russian. The colonel's plan showed Shep training at Star City for six weeks but specified lesson plans and lecture topics for fewer than four. Russian Space Agency officials insisted they couldn't sign training agreements because the managers who had such signature authority had not been hired yet. And the chief SM display designer, who had demonstrated some impressive laptop-based crew displays, told her NASA counterparts that she would "never give NASA the display software, no matter who" directed her. We had a long way to go.

On the positive side, we had an uplifting video-telecon with Andy Thomas, then flying aboard Mir, during a visit to the Mission Control Center in Korolev (the TsUP). Andy was in fine spirits three months into his tour as the last American aboard Mir but still eager to see some new faces. Later, walking with friends through Moscow's Ismailova Park, I found a graceful Soyuz booster poised on its launcher, a monument to Russia's historic achievements off the planet. Nearby we found that the former space exhibits hall had been transformed into a warehouse, housing row upon row of stacked cartons holding Japanese stereos, CD players, and televisions. But at the rear of the cavernous hall behind the electronics, we stumbled upon some of the former exhibits: full-scale replicas of Soviet planetary probes and satellites shunted haphazardly into a corner. Socialist glories had given way to consumer demand. Opposite the Soviet-era spacecraft sat a relic of an older Russian–American alliance: an Apollo command and service module. On loan here since the 1975 Apollo–Soyuz mission, the spacecraft sat neglected, its gold and silver exterior coated with a thick layer of brown dust. The Apollo seemed an apt symbol for the new Russia's attitude toward our joint space efforts.

In Moscow again in September 1998 for another Training Readiness Review at Star City, I took advantage of the meeting to visit with Hairballs Dan Bursch and Carl Walz. Their Star City quarters were adjacent to the NASA offices in the Prophylactorium, a crew training and quarantine facility built during the Apollo–Soyuz project. The "Prophy," like many Russian buildings, had decayed noticeably in the twenty years since its construction, but Dan and Carl's duplex was a great improvement over Norm Thagard's grim Star City apartment. NASA had purchased Canadian kit-built houses, shipped them to Star City, and paid Russian construction crews to assemble them on-site. Comfortably furnished and equipped with Western appliances and bathrooms, the quarters were convenient to the Star City classrooms and simulators.

Bill Shepherd's duplex had the same layout, but he had outdone himself by founding Shep's Bar in the basement, which became a Star City institution. Both astronauts and cosmonauts were welcome in this makeshift watering hole to chat, unwind after work, or screen a movie together. In the adjoining basement reached through a gap in the concrete foundation, Shep had installed a small but growing collection of exercise gear. The gym and Shep's Bar were the social centers of gravity for all the Expedition crews training in Star City, and it brought the cosmonauts and astronauts together in a way that the workplace seldom did.

At the conclusion of the September 1998 meetings in Moscow, many ISS problems remained unsolved: the lagging construction of the Service Module; delays in the US control software for the Node and Lab; the shortcomings in crew training at Star City and JSC; the difficulties in getting cooperation from Star City and Energia; and the woes of the cash-strapped Russian Space Agency. But the ISS partners had reached one important agreement: in late November 1998 a Proton would launch from Baikonur, taking the FGB to orbit. It was more than a year behind schedule, but the International Space Station was about to take flight.

MARK LEE HAD RETURNED from his fourth flight, STS-82, as a veteran of four space walks. After this second Hubble servicing mission, he had immediately gone to work solving some of the difficult EVA problems facing astronauts and ISS engineers. One of his tasks was to develop a practical plan for the space walks needed to install and outfit the US Laboratory, later named *Destiny*. In late spring 1997 Mark had asked me if I might be interested in joining him on some of the underwater EVA testing on the Lab. I didn't want to get my hopes up, but Mark seemed to be carrying around some good news, although he was not quite certain how to share it.

The chief solved the mystery a week later. Six months into my assignment as Tammy Jernigan's deputy for the ISS, I got a note from my STS-80 crew secretary, Tammy West, now working for the chief. "Colonel Cabana would like to see you in his office at your earliest convenience." It was June 3, 1997, barely half a year since we had run into that jammed hatch. Bob, I hoped, had found a way to remove the sting of that disappointment.

## SPACEWALKERS NAMED FOR SPACE STATION ASSEMBLY FLIGHTS

*From NASA Press Release 97-126, June 9, 1997*

A cadre of 14 Space Shuttle astronauts has begun intensive training in preparation for the spacewalks required for on-orbit construction of the International Space Station. "It is important for us to begin work now to train the crews who will support Space Station assembly flights," said David C. Leestma, director of Flight Crew Operations. "These crew members will be exceptionally busy preparing for some challenging and demanding tasks, from initial assembly through installation of the robotic arm and an airlock for station-based spacewalks. . . ."

The specific assignments for these US assembly flights, through August 1999, are:

| Flight | Crew Members | Planned Launch Date | Payload |
|---|---|---|---|
| STS-88 ISSA-2A | Jerry Ross and Jim Newman | July 9, 1998 | Node 1; pressurized mating adapter 1 and 2 |
| STS-92 ISSA-3A | Leroy Chiao, Jeff Wisoff, Michael Lopez-Alegria and Bill McArthur | January 1999 | Z1 integrated truss, pressurized mating adapter 3 |
| STS-97 ISSA-4A | Joe Tanner and Carlos Noriega | March 1999 | Photovoltaic module, P6 truss |
| STS-98 ISSA-5A | Mark Lee and Tom Jones | May 1999 | US Lab module |
| STS-99 ISSA-6A | Chris Hadfield and Robert Curbeam | June 1999 | Lab outfitting; remote manipulator system |
| STS-100 ISSA-7A | Mike Gernhardt and James Reilly | August 1999 | Joint airlock, high-pressure gas assembly |

"The assignment of these crew members is a critical element in our ability to bring together the elements of the Space Station on the ground and then successfully assemble them in orbit," said Randy Brinkley, Space Station program manager.

Grateful that Bob had found a way to get me another shot at an EVA, I asked him about Tammy's prospects. The chief assured me he was working on that.

Working daily on the details of the ISS program, I was in a great position to help prepare with Mark for ISS assembly flight 5A. STS-98 was scheduled to deliver the US Laboratory in May 1999, just after Shep's Expedition One crew arrived at the ISS. The voyage was still two years away, and I was always skeptical of ISS hardware schedules, but at least I knew my destination. When the Station finally got off the ground, I had an appointment with *Destiny*.

# 14

# DESTINY

*Every fallacy we detect can show us where we are at fault and guide us toward the truth.*
ROBERT H. GODDARD, ADDRESS AT HIS HIGH SCHOOL
GRADUATION, 1904

M ark Lee and I would be at the center of planning for ISS assembly flight 5A, the fifth American assembly flight in the original construction sequence. Flight 5A's sole purpose was to deliver *Destiny* to the ISS. It sounded simple enough, just a routine drop-off in orbit, but Laboratory installation was *the* critical step in Station assembly. The Lab was not only a research facility but also contained the computer, life-support, power, and thermal control systems that would bring the US "segment" (American-built portion of the Station) to full potential. It would become the Station's flight deck, the nerve center for operation of all major ISS systems. As experiments arrived on later shuttles, the Lab would assume its role as the center of scientific research aboard the outpost.

Activating the Laboratory required far more than just plugging it onto the front end of the Node, its designated berthing location. Berthing was just the first step. The 5A astronauts would conduct three space walks to connect the Lab to Station power, data, and cooling re- sources. Then our crew, working with Expedition One and both the

Houston and Moscow control centers, would activate the Lab's critical computers and systems.

Mark Lee and I had the job of executing the three EVAs required on the 5A (STS-98) mission. Our lead EVA officer was Kerri Knotts, a young aerospace engineer just four years out of West Virginia University. Kerri coordinated the task of transforming ISS assembly requirements into plans for the space walks, creating the tools and procedures we would need to get the job done outside. When we began working together in June 1997, she had already sketched out the major elements of the three space walks. Kerri was the coach, and Mark and I were her EVA team.

Our practice field was the Neutral Buoyancy Laboratory (NBL), a new facility for simulating space walks, located on the southeastern corner of Ellington Field about five miles north of the Johnson Space Center. Tammy Jernigan and I had done all of our EVA simulations under water in the WETF in Building 29. But even during my STS-80 training, it had become obvious that the WETF was too small for future training needs. The water tank could barely submerge a shuttle payload bay mock-up, and its twenty-five-foot depth meant that large mock-ups, like that of the Hubble Space Telescope, had to be cut in half to fit. The WETF was clearly outmoded for ISS training. The modules and trusses were simply too big to squeeze into the pool.

JSC took over a new but never-used factory from the defunct Freedom program and dug a twenty-foot-deep pit in the expansive floor. Erecting 20-foot-high steel walls around the pit created a huge 40-foot-deep water tank, holding 6.2 million gallons of water (enough to fill about 200 suburban swimming pools). At 202 feet long and 102 feet wide, the NBL is large enough to accommodate two simultaneous EVA training runs. And the mammoth pool can hold not only a shuttle cargo bay mock-up but also a substantial portion of the ISS pieced together just as it is in orbit.

The NBL's depth would limit astronauts to training runs of just a few hours without a decompression stop on the way up, so in the NBL, crew members breathe Nitrox, a combination of air and nitrous oxide. The gas mixture reduces the amount of nitrogen dissolved in the blood and permits longer, deeper dives without risking decompression sickness (the bends). To honor his commitment to the future of space exploration, the new NBL was named for the late Sonny Carter.

Kerri began putting Mark and me in the water to evaluate the techniques we might use to accomplish Lab installation and outfitting. Our first EVA would begin with our robotic arm operator shifting the Node's PMA-2 docking tunnel to a parking spot higher on the Z1 truss, a cube-shaped utility element, clearing the Node's forward berthing port for the Lab. After *Atlantis's* arm swung the Lab into place, we space walkers would wire in a set of backup heaters that would keep its critical equipment warm in the −200°F shadows of orbital night.

Our second EVA would tackle Lab activation. Mark and I would hook up a dozen or so umbilicals running from the Z1 truss, completing power, data, and cooling connections to the Lab. We would also install a grapple fixture on its hull that would later serve as the base for the ISS robotic arm.

The last EVA would see us assist in the transfer of the PMA-2 docking tunnel to the forward end of the newly installed Lab. We would reconnect the tunnel's power and data connections, making PMA-2 the ISS's front door for future shuttle dockings. Mark and I would round out our eighteen hours of space walks on flight 5A by unveiling the Lab's Earth observation window and transferring spare parts to the Station exterior.

None of the skills required was new to me, but unlike on STS-80, the three 5A space walks were not a test program. The Lab EVAs were central to the success of the mission. If we failed to get the job done, the Lab could be irreparably damaged and the Space Station's future research and control capabilities seriously degraded. Every time I went in the water, I was aware that our team simply had to find a way to get these tasks accomplished. Kerri's original plans were the starting point for a difficult and lengthy journey we were determined to see through to mission success.

Every month we would try out new concepts and ideas under water. Even for a space veteran, working in the vast NBL was an eye-popping experience. The waters of the 6-million-gallon tank were a clear aquamarine kept at a near-constant 85°F to keep our safety divers warm during their long shifts submerged. Hovering in my suit over the ISS modules and trusses forty feet below, their massive forms enveloped in the tropical blue of this indoor ocean, I could scarcely imagine the day when I would traverse this giant structure in orbit. Long hours spent working on the Lab, Node, and trusses, laid out horizontally on the floor of the pool, began to imprint a map of the Station in my mind; finding my way around on the vast complex soon became second nature.

An ISS submerged in the depths of the NBL was impressive, but we would one day have to work on the real thing, whose modules were being built at the George C. Marshall Space Flight Center in Huntsville, Alabama. Whenever we needed to check progress on the real Lab, Mark and I could grab a jet for the hour-long hop over to Huntsville. In Marshall's Building 4708, where Wernher von Braun's team had built the test versions of the Saturn V moon rocket's mammoth first stage, Boeing was assembling the first US station components. When I first saw them in 1997, the Node and Lab were hollow metallic cylinders, their three-eighth-inch-thick aluminum hulls painted in dark green anticorrosion primer. The Lab was twenty-eight feet long and fourteen feet wide, sized to just fit inside the shuttle's payload bay. Avionics, life-support, and research racks would narrow its interior aisle to a still-roomy eight feet. The aluminum hull would be armored with a gleaming silver shell of meteoroid and orbital debris shielding. It was hard to imagine this giant beer can could ever make it to space, but the Boeing technicians assured us that in two years this would be a fully outfitted orbital laboratory. Well, they had built Moon rockets on this floor. Perhaps a Lab for the Space Station wouldn't prove too much of a stretch.

Although a bit daunted by the complexity of our EVAs and the years of work still to be done, I was reassured by Mark's extensive experience. Forty-five years old when assigned to STS-98, Mark was an Air Force colonel, a former fighter pilot in the F-4 and F-16. With a master's degree in mechanical engineering from MIT, he had joined the astronaut corps with the Maggots in 1984. He had helped launch the *Magellan* radar mapping probe to Venus in 1989, served as payload commander on the Spacelab-J mission in 1992, test-flew the EVA rescue jetpack on STS-64 in 1994, and racked up three more space walks while visiting the Hubble Space Telescope early in 1997. One of the Office's most experienced space walkers, he was also blessed with the physical size and strength to work efficiently in the suit. Mark was Kerri Knotts's ace in the hole, an astronaut who could apply finesse (and muscle, if necessary) to almost any EVA task she could devise.

Kerri, Mark, and I tackled all the intricate details of the three 5A space walks, a more complex and delicate set of challenges than any I had encountered in space. One of my jobs on EVA 2 was to install the grapple fixture, the future foundation of the Station's robot arm, and make the wiring connections into the Lab's hull. Unfortunately for me, the work area on the Lab mock-up forced me to work heads down in the suit as I rehearsed the task. While the NBL provided our suits with neutral buoyancy so that we moved about much like we would in orbit, gravity still reigned, even under water. Working with my feet suspended from the shuttle arm for twenty minutes or more, I sagged into the suit's hard steel and fiberglass shoulders, blood rushing to my head. My eyes bulged, my head pounded, and the suit bit painfully into the tops of my shoulders. Grimacing with the effort, I would flex my feet hard into the boots to keep my shoulders off the metal arm bearings, but I could only take the punishment so long before crying "uncle" to the divers. After a short breather right-side up, I was back to head over heels, wishing all the while for real free fall. Teflon shoulder pads helped a little, but after each six-hour run, I would sport vivid bruises on my shoulders for days. The only sure relief was to take this job to orbit.

That would prove a little difficult without the rest of the crew. The planned launch date of May 1999 dictated that the Flight Crew Operations Directorate assign a crew soon. On June 23, 1998, I got a message from Ken Cockrell, "Taco," my STS-80 commander: "Needs to see you!" said the secretary's note.

He wanted to fill me in on some great news: our next chance to work together. Mark and I would be joined by Taco as commander, Mark Polansky as pilot, and Marsha Ivins as our robot arm operator. I knew from my continuing ISS work that the Service Module delays would push our launch back to the fall of 1999, but that would just give the five of us more time to get ready. STS-98 was a complex mission, and there were a thousand details about our ISS work still to be decided.

Mark Polansky (christened by Air Force buddies with the inevitable nickname "Roman"), had joined the Office in 1996 and would be making his first flight. Marsha was Mark Lee's 1984 classmate and had been working at JSC since 1974. A barnstorming Stearman biplane pilot and the Office expert on space photography, Marsha had flown four times—on STS-32, STS-46, STS-62, and STS-81—and done nearly everything an astronaut could do in space, including a *Mir* docking mission. She and I had worked together on the frustrating problem of where to stow cargo shipments in the limited space aboard the ISS. Marsha would be our arm operator, and I liked her determined approach to the difficult Lab berthing task: "Once I take it out of the cargo bay, I ain't puttin' it back."

Our formal crew announcement came in August 1998, and while I continued to hold down my ISS job for the Office, we forged ahead for the next year with work on the Lab and its three EVAs. Events in the ISS program continued to dominate our planning and rule our future.

### DIARY ENTRY, AUGUST 31, 1998:

The Russian financial collapse continues, now with the government tottering as well. . . . The fact that they are late with SM means that next April's [Service Module] launch will slip and so will my launch, probably til at least Jan. 2000.

The Russians have us over a barrel in a huge way. We need the SM for crew quarters and propulsion, life support and [guidance, navigation, and control]. But more significant is the necessity for at least 30 Progress launches for propellant and resupply. If the Russians go away, we will wind up having to replicate that capability at great expense. . . . But right now the Russians are not even buying the parts needed to keep their Progress production line open. Very worrisome for logistics in 1999, hence our worries over the launch sequence in 1999. When will *Mir* fall? When will we know whether the Russian government will ever supply the '98 funds to their space program, let alone the '99 money!?

A few weeks later I saw Randy Brinkley, a tough former colonel in the Marine Corps, holding his head in his hands as he listened to Russian officials during a Moscow Joint Program Review. I wondered whether a successful Station was even possible. Both ISS hardware and training schedules were lagging, and the Russians did not seem embarrassed that they were largely to blame. While the FGB was nearly ready for launch, the Service Module was hidden away within its Moscow factory. NASA officials were denied even a look at it during our September 1998 meeting in Moscow: Energia officials announced that because it was past 4 p.m., the factory was closed. Even Shep, the commander who would lead the first crew to the ISS, had trouble getting access to the SM for training. The visits were essential, since there was no simulator yet at Star City.

Routine crew access to a spacecraft was simply unprecedented in Russia. Astronauts swarmed over the shuttles and hardware at the Cape (the Office encouraged us to go there as frequently as possible), but cosmonauts had very little input to Russian vehicle design or testing. They flew the ships "as delivered." Thus, Russian engineers were naturally uncooperative when American astronauts started suggesting ways to make the FGB and SM switches and controls more "user-friendly." Occasional technical meetings to submit our suggestions to the Russian designers and trainers usually accomplished little. The Office began to make headway on Shep's behalf only as the astronaut cadre in Moscow gained experience. We had had a "Director for Operations, Russia" at Star City since the Phase I shuttle-*Mir* days, but the hardware was nearly an hour away in Moscow. The astronauts and engineers who were posted to Moscow made the contacts, identified the problems, and doggedly solved them. The "Russian Crusaders" got the job done, and they were invaluable to the chronically overworked Expedition crews.

ISS assembly began at last on November 20, 1998, when the FGB soared into orbit from Baikonur Cosmodrome atop a Proton booster. With the Service Module seriously behind schedule, the Russians had recommended delaying the FGB launch into 1999 (to avoid cutting into the FGB's fifteen-month orbital design life). But NASA knew an impatient Congress was watching, and Goldin, Abbey, and Brinkley wanted to keep this "first element launch" milestone in 1998. The race now was to get the rest of the assembly sequence moving.

On December 8, 1998, Bob Cabana's STS-88 crew aboard *Endeavour* caught up with the FGB, and my classmate, Nancy Currie, snared the twenty-one-ton module with the robot arm. Nancy, who had emplaced the Node atop the orbiter's docking tunnel the day before, now centered the FGB's hatch a few inches above the Node. With the alignment checked and double-checked, Bob fired the orbiter thrusters to shove the two modules together and engage latches linking the two. The joined FGB and Node formed the nucleus of the International Space Station, seventy-six feet tall and weighing about 80,000 pounds. The crew spent the next few days entering the Node, continuing down the tunnel into the FGB, and outfitting the interior. Jim Newman and Jerry Ross also performed space walks to connect electrical wiring between the two modules and install the Station's backup communications antennae. Watching the work from MCC, I shook my head in amazement:

the Hairballs were up there building the ISS. Not bad for a class that didn't even think to include the Station on its patch.

The nascent ISS would now wait patiently in orbit for the Service Module, originally scheduled to follow in the spring of 1999. The SM would dock to the combined FGB and Node, giving the ISS the capability to support permanent crews. A pressurized central core would run the length of the Station, with the Node at the prow, the FGB behind it, and the SM bringing up the rear. A pair of solar arrays on each Russian module would provide initial power, and the SM's thrusters would supply attitude control. This three-module train would stretch 118 feet from nose to tail.

But the wait for the SM grew ever longer as Russian government payments to its contractors failed to materialize. Incomplete and overdue, the Service Module became the symbol of the Russian Space Agency's inability to meet its ISS commitments. Even before the FGB was launched, the Russians had already announced an SM delay until the summer of 1999 at the earliest. Still saddled with *Mir*'s operating costs and unwilling to give up this famous symbol of its preeminence in long-duration flight, the Russians seemed none too anxious to transition to a venture in which they would only share the top billing. All NASA could do was press ahead with the Interim Control Module, hoping its open willingness to launch it might shake loose SM funding in Moscow.

By early 1999, just two months after the FGB launch, the Service Module had slipped again, into late September. Every month of delay cost US taxpayers $7 million in storage costs for the ISS equipment (the trusses, Lab, and airlock) stacked up at the Cape, waiting for assembly to resume. It would have been cheaper just to pay the Russians to finish the SM and get it into orbit, but Congress had barred NASA from transferring further funds to Moscow. We would wait more than a year for the Russians to scrape together the resources to ready the SM for launch.

The SM's arrival in orbit would break the assembly logjam and kick off a series of major construction milestones. With the SM in place, shuttle assembly flight 3A would deliver the Z1 truss to the top of the Node. The Z1 contained electrical and cooling lines, a high-gain communications antenna, and four control moment gyros to control the Station's orientation, or attitude. Arriving with the Z1 would be a third docking tunnel, PMA-3, occupying the Earth-facing berthing port on the Node.

Joined by Sergei Krikalev and Yuri Gidzenko, Shep would arrive on a Soyuz a few weeks later to become the Station's first permanent crew. The 4A shuttle crew would visit Expedition One and install the P6 truss, a 48-foot power tower supporting two huge solar arrays. The solar panels would span 240 feet, and their 9,600 square feet of photovoltaic cells would furnish about twenty-five kilowatts of power. Once the solar arrays arrived, the Expedition One crew and the ISS would be ready for the arrival of the first research module, the US Laboratory module *Destiny*.

Russian and American space station designers had followed two very different approaches to the problem of orbital assembly. Russian spacecraft from Salyut onward used automated docking techniques to berth Soyuz, Progress, and the half-dozen modules added to *Mir*. Electrical and plumbing connections were usually made automatically upon docking or hooked up later using cables strung through open hatches. Looking at the challenges of putting a million pounds of hardware together in orbit, American designers decided to take advantage of the shuttle's robotic and EVA capabilities. US modules and truss sections could be berthed via the robot arm, saving the weight and expense of thrusters and new automatic docking systems. Rather than try to cram all utility connections into the mating surfaces between modules or truss sections, US engineers decided to position them outside the Station's hull, where they could be connected by space-walking astronauts. Damaged umbilicals could then be reached easily for repair, and any toxic leaks of coolant, for example, would remain outside the crew spaces. Another safety benefit was keeping the power, data, and cooling connections free of the hatches so that they could be swiftly closed in an emergency.

There were disadvantages to this approach, to be sure. Precision robotic operations were demanding, sometimes at the very limits of the crew and the robot arm's capability. And multiple space walks were required to complete utility connections and outfit the module exteriors. Both robotic and EVA operations were risky, but involving the shuttle crews provided an extra reserve of troubleshooting and problem-solving capability.

Late in 1999, NASA shipped the new *Destiny* Lab to the Cape using the venerable "Super Guppy" transport, a converted 1950s-era Boeing Stratoliner purchased by NASA to move ISS modules from their manufacturing locations to the launch site. The turboprop Super Guppy looked like a bloated white whale with wings; its nose swung up and out of the way so that the outsized rocket and ISS components could slide into the cavernous fuselage. As progress continued on *Destiny's* construction, Roman and Marsha joined Mark and me in our EVA training out at the Neutral Buoyancy Lab. Roman took charge of our tools and checklists as in-cabin EVA coordinator, while Marsha handled the NBL's functional robot arm to practice moving us space-walkers about.

In summer 1999 the Service Module seemed no closer to launch than it had the previous year. Meanwhile, the two-module ISS continued to circle the Earth, husbanding its fuel and waiting for the next arrivals from Baikonur and the Cape. One of the visiting shuttle crews sent to maintain and outfit the Station included Tammy Jernigan. In early June 1999 she arrived at the ISS on STS-96, and the mission included her well-deserved first space walk. She and Dan Barry spent almost eight hours outside working on the Station's exterior, installing both US- and Russian-built construction cranes (the US version was the same one we had planned to test on our lost space walks). Tammy's EVA was sweet compensation for our bad luck on *Columbia,* and my own hopes rose with her successful emergence from the airlock.

The Lab's critical command and control software was still behind schedule, being debugged in Houston. Without those crucial computer routines, *Destiny* would be unable to communicate with and command the myriad systems aboard the ISS, everything from the US life-support components to the Russian propulsion system. Unlike the shuttle cockpit that was crammed with more than a thousand switches and circuit breakers, the Station would be controlled via laptop computers in the Lab. Without reliable software, the Lab would be an empty addition to the Station, a spare bedroom rather than the ISS nerve center. The Russians' complaint that their funding problems were shielding NASA from criticism about the US Lab's substantial delays had some merit. Until its software bugs were worked out, *Destiny* would stay on the ground.

For more than two years, first as deputy, then as chief of Space Station Operations for the Office, I had worked hard to make the Space Station a safe and productive outpost. My gripes with the assignment centered mainly on the Russians' inability to deliver training and hardware on time, and NASA's unwillingness to prod its major partner along more forcefully. Although the Lab had its problems, too, they were being worked out, and I was confident that when STS-98 finally launched, my own crew could handle anything space could throw at us. But now developments inside the Astronaut Office put our tightly knit team at risk.

Charlie Precourt, my Air Force Academy classmate and fellow Hairball, had replaced Taco as chief astronaut in 1998, just before the STS-98 crew announcement was made public. A few months later, Jim Wetherbee replaced Dave Leestma as Flight Crew Operations chief.

In July 1999, two years after we had begun working on the Lab together, Mark Lee and I were relaxing at Molly's Pub after another tough day in the NBL. What Mark told me now came like a bolt from the blue: Jim Wetherbee had asked Mark to remove himself—resign—from the STS-98 crew.

I was stunned. Had I heard him right? What possible reason could Wetherbee have for removing Mark? His leadership and hard work in preparing for flight 5A and the Lab EVAs had been not only superlative but, to me, indispensable.

Mark had a hard time explaining it. He shrugged his broad shoulders and looked off across the deck outside the crowded bar. "I think it has something to do with Jan and me getting divorced." Mark had been married to fellow astronaut Jan Davis, but their marriage had ended earlier that year. Wetherbee had cited only vague dissatisfaction with Mark's job performance as chief of the EVA branch for the Office— disagreements that Mark considered minor. Wetherbee and Charlie Precourt had not documented their concerns nor had they warned Mark that his flight was at risk. Wetherbee was now giving him just one option: step down, with the hope of assignment to a future flight.

Our crew was thunderstruck by the development. We knew of no problems or complaints about Mark's job performance. Marsha, Mark's classmate and friend of fifteen years, couldn't imagine how Wetherbee and Abbey could pull Mark after his efforts to plan the EVAs, the heart of our mission. Taco talked to Charlie, and I wrote a letter to him detailing the negative mission impact of pulling the 5A lead space-walker so close to the flight. Charlie couldn't help us. He told us his hands were tied, that it was up to Wetherbee.

By Labor Day, Mark had gotten nowhere in his efforts to change Wetherbee's mind, and a meeting with Abbey had been similarly unproductive. Our crew's efforts to win a review of the situation from NASA headquarters also went nowhere. The associate administrator at the Office of Space Flight told us that he preferred to let the responsible JSC chain of command handle the case.

Late one evening after this final bad news, I walked down with Taco to Building 4's deserted ground floor. Grabbing a Coke from the machine, I vented my feelings to my friend and two-time commander.

"Taco, what they're doing to Mark is inexcusable. I'm so disgusted with this whole mess that I'm tempted to just resign from the flight myself."

He sympathized, saying he, Marsha, and Roman were ready to do the same. But he was just as certain that throwing our badges on the table would be a futile gesture. Wetherbee would appoint a new crew, and our colleagues would have less than a year to prepare for the mission. It wasn't a realistic option.

The next day Jim Wetherbee called our crew in to Charlie's office. He officially notified us that he had decided to remove Mark from the crew. He wasn't prepared to share his reasons with us, "so as to protect Mark's privacy." If Mark agreed to the reassignment, he would have a shot at a later mission. If we didn't accept his decision, we would be reassigned as well. Wetherbee asked pointedly if we understood.

So far, the entire controversy had been kept under wraps. Later that week, Jim and Charlie announced the decision to the Astronaut Office as our crew sat stone-faced in the back of the conference room. Cady Coleman stood up to ask if her fellow astronauts might avoid future problems by learning what rules had been violated (a gutsy move on her part), but Charlie, again citing Mark's privacy, refused to share any details as to why they had chosen to remove him.

### ASTRONAUT PULLED FROM SHUTTLE FLIGHT

Houston Chronicle, *September 9, 1999*

In an unusual move, NASA pulled a key spacewalking astronaut from a high priority International Space Station construction flight next spring.

Astronaut Mark Lee was bumped from the five-member shuttle crew that is to deliver the bus-sized US research

laboratory to the new station in late May.... The space
agency ... did not disclose the reasons for its actions. Efforts
to reach Lee, 47, an Air Force colonel, were unsuccessful.

"Is this a disciplinary action? The answer is no," said NASA
spokesman Ed Campion.

With Wetherbee unwilling to discuss plainly what Mark had done
wrong, Mark could devise no strategy for making things right. He could
either leave the Astronaut Office or wait indefinitely for his superiors to
fulfill their vague promise to consider him for a future assignment.

Unfortunately the way Mark was being treated had its precedents.
Several astronauts who had made serious professional missteps either in
training or in flight had been (properly) disciplined, but they had received only official silence from the Office and Abbey as to their
prospects for redemption and a future spaceflight. Instead they had discovered by rumor or experience that a flight assignment would never
come their way again. As Bryan Burrough notes in his book *Dragonfly*,
Bob Cabana had tried to assign one such astronaut to a flight after a suitable cooling-off period, but Abbey turned down every crew nomination
that included him. "I put your name in . . . and it didn't go through,"
was all Cabana could say.

In Mark's case, he never got a straight answer on what he had done
to deserve removal, let alone whether he would fly again. Looking back
on the incident, I am still deeply disappointed in NASA's inability to
deal straightforwardly with the finest, most devoted group I've ever
worked with—the astronauts. Every astronaut, and especially those few
working under a cloud, merited a frank assessment of their actions and
their prospects for a future flight.

After Mark Lee agreed to step down, Wetherbee and Abbey never
gave him another flight assignment. He left NASA for private industry
in 2001.

WHEN WETHERBEE REMOVED MARK from the crew, I stepped into the
role of lead space-walker on the STS-98 mission, and not without some
trepidation. Already the head of the Space Station Operations branch
in the Astronaut Office, I now had responsibility for three EVAs, one of
them critical to the successful activation of the Lab. I had been practic-

ing for a space walk for the last eight years, and though I had been to vacuum in a space suit in free fall, I had never actually stepped outside. As Liz wryly noted, "You're NASA's secret weapon: the best-trained space-walker who's never done a space walk."

With Mark gone, Charlie Precourt had to come up with an astronaut to replace him. His pick was Bob Curbeam. "Beamer," a member of the 1995 astronaut class (the "Flying Escargots"), was already assigned to flight 6A, the ISS assembly mission next up after the Lab's arrival. A veteran of STS-85 in 1997, Beamer had been working on the 6A space walks with Chris Hadfield.

Beamer and I had never worked together. He and his family had arrived at JSC while I was in the middle of capcom duty, and my flight on STS-80 and his own mission the next year kept us from being more than nodding acquaintances. With his sly wit and one of the most sensitive BS detectors on the planet, Beamer blended easily into the crew. Now together we would have to pull off the difficult space walks surrounding Lab berthing and activation.

Beamer joined the crew when STS-98 was just nine months from its spring 2000 launch date. Kerri, Mark Lee, and I met immediately with Beamer to reshuffle the EVA plan and bring him up to speed on his space-walking tasks. Fortunately he was already familiar with the Station from his 6A training and could jump right into our NBL preparations. We would log more than 200 hours in the pool by the time we launched. The week after he joined the crew, Beamer was under water with me running through the assembly steps for EVA 1. With his ample strength and reach, he was also perfectly suited to take on Mark's space-walking tasks, including connecting the critical ammonia cooling lines linking the Lab to the Station's thermal control system.

Beamer reviewed our EVA plans after our first water runs together. He noticed that on the first EVA, we space-walkers largely busied ourselves with a series of secondary tasks, such as connecting backup heater cables to keep the inactive Lab from freezing. Why not jump right on the big job and attach the main power and cooling lines to *Destiny*? That would push the Lab's activation two days earlier and spare us the trouble of stringing unneeded heater cables. Mission Control and the two crews aboard the ISS—Shep's and ours—would get two extra days to deal with any unexpected problems cropping up as we turned on the Lab's computer, electrical, thermal control, and life-support systems. Within a few weeks our shuttle and Station flight control teams evaluated Beamer's proposal and found it made a lot of sense. Curbeam had made the first of his many contributions as a 5A team member.

The school bus–sized *Destiny* was the most complex of the ISS modules, thanks to its central control and research functions. While we worked on plans for getting it activated at the Station, Cape technicians were meticulously assembling its maze of internal systems. *Destiny* weighed about 32,000 pounds and featured hatches at both ends of its twenty-eight-foot length for connecting to other ISS modules.

Inspecting the Lab during one of our frequent visits to the Kennedy Space Center, I could easily see the reinforcing waffle pattern machined into the exterior of the high-strength aluminum alloy hull. Another obvious feature was the twenty-inch-diameter Earth observation window located at the center of the downward-facing hull section. Portions of the Lab were already being fitted with debris shield blankets made of Kevlar, similar to the material used in bulletproof vests. The final layer of micrometeoroid protection would come from those thin aluminum debris panels we had seen being fabricated in Huntsville.

Inside the Lab's once-empty hull, technicians were installing the trays of wiring and plumbing that would support both the control and research functions. Four standoffs—narrow utility spaces running the length of the module—were spaced evenly around the interior, marking the baseboards of the square center corridor. Six ISS standard payload racks installed side by side would form each of the four interior walls (labeled overhead, deck, port, and starboard). Each of these refrigerator-sized racks housed either research experiments or Lab systems such as power, life support, or computers. The modular racks could be removed and replaced as necessary; each came equipped with fluid and electrical connectors to support the equipment and experiments housed within. The entire Lab contained twenty-four rack locations—six on each aisle surface—with more subsystems built into the bulkheads surrounding the forward and aft hatches. Shep and his Expedition 1 crew would use many of the empty rack locations as temporary storage closets.

*Destiny* will eventually hold up to thirteen experiment racks dedicated to life sciences, materials research, Earth observation, and commercial applications. Because of weight restrictions, we would carry the Lab to orbit with just five racks, all housing essential support systems: thermal control, electrical power, computers, and air conditioning.

Those computers and guidance systems, though, would relieve the Service Module of day-to-day control of the Station. Successful Lab activation would shift operational responsibility for the ISS from Moscow to Houston's Mission Control Center.

That extensive capability made the new Lab the most expensive ISS component. The $1.4 billion module was one-of-a-kind: NASA couldn't afford a backup. Taco joked nervously that if we blew it, the five of us could turn over a lifetime of paychecks to NASA and still never come close to paying for *Destiny*.

Each of us played an essential role in our plan for delivering the Lab. Taco would lead the mission, piloting the shuttle through the crucial rendezvous and docking at the ISS. Mark Polansky would back him up as pilot and coordinate our space walks from inside the cabin. He would also fly the departure from the ISS. Beamer would, of course, join me on the space walks and handle *Atlantis*'s docking system. I would lead the EVAs, serve as the pilots' rendezvous assistant, and ride herd on our network of a half-dozen laptop computers. Another critical job, and perhaps the most pressure-packed, fell to Marsha. She was the flight engineer for launch and landing, but her big challenge was grappling *Destiny* from the payload bay and berthing it at the Station. I was the only other trained arm operator, but I would be outside during that crucial operation. Although we were all members of a team, it would fall to Marsha to put *Destiny* in place. Mission success depended on it.

Marsha had a reputation in the Astronaut Office for being a perfectionist, and she didn't suffer fools gladly. She spoke her own mind, pulling no punches, and she had left a trail of bruised and broken egos, managers and astronauts alike, in her pursuit of a better-run Space Station for the crews. Over the years I had lost count of the number of times she had delivered a scorching rebuke to another astronaut or JSC engineer who was too slow or ineffective. Her frank opinions and direct manner rubbed some in the Office the wrong way. The only thing that mattered to Marsha was whether you got the job done.

Her performance with the robot arm on STS-98 would be her biggest challenge as an astronaut. It was her fifth flight, but nothing she had experienced prepared her for maneuvering a large payload under such demanding circumstances. Her Lab berthing task was an intricately choreographed series of arm maneuvers, from grapple, to "unberth" from the cargo bay, to a half-flip of the Lab high above the payload bay, to its final delivery within the berthing envelope. The Lab nestled so snugly in the payload bay that she had only two inches of clearance to work with on either side. She had to bring the sixteen-ton module into precise alignment just four inches in front of the Node's forward hatch, yet during the berthing operation her direct view would be totally blocked by the PMA-3 docking tunnel outside the aft cabin windows. She would have to rely on TV cameras and computer readouts to accomplish the berthing maneuver. Marsha had serious doubts that the system—her brain and hands, the computers, the robot arm, and the berthing latches and bolts—could pull off the task. The job might be executed in weightlessness, but the responsibility for it weighed heavily on her slim shoulders.

After the mission, she told me that she felt physically sick to her stomach every time she practiced the berthing maneuvers. After each gut-churning session in one of the several robot arm simulators, she would collapse into her chair in our office and intone grimly: "We're doomed." Marsha had lifted the slogan from a popular candy commercial, echoing the lament of the newly introduced orange M&M over his abbreviated life expectancy in the palm of the consumer. The T-shirt she wore to training sessions featured the "I'm doomed" slogan, and the orange M&M itself became her mascot.

The surest way to build our collective confidence was to practice together at the NBL. Almost every month, Kerri would put us in the water for a week of EVA rehearsals, three six-hour runs in five days. Marsha and Roman, working topside, joined Beamer and me for each simulation. The back-to-back runs were tough on everyone's stamina, both physically and mentally, and we searched for some way to lighten the mood both under water and topside. For each run, our crew teamed up with Kerri and the rest of our EVA instructors to put together some kind of a "themed" lunch. Our EVA team soon became known as the NBL's resident caterers, serving up Tex-Mex, an Italian buffet, deli sandwiches, and—when Taco set up his propane cooker in the parking lot—boiled shrimp and even a deep-fried turkey. Beamer and I had a complaint, though: while he and I contributed food, we were submerged in our suits at lunch hour, sustained for six hours by nothing but sips of water from our in-helmet drink bags. We got little sympathy. During one "working lunch," a utility diver handed us a waterproof photo of our crewmates topside, tearing with relish into Taco's golden-brown turkey. We were forced to settle for leftovers.

In both neutral buoyancy training and in flight, the most important factor in working comfortably in the space suit was glove fit. Pressurized against a vacuum (or the water) like the rest of the suit, gloves become stiff, like miniature balloons, and hours of hand-intensive work in them can turn forearm muscles to jelly. Like other components of the shuttle EVA suit, the gloves came in standard sizes, mixed and matched to fit each astronaut. But hands differ so subtly from individual to individual that fine adjustments in finger length and palm size were necessary, even when hand size was roughly similar. For the Space Station EVAs, NASA suit contractor Dover ILC had developed a more flexible glove that incorporated a Russian wrist-joint design. The new, more flexible Phase VI gloves fit better and eased the EVA workload noticeably during hours of pressurized operation.

A new experience for Bob and me was a series of classes held in JSC's Virtual Reality Laboratory that were aimed at preparing us for that least likely but most serious EVA event: coming adrift from the Space Station. I would always be tethered to the shuttle or Station by a 50-foot, 2,000-pound-test steel tether, and I usually had yet another short tether anchoring me to the work site, but if I did drift free, the docked shuttle would be unable to retrieve me. The solution was the Simplified Aid for EVA Rescue (SAFER), a nitrogen-powered jetpack that enabled a lost crewman to pilot his way back to the Station and safety. In the Virtual Reality Lab, we would don a set of 3-D television goggles and practice SAFER flying until we got its handling characteristics down pat.

**DIARY ENTRY, AUGUST 10, 1999:**

The fun part is when the instructor tosses you off the Space Station or docked orbiter, tumbling you end over end, and then asks you to recover and fly back. First, engage attitude hold and stop rotating. Then, select rotation control and spin in place to find a view of the station and orbiter. Next, select translation control and thrust toward the station. About half a dozen clicks of the RHC sends you perceptibly toward the station. On the way back to the ISS you correct trajectory by changing your pointing (attitude), then thrusting in that new direction.

Alternatively, you can sidestep or thrust up or down on your way in. As you close within grappling distance, your suited hands appear in the virtual reality view and enable you to grasp the handrails. Voilà! Great fun...the equivalent of a space-age video game. I almost wish I had the chance to use this capability.

In the same way, Marsha and I used virtual reality to practice her first robot arm task: parking the PMA-2 docking tunnel up out of the way on the Z1 truss. The space helmet visibility and 3-D effects were uncanny, with a realism that neither the NBL nor the shuttle simulator could match, enabling us to rehearse the space-walking assist I would provide her during the PMA berthing.

In the end, though, the success of our EVA work would come down to our underwater rehearsals and the space-walking skills acquired there through long hours of practice. For Beamer, Roman, and me, it was practice, practice, practice. With Kerri teaching at first, then observing and critiquing, we repeatedly ran through each space walk so thoroughly that by launch day we had put in twelve hours in the pool for every hour that we planned to spend outside at vacuum. Some days, everything clicked. More often, each run revealed a problem that we had to confront as a team and solve. A typical experience:

### DIARY ENTRY, AUGUST 2–4, 2000:

EVA I...was the first time we'd tried a new timeline for the EVA, trying to squeeze in the connection of Z1 to Lab umbilicals at the end.... The result was a busier timeline than we've been used to, but it still started off well. Then a number of things wound up hurting me...at the end of the run. First, I did the connection of the launch-to-activation heater cable from the Lab to Node 1. Pretty simple, but it then involves wire-tying the cable to handrails on the Lab and Node endcones. That took about 45 minutes. Then, for only the second time, I did the starboard umbilical connections while Bob was completing the port and fluid umbilicals. This job had me heads-down in a foot restraint in a tight spot near the Z1 tray. Every connection involved putting a heavy switch box, the CID [circuit interrupt device] into position against gravity while trying to mate the electrical connector in a particular orientation. Very hand-intensive and tiring, especially hanging upside down. The longer the work took, the less efficient I got. Soon I was very tired and frustrated at not being able to complete the work quickly. Had a hard time wiring the CIDs to the tray and then getting them out of Marsha's arm path when she moved the PMA away at the end of the mission. And I couldn't reach one particular connector, so I had to get out [of the foot restraint] and free-float it into position from the Lab side, meanwhile tangling up in the other wire connections already mated. It was inefficient and sloppy work. But the alternative was more heads-down time, which I just couldn't take. By the time I finished that job, I was bushed. It took quite a while to finish up the Velcro and get back into the airlock. In retrospect, I should have just gone upside down and finished the job quickly.

My fingers are still sore as I type this on the 4th. The first EVA run was one of the most tiring I've been on in a while. That last hour nearly did me in. I had the physical strength to work the jobs, but my endurance was tested, to be sure. My hands never gave out, but I had the starch taken out of me by the upside-down time. Another sore spot was a kink in the LCVG [long underwear] cooling tubes that bit into my arm and abraded some skin off my arm near my elbow. It was raw when I came out of the run.

One last trial was a literal pain in the neck. I was working in Roman's LCVG in case mine malfunctioned in orbit, and the shoulder pads were transferred over from my old one. But the pad on the right chafed my neck, and it soon became a major irritant, biting painfully whenever I turned my head. It was yet another reason I felt so wrung out at the end of each practice EVA.

All the while, we pressed ahead with shuttle training. I would be on the middeck for launch, but on entry I would join the pilots and Marsha upstairs. Our sessions in the simulator were as exciting and trying as ever. In one of our entry simulations, I joined Taco, Roman, and Marsha in the motion base shuttle simulator; we were about 50,000 feet up, heading for KSC, when our instructors failed two of the orbiter's three inertial measurement units (IMUs). Without these stable guidance platforms that sense the orbiter's flight attitude, the shuttle can't maintain a controlled glide. With two of them down, retaining the third IMU was crucial.

Then one of our four general-purpose computers failed, taking with it the last IMU. Mission Control immediately advised us to engage the backup flight system (BFS), our fifth flight computer, to restore communications to the IMU. Taco mashed the BFS button, lights flashed, flight instruments buzzed, and we heaved a sigh of relief at seeing the GPC and the third IMU communicating again. As we slowed through Mach 3, analyzing the next failure, the orbiter abruptly rolled off on one wing and departed controlled flight. The cab of the motion base lurched to a stop, the scene out the front windows froze, and the cockpit got very quiet. *Atlantis* had just died, along with the four of us.

The four of us sat there morosely, trying to figure out what went wrong. There it was. The open checklist on my lap contained a step in the "post-BFS engage" procedure: "inhibit air data" to the guidance, navigation, and control software. By failing to do so, we had allowed a failed air data probe to feed corrupted information to our last computer. The bad probe data had locked up the guidance software. Missing just one step in the checklist had sent us tumbling out of control.

I had read the checklist step but hadn't pushed the pilots to carry it out; I was a bit rusty on the procedure, and the critical inhibit step didn't jump off the page at me. The pilots were too busy flying and dealing with our bad probe to notice the misstep, and Marsha had been swamped by yet other malfunctions. All of us let this one get by us, and it bit us—hard. We had more work to do on both our proficiency and our teamwork.

One of the best places to escape the pressures of the training flow— the most intense I had experienced—was the astronaut gym. We often ended our pressure-packed days at the gym for an hour or so of strength training, following a fitness program developed specifically for us by our fitness coach. For Beamer and me, peak physical conditioning was all-important; it might provide the vital margin between success and failure on our space walks. Amid the free weights and exercise machines, we worked off the day's frustrations and found a few rare minutes to swap ideas and mull over how we could best get the job done. At the end of each session, we were a little stronger, and a little closer as crewmates and friends.

Swept along by our relentless training schedule, we tried to stay aware of those key events in the Station program that dictated our eventual launch date. On July 11, 2000, more than eighteen months after the launch of the FGB and well beyond the expected lifetime of its avionics, the Russians had finally launched the Service Module. After all the delays and foot-dragging from our partners in Moscow, I had scarcely believed it would fly.

The Russians had dynamited the logjam. On July 25, 2000, the SM executed a successful docking with the rest of the Station. With the SM's living quarters, guidance systems, twin solar panels, and capable thrusters, the three-module ISS could support a permanent astronaut crew. I wrote in my diary that there was no "big obstacle in the way of our flying early this winter . . . the Russians seem to be holding up their end."

Two shuttle missions followed in rapid succession. The STS-106 crew, with my old crewmate Terry Wilcutt commanding, launched on September 8, 2000. Terry and his crew docked with the ISS, laid in supplies for Expedition One, and readied the interior of the Station for their arrival. Then further construction got under way. The STS-92 crew launched on October 11, delivering the Z1 truss and PMA-3, the third docking tunnel, to the ISS. The crew berthed Z1 on the Node's zenith port, where it would serve as the foundation for the solar array tower to come next. Hairballs Leroy Chiao, Bill McArthur, and Jeff Wisoff joined Mike Lopez-Alegria in a series of space walks to berth the docking tunnel, connect umbilicals, and deploy the big $K_u$ communications antenna. Test-flying the SAFER jetpacks, Jeff and Mike did in space what the rest of us could only practice in virtual reality.

With a rush, all obstacles to the first expedition had fallen. Bill Shepherd, Sergei Krikalev, and Yuri Gidzenko climbed the steps to the gantry elevator at Baikonur on October 31, 2000. Their Soyuz blasted from the bleak grasslands of Kazakhstan and docked at the ISS less than three days later. The trio opened the hatch into the SM and became the first crew to take up residence aboard the Station. STS-97 on *Endeavour* docked at the outpost on December 2, 2000, and erected the P6 truss atop Z1. When the shuttle departed, a pair of solar arrays stretching more than 240 feet from tip to tip fed a steady stream of electricity down into the ISS. Most of that power would go to *Destiny*. Its delivery crew was nearly ready.

Over frozen margaritas, our favorite remedy for muscle soreness after an NBL run, we reviewed the final few weeks of training before our January 18 launch date. Shuttle training—rendezvous, ascent and entry, emergency escape, integrated simulations—would be finished just before Christmas, except for a few refreshers. We would wrap up our water training in the first week of the new year. The pending countdown rehearsal and quarantine were for us a welcome chance to get off the training treadmill.

But even before our abbreviated Christmas break, there was bad news from the Cape. Technicians examining *Endeavour*'s left-hand solid rocket booster recovered from the sea for refurbishment had discovered that one of the separation charges on the lower external tank attachment strut had failed to fire. The backup charge had worked properly, shearing the strut and enabling the booster to drop free. Had the booster not separated cleanly, it might have torn away and collided with the tank or orbiter, destroying the vehicle.

Technicians soon pinned down the reason for the pyrotechnic failure: a worn wiring connector in the cables carrying signals from the orbiter computers out to the SRB. The reusable cables were inspected before stacking the boosters for flight, but despite passing electrical checks the defect in this one had slipped through. Now the booster cables on *Atlantis*, our ship, were suspect. Inspections on our boosters in the Vehicle Assembly Building would delay our launch a few days; our countdown rehearsal slipped past Christmas into the first week of 2001.

When we arrived at the Cape to perform our final *Destiny* inspections and strap in for our mock countdown, we found our reputation had preceded us. Earlier in the fall, we had gathered in the Building 8 studio for our traditional crew portrait. Taco, Marsha, and Roman wore their orange launch-and-entry outfits, while Beamer and I looked *very* special in our white EVA space suits. But Marsha had dreamed up a rather unconventional pose for a spoof portrait—one she thought might prove ideal for NASA's spaceflight safety campaign. A typical agency safety poster might feature an astronaut holding her cute daughter, a reminder to workers that real people, real families, staked their lives on the dedication of each and every employee. Marsha wanted to take this traditional NASA workforce safety pitch to a new level. Setting the fashion pace in a skin-tight black leather ensemble, she led us back into the studio, looking like we had shoved our way right off some gritty downtown street. In black jeans and shirts, and tattoos painted on by NASA graphics people (who ran a body-painting sideline), we assumed what we hoped was an intimidating pose. Marsha's graphics colleague, Sean Collins, did the poster layout, and together we put out the word: "Think Safety—You Got A Problem With That?"

Taco somehow got the poster approved personally by George Abbey, and it hit the streets at JSC late in the fall. The posters flew off the shelves. By the time we arrived at the Cape for our countdown rehearsal, contractors and NASA safety managers had printed up hundreds more, and according to the *Orlando Sentinel*, demand "quickly skyrocketed to supermodel-like heights," adding that we had posed for "arguably the most unusual astronaut crew photo ever."

Perhaps shuttle managers had our poster's message in mind as they mulled over our booster cabling problem. Finishing our first four days in quarantine at JSC, we drove to Ellington Field on January 15 for our T-38 trip to the Cape. While filing our flight plan, we got a call from FCOD deputy chief Steve Hawley in Florida reporting that the mission management team was still worried about possible suspect cables on our SRBs. Cape technicians had found four more booster cables out of the hundreds in inventory that failed to pass electrical checks. These worn cables were similar to those on *Atlantis* that carried critical steering and firing commands to the booster rocket nozzles and explosive separation charges. Hawley called to warn us: "They're still discussing things . . . don't take off 'til I give you the thumbs up." We watched the clock in the Ellington operations office; Hawley's next update told us that after more discussion it was likely the team might delay us a day or two. But even that call didn't prepare us for the final decision that day. "It's a roll-back," Hawley announced. "You're scrubbed for at least two weeks."

Since the suspect cables lay in booster wiring trays that were inaccessible to technicians at the pad, the shuttle program manager, Ron Dittemore, ordered the stack rolled back to the VAB for the necessary inspections. "I guard against the phenomenon of 'Go Fever' like it was the plague," he noted, justifiably calling the launch delay a necessary precaution. He had our vote.

When we had parked our cars at Ellington that day, we had been a half hour from takeoff for the Cape in our T-38s. Now our two-hour delay had turned into two weeks. But there were a few benefits. Our new launch time, dictated by the timing of our ISS rendezvous, meant a much easier sleep shift for us in quarantine. Our EVAs would now occur during Houston prime time instead of the middle of the night, a boon for our flight control team. And we all could use the extra rest.

Our new date was February 7, 2001. We would be the first human space launch of the new millennium. We took a few days off, got in an extra sim or two, and reviewed our space-walk plan for the umpteenth time. I had the EVA steps down so well I could almost do them in my sleep. It was time to go and visit Shep, Yuri, and Sergei at their new address.

The STS-98 crew reentered quarantine on January 31, 2001. It was a familiar routine out at the quiet JSC crew quarters: meals with spouses, daily exercise, and time for a crew movie or two. Roman's favorite was *So I Married an Axe-Murderer*. I remember wondering what his fiancée, space food specialist Lisa Ristow, thought of his film preferences.

The weather was dicey when we flew to KSC on the afternoon of February 4. Storms along our route and low ceilings at the Cape meant there would be no grand flyby of the launch pad on this trip. Our four T-38s ventured out over the Gulf of Mexico in single file as we calculated our fuel to the last pound, hedging against a socked-in Cape diverting us into the Florida panhandle. But an instrument approach guided us in to the shuttle runway, and we taxied up to our families waiting on the ramp. It was the first time our schedules had enabled Liz and I to enjoy the traditional crew arrival at the Cape, and it was a big moment for our family. Annie and Bryce were there, too, and I could get as close as ten feet to them, the limits defined by a big yellow taxi stripe on the concrete. No tears this time, just big smiles and the expectation of a perfect launch on Wednesday evening.

Training and preparation followed us even to the Cape. Marsha ran robot arm simulations on a computer workstation she had set up in crew quarters, and we hit the checklists and filled page after page in our crew notebooks between relaxing visits to the Beach House. I took time to tuck away the scripture readings for the two Sundays I would be in orbit and planned to take Christ with me in the form of the Eucharist.

The afternoon before launch the crew met Liz; Beamer's mother; Mark's fiancée, Lisa; and friends of Taco and Marsha for our tour of *Atlantis* on the launch pad. Liz and I climbed the work platforms around the shuttle, still protected within the rotating service structure, walking hand in hand around the orbiter's nose and crew cabin. In the White Room, Liz took a peek into my home for the next two weeks. My launch seat was visible on the middeck just to the left of the hatch. I would be downstairs by myself this time. For Liz, seeing the shuttle up close was sobering. She had no choice but to trust this complex and delicate machine, devised by fallible human hands, to take me to the Station and bring me home again safely.

We said our good-byes in the privacy of the Beach House, then went outside on the deck to wait for Joe Tanner, her astronaut escort. When Joe pulled up, Liz gave me a brisk kiss, said good-bye, turned and walked away. Liz told me later that on the ride back to Cocoa Beach, she reassured Joe: "I just want you to know . . . I gave Tom a nicer good-bye earlier."

Later that evening, Marsha and I ran through our last inspection of the orbiter cabin. Crawling through *Atlantis* in our bunny suits, we checked our cabin wiring diagrams one last time, tracing power, computer, audio and video cables from one corner of the cockpit to another. Emerging into the White Room, we checked our watches and hurried across the swing arm away from the glare of the xenons. Looking up from the shadows cloaking the gantry, past the gleaming boosters and the huge bulk of the ET, we spotted the Station hurrying along on the northeasterly leg of its orbit, carrying Shep, Sergei, and Yuri in their thirteenth week aloft. In less than a minute, their bright star was gone, lost out over the Atlantic. I looked over at Marsha with a shake of the head. "How are we *ever* going to catch up to that thing?" Our eyes turned in unison toward the sleek rocketship behind us.

# 15

# RENDEZVOUS

*I think Isaac Newton is doing most of the driving now.*
BILL ANDERS, RETURNING TO EARTH ON *APOLLO 8*, 1968

Greg "Ray J" Johnson hovered over my right shoulder, checking my shoulder straps and tucking communications cords under my parachute harness. Trussed up like a Christmas turkey, I lay on my back, cinched firmly into the lone seat on the middeck. I couldn't wiggle more than a half inch in any direction. Clad head to toe in his white bunny suit, Ray J slapped me on the arm and pronounced himself satisfied with my straps. Standing on the temporary floor panels in the topsy-turvy world of *Atlantis's* now-vertical middeck, he would be the last person on Earth to see our crew that evening.

Ray J shuffled around to face me and leaned in over my seat, crouching under the storage lockers two feet from my face. "Tammy wanted you to have this," he grinned. Puzzled, even suspicious, I accepted a small plastic bag, and opening it, removed a walnut. Attached to the nut by a length of pink yarn was a purple tag about the size of a business card. A hand-lettered note read: "Tom's Right Nut—An EVA good luck charm—from the 5A crew."

The note's author was Marsha. She had written it two years ago, when she had delivered the nut to Tammy, strapped in for her STS-96 launch. My old EVA partner had now returned the good wishes. I laughed with Ray J, who knew the history here. I turned the card over:

> Dear EV1—
>
> It takes cojones *to fly in space. So I'm returning this* awesome *good luck charm. Can't wait to see you egress the airlock.*
>
> —Tammy

I velcroed the bag onto a locker, shook Ray J's hand, and listened as he cleared the cabin and moved to the orbiter hatch. The five of us chorused our thanks to him and the close-out crew before they disconnected their headsets and withdrew. Alone in the middeck, perched near the forward left corner of the cabin against a metal storage cabinet, galley, and lockers, I listened to the whir of fans and the calm voices of the launch team ticking off *Atlantis's* countdown checklist.

This was my first launch alone downstairs. Although linked to the flight deck by intercom, I was isolated on a middeck crowded with ISS cargo for Shep's crew. The White Room team completed their hatch checks; we could hear them signing off with launch control and clearing the pad. *Atlantis* was less than an hour from launch.

While I waited on the middeck, Liz and the kids had arrived at the LCC with Joanie Cockrell, Julie Curbeam, and their children. The kids were soon busy sketching their version of the STS-98 mission patch on a large whiteboard in another room while the wives settled in for the last ninety minutes of the count. They appreciated the NASA Security people keeping away reporters and curious employees, but a few visitors were welcome. Mike Mott, a Boeing executive and old friend of Taco and Joanie, dropped by, soon joined by Andy Allen, a former astronaut now with United Space Alliance. The wives were happy to see them, and their conversation proved a pleasant distraction.

The next visitor was a stranger, but Security couldn't turn him away. NASA administrator Dan Goldin was the kind of VIP visitor who the wives would have preferred to see *after* the launch, but they put on their happy faces to entertain him for a few minutes with the help of Mike and Andy.

The small group exchanged introductions, then Dan began a rambling speech. Liz expected NASA's top official to express appreciation for their support of the crew and optimism for an on-time launch. Instead Goldin's remarks took an unusual direction. "We've tried to make this vehicle as safe as possible." Liz looked at him with disbelief as he continued. "We've tried to make this launch as safe as possible." What could he possibly be thinking? How was Dan going to salvage this one? He followed his thoughts to their chilling conclusion: "But after all . . . this is space." Goldin pursed his lips, raised his eyebrows, and shrugged. The three wives were stunned. As he left the room it was all Liz could do to voice a polite good-bye.

An hour before scheduled liftoff, the launch team resumed the countdown and brought us out of the scheduled hold at T-20 minutes. The count proceeded to T-9, holding there in readiness for the opening of the launch window just before 6:00 p.m. No problems so far; it had been a very clean count. It was so quiet aboard that I pulled out a peanut butter and jelly sandwich, polishing off the snack and my antinausea pill with some cold water from a drink pouch stashed near the galley. At 5:46 p.m. we got the news that our transatlantic abort runways, Ben-Guerir in Morocco and Zaragosa in Spain, were both go. The sun dipped toward the scrub palmetto and orange groves lining the darkening shores to the west.

At six o'clock, just before the final "go/no go" poll, the launch team reported that one of the multiplexer-demultiplexer (MDM) circuit cards aboard the orbiter was acting up. Back in the orbiter aft compartment, the "OA1" MDM's number 6 circuit card wasn't relaying data correctly. At the four-minute mark, if the card didn't pass along the correct APU turbine speed, the ground launch sequencer (GLS) would halt the countdown. We wouldn't know until that turbine speed measurement became active at APU startup. Upstairs just behind the pilots, Marsha groaned at the ongoing discussion: "Come on! Don't scrub for instrumentation. Get on with it." After three hours on our backs, none of us relished the thought of crawling out of Atlantis and starting over the next day. But we had few options beyond listening and hoping that MDM would come through.

While the engineers discussed the problem over the firing loop, the launch director, Mike Leinbach, began the formal poll of his team. One by one, the console positions checked in with their status. The litany reached the Supervisor, Range Operations (SRO).

"SRO?" Leinbach queried.

"SRO is go. You have a 'range clear to launch.'" That call always made the hairs stand up on the back of my neck. SRO's range safety engineers had their collective fingers on the destruct button.

The quality control guys were next. "Safety and Mission Assurance?"

"SMA is go!"

"International Space Station launch manager?" It was his Laboratory we were committing to flight.

"We are go to proceed."

"Range weather?" The sky outside remained clear in the soft light of dusk.

"Roger, sir, we have no constraints to launch."

Leinbach had only one dissenter, the Instrumentation console operator, awaiting good telemetry from card 6. But for that check we would have to wait for APU start-up.

"Ops Manager, Launch Director. Verify no constraints to launch." Leinbach was calling Jim Halsell, my astronaut classmate and manager, Shuttle Launch Integration. It was Jim's call as to whether we could proceed, and it was reassuring to know that a Hairball was in charge. He and the rest of the management team had followed the discussion from their console. Jim spoke:

"The Mission Management Team understands the issue regarding the fault on OA1 card 6. We understand that we are go for launch at this time. We concur. You are 'go' to proceed with the countdown." If Jim was voting to launch, that was good enough for me, Taco, and our three crewmates.

Leinbach moved briskly. "Copy that, sir. The worst case is that GLS will catch us and hold at four minutes. *Atlantis*, Launch Director."

"Go ahead." Taco's voice was taut.

"Okay, Taco, we've got a good day to go fly. We'll get this turbine speed issue behind us and you'll be on your way. So we wish you luck as you deliver the heart and soul of the International Space Station. And have fun."

Taco's reply crackled over the intercom and out to the launch control center. "On behalf of the crew, thanks to the KSC team and the shuttle and Station programs. Let's get this vehicle off the MLP."

I pictured *Atlantis* booming off the mobile launch platform as Leinbach answered, "Copy that. That's the plan. NTD, with that, you're clear to proceed."

I gave an automatic tug on my seat harness and made sure I was centered up on the parachute and seat back. Steve Altemus, the NASA test director (NTD), came on the loop.

"Attention all stations! The countdown clock will resume at T-minus-nine-minutes momentarily." Leinbach had just cocked the hammer on this big cannon.

The calm voice of Janiene Pape, the GLS engineer at the Integration console, answered instantly: "I copy. Three . . . two . . . one . . . T-minus-nine-minutes and counting. GLS autosequence has been initiated." We were now in the hands of the launch computer.

Three miles away at the Banana Creek viewing site, our families and friends in the bleachers watched the big countdown clock start its final run to liftoff. A hush fell over the crowd as the opening drum roll of "The Star-Spangled Banner" brought hundreds to their feet. *Atlantis* glittered in the floodlights across the calm water as the crowd sang the anthem with spirit; my brothers reported later that there wasn't a dry eye in the stands.

Our chatter in the cockpit had ceased. I heard only the muffled, vacuum-cleaner whine of the big cabin fan behind the polished aluminum floor panels. Each was alone with his thoughts. *It all comes down to this.*

We heard the familiar voice of LeRoy Cain, our ascent flight director back in Houston, confirming the liquid oxygen drain-back numbers for the external tank. Nearly a thousand miles away, his team was ready to guide us into orbit.

Jim Taylor, the orbiter test conductor, took over for the last few checks with our crew. "PLT, OTC, connect essential buses to fuel cells."

"That is in work." Roman put *Atlantis* on internal power. Our rookie's voice was solid, confident. He glanced out his side window. "Hey, you can see the full Moon rising over the ocean. It's getting dark outside."

The swing arm just outside the cabin lurched and began its retraction toward the launch tower. As it cleared the hatch, a remote camera in the White Room caught Taco's wave to our friends and loved ones, a last gesture of thanks before things got very busy. The quiet waters surrounding the pad were a sheet of liquid silver under the twilight sky.

"PLT, perform APU prestart."

"APU prestart is in work." Roman reached to his right and flipped the switches arming the auxiliary power units. "APU prestart is complete. Two gray talkbacks and one barber pole." Two APU indicators were normal—gray. The third was showing an intermediate status, but we had been briefed to expect that. No surprises. We waited for the five-minute mark and the news from card 6.

"PLT, OTC, start APUs." Again Roman reached down near his right knee, taking the three APU start switches to "Run." Hydrazine sprayed onto catalysts back in *Atlantis*'s aft and flashed into hot gas. The APU turbine blades caught the energy from the searing exhaust. I heard the whine far below me as the three power units spun up to operating speed, pressurizing our hydraulic systems.

"OTC, PLT. APU start complete. We've got three fine-running APUs."

"Excellent," came the reply from the LCC.

Taco, Roman, and Marsha could see three good turbine speed read-outs on their displays. "Looks like we have all the data we need," I added. Card 6 had delivered, and the GLS, tracking over a thousand separate measurements on the vehicle and launch pad, was still happy. "T-minus-four-minutes and counting."

Chilled water gurgled through my long underwear in my snug cubby-hole at the upper left corner of *Atlantis*'s middeck. Acutely sensitive to every vibration and movement, I felt *Atlantis* shudder as the hydraulic actuators drove the flight control surfaces, the elevons, rudder, and body flap, through their range of motion. Now the main engine bells swiveled in unison, limbering up for ignition. We felt the orbiter sway along with the nozzles. Our machine was alive!

"Three minutes and counting."

Roman keyed the mike. "OTC, PLT, the Caution and Warning memory is cleared. There were no unexpected errors." He had the fault summary display ready to register any ascent failures.

"Copy, PLT. All flight crew members, close and lock your visors and initiate $O_2$ flow."

As Taco echoed the call, I pulled down my clear visor and snapped it closed. Flipping the lever on the suit, I took a tentative breath; each inhalation now brought a whiff of cool oxygen into my helmet. We checked in on intercom: Taco, Roman, Beamer, Marsha, and finally, me. It was our verbal equivalent of a final handshake.

"One minute, thirty seconds." Outside, the pad microphones picked up the steady chuffing of the APUs' rhythmic exhaust, keeping time with the countdown clock.

Silence on the flight deck. Thirty seconds. The ground launch sequencer handed off the countdown to *Atlantis*'s own onboard computers a foot above Taco's helmet. The four general-purpose computers marched in unison toward main engine ignition. We were in their hands now . . . and God's. I had never felt so alive.

"Twenty seconds," came the subdued call over the radio. At eleven seconds, the two attitude indicators in front of the pilots snapped to attention as the computers started the navigation routines for launch. "*Nav init*," said Taco quietly.

Now events tumbled together too swiftly to note each with detachment. Over the firing loop came the final call from the LCC: "Go for main engine start." At six seconds to go our spaceship trembled with the sudden release of stored chemical energy. A vivid image of valves slamming open, of freezing propellants rushing into turbopumps, flashed through my mind as *Atlantis*'s engines rumbled to life. The initial tremor built to a roar, shaking my seat, the cabin, the entire stack. Metal rattled on metal in the middeck as the three motors muscled to full thrust.

Above the racket came Roman's call, "Three at a hundred!" A split-second later—BOOM! The boosters crashed into life, shaking the stack with a bone-jarring impact. My seatback gave me a brutal shove; I barely got the stopwatch on my kneeboard started. The vehicle lurched off the pad, the SRB nozzles 120 feet below swiveling to balance our four and a half million pounds on a pillar of white-hot fire. We were off just two minutes late, at thirteen minutes and two seconds past six, February 7, 2001.

"Tower clear," Taco called, getting an eyeful as the gantry plummeted from view. *Atlantis* surged upward, the bang and rattle settling into a more tolerable yet still pronounced shaking. *Atlantis* began to rotate onto the proper track to chase the ISS up the East Coast. I felt the orbiter pirouette gracefully, swing past the mark, then gently correct the overshoot. Heads down now, we arced out over the Atlantic; my body slid tightly against my shoulder straps and stayed there.

"Houston, *Atlantis*. Roll program," Taco announced.

"Roger, roll, *Atlantis*." The reassuring voice of Scott "Scooter" Altman came back to us from Mission Control. His timing was good, for just then a rising howl began to penetrate the cabin walls as if some tortured spirit clung to the hull outside. The low moan built to a blended roar of pure power, the slipstream blasting past the cabin walls to join the full-throated scream of the SRBs. Ten tons of rocket fuel per second burned furiously and shot from their twin nozzles. The sound seemed to surround and even surge through me, a challenge from *Atlantis* to hang on, if I dared.

The vibration of first stage pounded us as we blew through the sound barrier at forty-five seconds. I watched the clock to anticipate each milestone as we soared upward. Roman confirmed the throttle-back of the main engines as we rattled through maximum dynamic pressure, shock waves peeling back from *Atlantis* and rippling the brilliant plume of her exhaust. We rumbled upward at a g and a half or so, the acceleration still trying to tug me off the top of my seat.

*Atlantis* surged forward as the three main engines returned to full power, squeezing me back into my parachute with two and half g's of acceleration. "Three at 104," Roman confirmed. Full throttle now at seventy seconds.

"*Atlantis*, go at throttle up," Scooter called.

"Go at throttle up," Taco came back, his voice straining against the g's. Above us, the sky had turned black, but our climb had taken us up out of Earth's shadow. The sun, once set, had risen again in the west. *Atlantis* was again bathed in brilliant sunshine streaming in through Roman's window.

The g pressure began to ease off as the boosters exhausted their solid fuel. I kept my eyes glued to the stopwatch, waiting for two minutes and solid rocket motor separation. Weight dropped to near normal, the vibration quieting as we seemed to coast upward.

Upstairs, Roman saw the forward reaction control system (RCS) jets fire just before separation, the thrusters flashing outside like spotlights penetrating a snowstorm. Then an instantaneous flare of orange-yellow wrapped the cockpit in fire.

BANG! Explosive bolts sheared the 150-foot booster casings from the external tank. Those cable inspections had done the job, and the separation motors washed the cockpit windows in the glare of their exhaust. The fireball was gone in a split second, replaced by black space.

The ride smoothed instantly, the difference between night and day. It was as if we had stopped short, with *Atlantis* hanging in space. I swung open the faceplate with relief and took a deep breath of cabin air.

Hey, we're through two minutes twenty seconds...what a ride!!!
Man, the boosters just came off...feel light in my seat, we've got
about 1 g on my back. Everything's going great, two minutes thirty
seconds.

Still at full throttle, the main engines shivered *Atlantis* with a faint,
pulsing vibration. Like a taut violin string, the orbiter sang with the different frequencies of the spinning turbomachinery and roaring engines
aft. This throbbing ascent was quite different from the smoothness of
my two rides on *Endeavour*.

*Atlantis* had fired up her two orbital maneuvering system engines just
beneath the tail to boost our thrust and get rid of some now-excess propellant. The increased push was imperceptible, just 12,000 pounds compared to more than a million from the main engines. Ice crystals or
perhaps flakes of loose foam from the external tank flashed in the sunlight, flicked by the cockpit windows, and vanished. Taco watched the
debris drift off to the side and hang with the stack a while, seemingly de-
fying our Mach 10 pace upward.

Five minutes. With the stopwatch and cabin altimeter my only instruments, I relied on Roman for our Mach number: 11.4 now, with
1.6 g's.

Five minutes twenty seconds (gasping). Ah, this is great. You can feel
the surging of the engines. There's a slight vibration, just enough to
jiggle the tether that I'm holding onto the tape recorder with. But
not a shaking, just a vibration. More so than *Columbia,* which was really rock smooth. This is like a low rumble.

The main engines swiveled slightly, and *Atlantis* nimbly rolled 180
degrees around the tank's long axis. Sitting high above the roll axis, the
pilots clearly sensed the rotation as we swung heads-up.

The engines had been sucking fuel and oxidizer out of the external
tank at the rate of a thousand gallons per second, yet the vehicle's
weight had kept our acceleration relatively low. At just over six minutes,
we were only a bit more than halfway to orbital velocity. But with the
tank growing lighter every second, our acceleration was reaching past
2 g's. In the next two minutes, the last quarter of the ascent, we would
nearly double our velocity. *Atlantis* was sprinting for orbit.

(Breathing hard) OK, Mach nineteen. Good, good g's. Single engine press, 104. [We could make orbit with a single engine at full throttle.] Seven minutes. Oh, it's hard to breathe. Yeah, I can feel it. Ah, awesome, awesome! Getting a good squeeze at 3 g's.

The racing shuttle pressed through Mach 19. With *Atlantis* at 3 g's, the computers throttled the engines back to stay within the stack's structural limits. My body now weighed about 480 pounds. I found it tough to breathe, and someone joked again about that gorilla on his chest.

To Roman, on his first ride to orbit, the entire ascent was like an out-of-body experience, full of astonishment, wonder, and exhilaration. The acceleration was "the big shove that kept on shoving." I welcomed the giant hand pressing down on us, for every second I was pancaked into my seat was a second closer to the right MECO velocity and a successful rendezvous with the Station. Speed was all!

Beamer marveled at the sight of the engine plume creeping up the side of the stack, flickering outside the overhead windows. Tucked in my niche downstairs, I could only imagine our ship wrapped in the expanding fingers of superheated steam flung from the engine bells. To be part of such a physics demonstration, to ride a spaceship clawing for the necessary velocity with its last few tons of fuel, to be alive at the center of it all, was worth all the months of training and worry. I smiled—couldn't help it—at the image of my dad hanging on with me. He had been there through every launch.

At MECO, main engine cutoff, an amazing reorientation took place. Shoved all the way to orbit on my back, knees up, I suddenly sensed that the middeck was "rotating" to a normal 1 g orientation. With the thrust vector gone, my brain simply put the middeck back to right. I once again felt I was sitting straight up in a chair with the locker wall no longer overhead but in *front* of me. The 90-degree mental transformation took place smoothly and instantaneously.

About ten seconds after MECO, a loud, jarring clang shot through the middeck as pyrotechnics severed the attachment fitting for the external tank a few feet beneath my boots. Next, what sounded like multiple cannon blasts from the primary RCS jets nudged *Atlantis* away from the now-empty tank. We glided up and ahead of the ET as I prepared to unstrap, now in blessed free fall. I slapped the recorder onto some locker

Velcro. Outfitting the middeck for orbit was my big job in the five hours
before sleep.

I was soon immersed in the blur of post insertion work on the middeck. These hours demanded all the concentration and efficiency I
could muster even as we began our physiological adaptation to weightlessness. Stuffing my gloves, kneeboard, and helmet into a ditty bag tied
to the galley, I released seat straps and parachute buckles, then floated
very tentatively out of my seat. Yes, that initial awkwardness was back.
Still encumbered by the bulky LES and chute harness, I reached down
and inserted the safing pin into the cover over the hatch jettison handle, disarming the explosives designed to blow open the side hatch.
Thank God we hadn't needed that bail-out option today.

I floated carefully up the ladder to the flight deck and was rewarded
with a big smile from Marsha. I gave her a hug and clapped Taco on the
shoulder: "This is a damn fine spaceship you're commanding," I told
him. I congratulated Beamer and Roman, then peered out the windows.
Earth was invisible: we had flown eastward into orbital night, and our
world consisted only of the brightly lit cockpit. I headed back down to
start the post insertion checklist.

For the next four hours *Atlantis*'s cabin was a bustle of concerted activity as we converted our rocket ship into an orbital home. I was in
charge of the middeck. On my kneeboard cue card I had typed up a master list of all my duties during the next three hours:

+ Hand-held mike and gray tape to the flight deck.
+ Help Beamer out of his suit.
+ Doff my own launch and entry suit; change into fresh clothes.
+ Unstow sleeping bags.
+ Rig stowage net at right front of forward lockers.
+ Activate Waste Control System (the toilet).
+ Rig emergency eyewash kit near galley.
+ Unpack personal toiletry kits and food trays.
+ Help Marsha out of her suit.
+ Set up laptop computers on the flight deck and middeck.
+ Stow seat pads, headrests, and parachutes.
+ Stow launch and entry suit bags behind net.
+ Stow parachutes in bags and float behind net.
+ Get Roman and Taco out of their suits.

+ Install emergency breathing masks and cabin air cleaner.
+ Dismount the escape pole and strap to middeck ceiling.
+ Roll up the "trampoline," the fabric partition covering the airlock hatch, and stow.
+ Open airlock and connect air ducts, activate lighting.

The work sounds mundane, and much of it is mere housekeeping, yet it has a special choreography all its own. The challenge is to put these dozens of steps into just the right sequence so as to minimize wasted effort. For example, I needed to get the sleeping bags out of the way so that I could rig a stowage net to hold all our bundled suits. In the same way, it was important to activate the toilet promptly so that it was ready for crew members as they floated downstairs to peel off their LESs. In the mock-ups and simulator at JSC, we had spent hours rehearsing these moves for maximum efficiency.

My first headache on the mission was setting up the laptop computer network. We had six laptops onboard, and four had to be up and running before we started our sleep period five hours after launch. First up were two on the flight deck that used data fed from our GPCs to display critical rendezvous guidance, and, of more general use, a world map showing our ground-track and present position. The moving map set up atop the dashboard cleared up those pesky "What continent are we over now?" quandaries.

Surrounded on the middeck by a web of Ethernet and power cables linking two more laptops, a wireless router, and an inkjet printer, I wrestled with one machine that stubbornly refused to boot up. Even with my checklist and MCC's help, I spent a frustrating hour trying to bring it to life. Why should a machine that checked out perfectly on the ground be "toast" just eight and a half minutes after leaving the launch pad? Perhaps the launch vibration and a sustained 3 g's had something to do with it.

Thanks to the computer glitch and all the unpacking chores, we didn't get to bed that night until about six hours after liftoff, more than an hour late. Still, we had had a fine day from strap-in to lights-out in orbit. The Lab was on its way, and *Atlantis* swept toward our rendezvous forty-eight hours ahead. I clipped my sleeping bag near the floor among the ISS cargo pallets sprouting from the middle of the deck. The whine of the cabin air circulation fan just under the floor made it a noisy bedroom, but earplugs and fatigue helped overcome any restlessness. Our nightlight was a thin line of sunshine seeping in around the hatch window cover. Folding my arms inside my sleeping bag, I realized I had hardly looked out a window at the Earth, a regret soon lost in a dreamless, exhausted sleep.

Over the middeck speaker in the semidarkness came the jazz strains of "Where You At?" a tune performed by our pilot's uncle. Capcom Ellen Ochoa had a message for Roman: "Welcome to your first full day in orbit!" A groggy but pleased Roman answered: "Thanks, Ellen. It feels like our natural habitat, and we're having a blast."

I felt few signs of space sickness, just a nagging headache and a little light feeling in the stomach. Just after breakfast, I managed to fix our balky laptop by replacing its hard drive with a spare. Beamer, Roman, and I then joined forces in the airlock, getting our three suits, which included Roman's spare, ready for the space walks two days hence. We unpacked each suit from its airlock storage rack, applied power, and checked its oxygen supplies and radios. I managed to stump Mario Runco, our capcom, and my EVA colleagues in MCC for the better part of thirty minutes when I mistakenly flipped an oxygen supply lever in the backpack in the wrong direction, giving them an unreliable pressure reading. As Mario, Kerri, and the suit engineers puzzled over the reading, I sifted through a few bad memories about canceled space walks. But before the discussion had gone very far, I decided to double-check the proper direction to flip that lever. Sure enough, I had been pushing in the wrong direction. MCC graciously accepted my apology.

The SAFER jet backpacks were next. No virtual reality here: these two units were fully functional. Each SAFER was the size of a large tackle box and was designed to fit beneath our suit backpacks. A high-pressure nitrogen tank at 8,000 psi fed cold gas to twenty-four thrusters controlled by a simple joystick, tucked away until needed.

Mission Control sent up our first full batch of e-mail after dinner. Marsha called us to the middeck with a shout: "Hey, you guys! Come take a look at this!" The mail included an image of our liftoff taken from near the launch control center. The setting sun just below the western horizon caught *Atlantis*'s rocket plume and painted it in colors from soft grayish pink to brilliant white. Teased by high-altitude winds into fantastic billows, the plume in turn had cast shadows far out over

the Atlantic. Narrowed by perspective, these deep indigo slivers of evening lanced precisely onto the rising face of the full moon.

An hour into our sleep period, I had finally finished setting up the
last pair of computers for the next morning's rendezvous and docking. I
took a few minutes before sleep to mentally unwind and look outside.

Floating in the commander's window, I watched some segmented
cloud formations drifting below, their polygonal borders fitting together
like a tiled mosaic. Outlined by thin lanes of clear blue ocean, the cloud
pattern may have been caused by cold water welling up to a warmer surface. Every glance out Atlantis's windows yielded another surprise, another quick intake of breath at Earth's complex beauty.

We awoke on flight day 3 to the first of our rendezvous maneuvers,
the series of rocket burns designed to catch the Station and position us
for a docking approach. All five of us had to work closely with Shep's
team and both Houston and Moscow to bring off a successful rendezvous. Guided first with ground-tracking data and later with our orbiter's sensors, we would repeat the techniques perfected on previous
missions to Mir and the ISS.

Our onboard bible for the approach was the rendezvous checklist, a
carefully crafted plan that meshed the efforts of two space crews and
both control centers. Taco, Roman, and I had spent dozens of hours
practicing this drill in the shuttle engineering simulator. The SES featured a planetarium-style dome that projected a huge, impressively detailed image of the Station over our shuttle cockpit. Our instructor had
been Val Murdock, and by the time we headed to the Cape she had me
convinced that every rendezvous must have a minimum of three sensor
or orbiter computer failures.

We began our final chase some ten miles lower and thirty miles behind the Station. With Atlantis closing steadily on that inside track, we
executed the first of seven burns that slowed our rate of closure and
raised our perigee enough to bring us to the terminal initiation, or $T_i$,
burn. That critical firing just 50,000 feet behind the ISS would bring us
to the entrance to our narrow docking corridor 800 feet beneath the
Station.

Five of us crowded onto the flight deck for this delicate final pursuit
phase. Loosely belted into the front seats or hovering aft beneath the
windows, we all had a job to do. Taco and Roman handled the autopilot

and thruster firings and monitored the orbiter's rendezvous software. Beamer took charge of the orbiter docking system, working the panel at the center rear of the cockpit just beneath the overhead windows. Marsha managed the TV system (the pilots' key visual sensor during the approach) and monitored our closing speed and distance with a laser rangefinder. I hovered over our four rendezvous laptop computers, which digested radar and laser sensor data and displayed it graphically to the pilots. My impression from rendezvous simulations was that there was always too little or too much information: either our sensors were down and we had to feel our way in manually, or so much tracking data came in that it was tough telling which numbers should be trusted. We would see the pendulum swing in both directions as *Atlantis* closed with the ISS.

Shortly after our initial burn, we rotated *Atlantis* to point our startrackers, radar antenna, and overhead windows at the ISS. From forty-seven miles out we spotted the gleaming star of Space Station *Alpha*, Shep's radio call sign for the ISS, hovering above the eastern horizon. "Tally-ho on the Station," Taco called to Houston, and we began a looping swing toward Shep and company that would end in the $T_i$ burn.

At twenty-eight miles our radar locked on to the ISS and began feeding tracking data to *Atlantis*'s computers. Closing to eight miles behind and 1,200 feet below the ISS, Taco and Roman set up the $T_i$ engine firing, and a few seconds of thrusting from our two OMS engines inserted us perfectly into the trajectory needed for final approach, just ninety minutes away.

We hurried through four small thruster firings to fine-tune our arrival beneath the ISS. At about 4,000 feet out, our payload bay pointed directly at the Station, we caught sight of the ISS solar arrays reflecting the glare of the rising sun. We were talking directly now to Shep, Yuri, and Sergei over a short-range VHF radio set up in the middeck. The growing Station resembled a delicate insect drifting against the blackness of space.

Roman remembered how in the simulator he had been impatient at the deliberate pace of our final approach: "Jeez, it's going to be another forty-five minutes until we dock!" Now he sensed a sort of time compression, with events aboard moving much faster than normal. As we closed to within 3,000 feet, a stream of laser pulses from the trajectory control system (TCS) found one of the reflectors dotting the Station's hull and locked the laser tracker onto the ISS, providing us with very accurate information on range, angle, and closing rate. The data went straight to a laptop display showing our progress up the approach corridor beneath the outpost. The pilots now used that picture to make thruster corrections as the flying tolerances grew tighter and tighter.

Taco had left his commander's seat and was now flying *Atlantis* from his post beneath the starboard overhead window. Peering up through a circular sight-ring and crosshairs, he fed in thruster pulses to nudge *Atlantis* into the final docking corridor, 600 feet out. Using his right-hand joystick, the rotational hand controller (RHC), Taco could pivot the orbiter in any direction around its center of mass. But because *Atlantis's* computers could handle that chore more easily, he usually left attitude control in the hands of the autopilot.

The speed and alignment of our approach, however, was strictly a hands-on flying task, accomplished with the left-hand joystick, the translational hand controller (THC). By nudging the THC left, right, up, down, in, or out, he could command thruster pulses to keep *Atlantis* aligned with the Station's PMA-3 docking port, projecting downward from Node 1. Roman and I fed Taco a steady stream of verbal cues on our range and closure rate (called R-dot). The Station swelled impressively in size, its three sets of solar arrays spreading from its central core, which stretched more than 100 feet from Node 1 to Soyuz.

Even 600 feet away, I was getting nervous. The golden solar arrays and silvery white hull seemed motionless against the black sky, but I couldn't dismiss the fact that both ships were racing around the globe at five miles per second. Each of us concentrated on our cockpit tasks, and there was little superfluous chatter. The quiet was punctuated by the occasional thud of a thruster firing and the clicking shutters and whirring motor drives of our cameras. The Station was an irresistible photography target; it was etched so cleanly on the black sky above that Marsha said later, "It was if someone had squeegeed your eyes." Although we knew Shep was watching our approach from the Service Module, no faces were yet visible in the windows.

It was time to swap ends. *Atlantis*, still headed nose first along our direction of flight, would bury her tail in the Station's belly if we docked in this attitude, and we needed the payload bay and Lab out in front of the Station for berthing. So at 600 feet, Taco initiated a planned, computer-aided yaw maneuver to swing our nose around 180 degrees. Roman, who had analyzed our performance in the simulator to determine just the right range and position to begin the maneuver, timed the tail-forward swing perfectly. Above us, the Station pivoted gracefully in our windows, mirroring the orbiter's actual motion. Three hundred feet now, closing as planned at three-tenths of a foot per second.

Floating in the overhead window with her laser range-finder, a modified version of a state trooper's speed gun, Marsha called out distance and closing rate. Beamer had the docking system in the green, ready for contact. Roman was now backing up Taco, monitoring the approach and occasionally squeezing off a photo. I called the R-dot to Taco every ten feet or so, the checklist going just like clockwork.

*Atlantis* glided up the corridor, closing the range steadily. At 170 feet, Taco fired several thruster pulses and brought us to a temporary halt, a chance for everyone in orbit and on the ground to take a deep breath before pressing in for docking. Expedition One called that they were ready, and Moscow confirmed that its systems aboard the ISS were go. Houston agreed: "*Atlantis*, you're go for docking." With two quick THC pulses, Taco started us up toward our meeting with the International Space Station.

The ascent proceeded at a slow, steady two-tenths of a foot per second. At two feet every ten seconds, it would take us nearly fifteen minutes before our docking ring contacted the Station's. At 100 feet the radar echoes grew too noisy for accurate ranging, so we switched solely to laser data. Marsha fed us a steady diet of range calls: "Seventy-eight feet, R-dot point one two." "TCS agrees," I chimed in; the payload bay laser was still solidly locked on ISS.

In the darkness of orbital night, the Station hovered in the wan glow of our payload bay floodlights, its feathery solar arrays fading into the inky gloom. The metal docking ring on PMA-3 gleamed in the floodlights as our own matching port stood ready just outside our aft windows. Earth was forgotten. All eyes were focused on the Station as its bulk slowly hove into the circle of light, like a shipwreck emerging from the gloom of the deep ocean. *Atlantis* weighed about 120 tons, the ISS about 100, and the two vehicles seemed to squeeze the vacuum between them as they closed the remaining distance.

To Taco, the final approach seemed like "a dance between two clumsy hippopotami." Inside 100 feet, he let the R-dot slow to a tenth of a foot per second. We were within one shuttle length of the Station. Houston was quiet; responsibility for the docking was now in our hands.

At thirty-five feet, the docking ring above seemed close enough to touch, and Taco slowed our approach. Our lights clearly illuminated the docking target mounted on the Station's tightly sealed hatch. Roman and I crowded up to the TV monitors and stared hard at the zoomed-in image of the target. The alignment cross was neatly centered in the bull's-eye, and we told Taco that the approach errors were insignificant.

"Houston," he called. "We don't see a fly-out required. We're pressing in." Mario Runco, the capcom, answered promptly: "We concur."

Across thirty feet of emptiness, Taco made a final call to Shep and his crew: "*Alpha, Atlantis,* here we come."

With a barely perceptible pulse from the thrusters, Taco nudged us upward at the desired tenth of a foot per second. Marsha started the range calls again. On her *Mir* mission, the orbiter docking ring was set farther aft in the payload bay, and that approach was "much less excruciating" than this one, where the controlled collision would happen just outside our windows. The two docking rings glided toward impact.

At fifteen feet, she began reading ranges directly off the template on the TV monitor. Taco's margin for error was just 0.03 feet per second. Too fast and he might bounce the orbiter's docking ring off the Station. Too slow and the capture latches might not snap firmly home.

"Fourteen feet, point one two."

"Twelve feet, point one one."

Taco's eyes were fixed on the target above him. Drifting off-center just three inches would put us outside the docking envelope and force an abort.

"Ten feet, zero point nine on the R-dot," I called, glancing at the laptop. My voice went up a notch in both volume and pitch.

Up in the commander's seat, Roman powered up the firing circuit for the automatic thruster sequence that Taco would trigger two inches from contact. "PCT is armed," he announced. The computer-controlled shove would bang the two rings together with enough force to guarantee capture. The last few feet came in a rush. Taco kept the closure rate steady, hovering around 0.1 foot per second, and kept the target cross centered with an occasional push on the THC.

"Eight feet, point one zero."

"Six feet, point one one."

"Five feet, point one one."

"Three feet, point zero nine."

From the cockpit, the Station seemed to descend on us like a giant industrial press, an unstoppable mass bent on ramming straight through the cargo bay. I tensed for the impact even as I called out the distance remaining.

"Eighteen inches, R-dot is good."

"One foot, petal overlap." The three metal alignment vanes atop each docking ring swept past each other toward impact.

"Six inches . . . two inches!"

Just before the rings slammed together, Taco punched the PCT button on the autopilot panel in front of him. *Atlantis* shook with thruster firings as the shock absorbers beneath our docking ring compressed with the force of the impact.

"Capture!" called Beamer as two blue lights flashed on his docking panel. The two spaceships bobbed gently, held lightly together by the mechanical latches on the docking petals. Damping springs quickly brought the relative motion to a halt.

"Capture confirmed, Houston," radioed Taco.

"Nice job, nice approach," replied Mario.

Handshakes and big smiles filled the cockpit for a few seconds. The sunrise glowed first a pure blue, then a rosy pink, then silver-white across the Russian solar arrays visible out front. In the harsh sunlight now washing over us, the entire Station came into sharp focus. Having verified a good docking, Beamer commanded the two rings to retract and lock together into an airtight embrace. The two docking tunnels now formed a slender vestibule between a pair of sealed hatches: one on *Atlantis*, another above in PMA-3.

Working with me in the airlock, Beamer quickly equalized the pressure between our cabin and the tunnel vestibule just above. Through the upper hatch window, we could see Shep, Sergei, and Yuri opening the Station hatch on their side.

We were ready for a reunion. Taco swam up through the airlock and joined us below the hatch. He peered up through the port, giving Expedition One a quick wave.

"Everything ready, Tom?" Taco grinned; I should have seen it coming. "Let's see how you do opening up *this* hatch."

I cranked the handle smoothly through a full circle. A quick tug, and the hatch came off the seals, mingling the atmospheres of the two spaceships. With Beamer's help on the stiff hinges, we swung the hatch down and flush against the airlock wall. Just like that, the door was open, and the five of us floated aboard *Alpha*.

The last time I had been inside PMA-3, it was bolted firmly to a work stand on the floor of the Space Station Processing Facility at the Cape. The tunnel's cream-colored walls widened into Node 1's nadir docking port. Taco, a body length above me, rose into the burly embrace of a grinning ex-SEAL and two exuberant Russians. Drifting after him through the open hatch, I entered the roomy world of the International Space Station.

Shep, Yuri, and Sergei met us in their "foyer"; for a few minutes the Node, its bulkhead trim a cheerful salmon color, was a scene of friendly chaos as we traded hugs, handshakes, and huge grins with our colleagues. The Expedition One crew had been on their own since *Endeavour*'s astronauts had departed nearly two months before, and they were genuinely glad to see us.

The ISS surprised me: it wasn't some austere outpost clinging to the very edge of human existence. It was instead a home port in space. The Node's roomy interior, warmer temperature, and softer lighting made it distinctly cozy compared with the shuttle's cramped, sterile cabin. Forward was the closed hatch leading to PMA-2, the forward docking tunnel that I would help Marsha move the next day. In twenty-four hours it would be the door to *Destiny*. Above us was the dead-end vestibule leading to the Z1 truss; left and right were hatch openings that served as storage bays for white cargo bags and stray pieces of Station equipment. Aft, past the salmon bulkhead trim, PMA-1 led to the FGB and the Service Module. Shep formally welcomed us aboard with a brief speech over the radio to both control centers and a ceremonial ringing of the ship's bell he had installed above the port hatch. For the next week the eight of us would form one crew, working together to berth the Lab and make it a functional part of the Station.

We moved on to the Service Module for a quick safety briefing and tour, then our hosts returned with us to *Atlantis*. Over the next few hours we inventoried and turned over to Shep's crew a half-dozen suitcase-sized bags of priority cargo. Included was a locker full of fresh foods and snacks, a welcome break from the freeze-dried, canned, and thermostabilized fare they had been eating for more than three months now. Four hours after boarding the ISS, we were back in *Atlantis*, closing the hatch to prepare for our EVA the next day.

Taco and Roman sent air whistling overboard through the airlock depress valve, lowering the cabin pressure to 10.2 psi. Just as on STS-80, the reduction from sea-level pressure would reduce the nitrogen content of our blood, lessening our chances of getting the bends and shortening our preparation time in the space suit the next morning. To further reduce our blood nitrogen levels, Beamer and I donned portable breathing masks to begin flushing that unwanted gas from our lungs. Looking like a couple of wayward scuba divers, we breathed pure oxygen while readying the tools for the first EVA. We quickly located nearly every item on the checklist, but three critical electrical components, called loopbacks, were missing.

For months we had trained underwater to install these soda can–sized plugs on the Lab's hull, where they would shunt electrical power or data to the proper circuits. These had to be in place for Lab activation after EVA 1. As we systematically searched our lockers again for the loopbacks, Marsha checked her inventory sheets. No loopbacks were listed, but she did find several items called 1553 bus terminators.

No doubt at some point in our training we had been exposed to the proper technical name for the loopbacks, but none of us had ever used that term. We quickly suspected that 1553 bus terminators were in fact the missing loopbacks. A quick call to Houston confirmed our suspicion. Unfortunately the lost components had been packed in a cargo bag that we had just transferred to the Station. They were now on the wrong side of a sealed hatch.

The pressure differential between the two vehicles prevented us from just opening hatches and grabbing the parts. My first thought was that we would have to repressurize either our cabin or the Station's docking tunnel, wasting precious breathing gas in the process. I had just made the call to Houston admitting our mistake when Sergei voiced a suggestion. Because our missing parts were only the size of a six-pack, he could place them into the foot-wide vestibule between the orbiter and Station hatches; it would take just a few liters of air to equalize the pressure there and retrieve them.

Sergei's solution soon had us back in business. As I grabbed the loopbacks from the vestibule, I met his gaze through the tiny porthole in the Russian-built Station hatch. My nod and wave of thanks were returned with an easy smile. The irony of having a former communist rescue our first EVA from an embarrassing foul-up wasn't lost on this former Cold Warrior. This partnership might yet work.

One final night of waiting and worrying remained. Restless in my sleeping bag, I fretted over the daunting work ahead of us on tomorrow's EVA. Outside in the cargo bay, *Destiny* awaited my first space walk.

# 16
# STEPPING OUT

*My God, the stars are everywhere.*

MICHAEL COLLINS, *GEMINI 10*, 1966

Roman was in position on the flight deck, having sealed the airlock inner hatch behind us. He stood by on intercom that day, February 10, 2001, to help us march through the airlock depressurization checklist. Crammed with our tools into the cylindrical five-by-seven-foot airlock just aft of the crew cabin, Beamer and I had about as much maneuvering room as two elephants in a phone booth.

I was adrift. Buttoned up tight into my space suit, I floated easily inside its protective envelope, close against its inner, airtight surface. The fit, though snug, produced not a single pressure point. Weightless inside *Atlantis*'s airlock, I ran my eyes over the suit gauges, then rested my gaze on the bare aluminum floor a few inches outside my faceplate. Beamer and I were about to embark on our first space walk, yet for the moment we were relaxed. I rested, waiting for the end of our oxygen prebreathe.

Our radio headsets crackled with conversation as Roman checked off the remaining preliminaries to the space walk with Mission Control. Through the helmet shell, from the world outside the space suit, came a muted, sporadic tinkling sound, the result of minor collisions between our drifting tools and the airlock walls. My view was limited to a few square feet of the airlock floor, the top of my head pointed toward the outer hatch. Beamer hovered two feet above me, wedged into the top of the airlock, his knees bent downward and nearly touching my suit's life-support backpack.

Breathing pure oxygen for the last hour, we were nearly finished flushing the nitrogen from our bloodstream and out through our lungs, an essential step before we lowered the space-suit pressure to 4.3 psi. Breathing pure oxygen at that pressure, our blood would be fully charged with the gas, yet the suit would be flexible enough to permit us to work effectively in the vacuum outside.

In a few minutes we would vent the airlock atmosphere overboard. It's a big step on any mission, but it seemed at that point like just another step in the checklist. We thought, instead, of the work ahead, the job we had trained for together those past eighteen months. I wondered about our true chances for success.

The rest of the crew was ready, too. Marsha already had *Atlantis's* robot arm locked to the grapple fixture on PMA-2, the forward Space Station docking tunnel. She and Taco had just commanded the release of the Node's motor-driven bolts that had held the tunnel in place for over two years. With a go from MCC, Marsha backed the tunnel carefully away from the Station. As she eased it forward of the Node, Beamer and I heard over the radio the words of Ray Price's familiar country tune: "Please release me, let me go. . . ." The music, from Marsha's CD player, was her way to relieve a bit of the tension. While she hoisted PMA-2 out and up from the Node, Beamer and I completed the last fifteen minutes of our prebreathe. When we exited, she would have the arm and PMA-2 in position for our work together.

Our preparation checklist at last complete, Beamer and I were ready. Up from Houston came the go to proceed. Roman stepped off smartly with the airlock depress procedure. I exhaled deliberately and took a deep breath: my second attempt at an EVA was under way.

"Airlock depress valve to 5," said Roman, and I heard the excitement in his voice. *We are actually going to do this!* I forced myself to stay calm, fearing the same letdown I had experienced in *Columbia's* airlock four years ago. *I'll believe it when I actually float out the door.* With the help of an airlock handrail, I twisted back to my right and rear, put a gloved hand on the black depress valve knob, and rotated it clockwise.

"Depress valve to 5," I confirmed, but my voice was nearly drowned out by the roar of air escaping through the valve and through the shuttle's thin aluminum hull. In the dim fluorescent light of the airlock, I saw the pressure unwinding toward zero on my suit's chest-mounted display and control module (DCM). The noise began to diminish as the thinning air lost its ability to transmit sound.

As the air swirled out of the chamber, I could feel the suit stiffen with the growing pressure differential between its interior and the near-vacuum outside. Like an inflating balloon, the legs, arms, and gloves grew more rigid. That was a good sign: my little spaceship was holding pressure. The bad news was that for the next seven hours I would be constantly working to bend that space suit to my will and to the job at hand.

When the pressure dropped to 5 psi, Roman made the usual checklist call to close the airlock depress valve, giving the suit circuitry a chance to perform its automatic leak check. I reached out with stiffened fingers to the valve knob—shaped just like the one on my kitchen range at home—and deliberately twisted it to the "0" position.

*Something's wrong.* The roar of escaping air didn't stop but immediately grew louder. The digital readout on the DCM showed the airlock pressure still dropping. My mind raced, sorting through possibilities, but I couldn't put a finger on anything. *The depress is still continuing! Why?* Beamer cut through my confusion with a call over the intercom: "The depress valve—it's not *closed*—it's on zero." His voice carried barely a hint of reproval; he was just calmly stating a fact. I whipped a glance back at the valve and cursed. Even before looking, I knew I had botched it. Instead of taking the valve to CL, I had twisted it to the zero mark on the dial. You might think that zero meant no air flow, but the valve's zero position drops the pressure rapidly to just that value through the biggest vent in the airlock! Not surprisingly, the pressure was now down to 4 psi, heading for vacuum, and we had missed the automatic suit safety check. I wrenched the valve closed; the whistle of escaping cabin air stopped abruptly, leaving us floating in silence. I fumed inside my helmet at my mistake. While we waited out a two-minute manual leak check, I took several deep breaths and forced myself to relax. *Concentrate on doing each step right.* Grateful to Beamer for his presence of mind, I took some solace in the fact that all I had cost us was a couple of minutes' delay in getting out the hatch.

Nevertheless, I couldn't afford to be careless outside. Once we took *Destiny* off shuttle electrical power and Marsha hoisted it out of the bay for berthing, we would have only thirty-six hours or so to complete the installation; any longer and the chilling cold of deep space would freeze the Lab's water coolant, rupture the thermal heat exchangers, and turn the module into a $1.4-billion fixer-upper.

Leak check completed, I twisted the depress valve to the now-famous zero position once more, releasing the remaining air into the void. As the internal suit pressure stabilized at 4.3 psi, the rush of oxygen from the suit fan dropped to a background whisper: there were now fewer molecules to carry the sound. *Comforting thought.* My voice took on a lower, rougher pitch, the result of the reduced gas density passing through my larynx. Now the moment of truth: I was ready to tackle the hatch a few inches from my helmet. Cranking the hatch handle clockwise, I grinned as the hatch rim came away easily from the door seals. A few seconds later, overcoming some stiff hinges, I had the circular door folded down onto the airlock floor. All that remained between me and *real* space was the flimsy thermal cover protecting the airlock from the orbital extremes of heat and cold.

Fastened around the rim by a generous band of that space age marvel, Velcro, the cover took a surprising amount of effort to open. During our Neutral Buoyancy Lab training, the waterlogged cover had floated effortlessly out of the way with a single push, and in tests at the Cape it had given me little trouble. But this Velcro proved surprisingly stubborn, so I jammed the extended fingers of my right hand between the cover and the bulkhead and shoved hard. Opening a hand-sized slit in the circumferential ring, I karate-chopped my way around the rim of the hatch cover. One last swat outward, and the thermal cover popped open and floated down out of the way. Brilliant sunlight flooded into the airlock.

Stowing the suit connection umbilicals that brought us oxygen, power, and cooling water from *Atlantis*, we flipped on our suit water supply switches. Now our cooling came from a backpack evaporator that chilled the water circulating through our long-john-like LCVGs. Already warm from wrestling with that Velcro, I twisted the control dial on my DCM to drop my suit temperature.

Still tethered to the inside of the airlock, I rolled to my right and turned to face Beamer above me. Just as I had practiced under water, I eased the suit and backpack gently through the thirty-inch-wide hatch. A gentle pull on the outside handrail brought me upright—I was through! I stopped outside the airlock for a moment, exulting in the sensation of free fall. Twisting slightly to the left and right, I discovered that moving the suit took almost no effort at all. After waiting and working nearly eleven years for this moment, I was free—and perched on the edge of the cosmos.

My gold-plated visor filtered the bright sunlight streaming into the payload bay to a creamy yellow-white. I felt the warmth on the back of my legs and arms, surprised that the sun's heat penetrated the suit so easily. Before donning my safety tether reel, I pivoted easily to my left and glimpsed the brilliant blue of Earth peeking just above the port edge of the payload bay.

Two hundred forty miles below and perhaps half a world away, Liz was listening for my first words outside the airlock. She remembered the disappointment of those lost space walks, too, and I wished she could see how all those nights away at the Cape, the long days in the pool, the simulator, and the mock-ups now faded in memory, swept away in this glorious light from above. I called back to her: "It's like opening the door on a new life."

Marsha was waiting, and the meter was running on our limited suit resources. I quickly hooked Beamer and myself into our safety tethers—fifty-foot stainless steel cables—reeling out from spring-loaded spools bobbing at our waists. It was a delight scuttling about the payload bay's handrails, and I was relieved to notice no disorientation at floating about in the suit. The underwater training back at the NBL had prepared us well. We freed ourselves from the three-foot-long Kevlar fabric tethers tying us to the airlock, clipped them to our waists, and began our work in earnest.

Our space walk that day was part of a carefully planned sequence of berthing events. With the PMA-2 docking tunnel clear of the Node's front door, I would help Marsha park the tunnel onto a temporary berthing ring one story up on the Station's Z1 truss, clearing the way for Destiny's arrival. Beamer would unplug the Lab's electrical connections to Atlantis and peel off the protective covers over the module's aft hatch seals. Marsha would then grapple Destiny and lift it ever-so-carefully straight out of the payload bay. The job was like backing a Lamborghini out of the garage with your eyes closed, knowing that if you dented a fender, the nearest body shop was 240 miles away—straight down. With the Lab clear of the bay, she would flip it 180 degrees and carefully align its aft hatch with the Node. Finally, she and Taco would command the mechanical capture of the Lab via motor-driven latches and bolts, sealing the hatches between the two modules. Beamer and I would then be cleared to mate the vital electrical and fluid connectors to complete the berthing and start the activation sequence.

Pulling tools deliberately from their stowage compartment next to the airlock, I noticed how easy it was to get 375 pounds of astronaut and space suit moving too fast. Without the familiar drag from the water, I felt a bit like a balloon, vulnerable to the slightest misstep with glove or boot. That tentative sensation lasted for about fifteen minutes, gradually fading as I became comfortable with the suit's inertia. Twenty minutes into the EVA, I left the cradling walls of the cargo bay for the first time, climbing the orbiter's docking tunnel up toward PMA-3 and Node 1. I took my time, testing the reaction of the suit to my hand-over-hand progress toward the Station. I was hand-flying my own independent space vehicle, although this one had a *very* tight cockpit.

As I climbed toward the Node, I got a peripheral glimpse of the black sky and the golden solar arrays spreading above, but Earth remained hidden behind *Atlantis*'s fuselage below. I was a bit reluctant to take in the full view of Earth or the Station, having heard enough stories about spatial disorientation from my colleagues to know that I didn't want to risk tumbling my gyros just as I got to work. Jeff Wisoff, my STS-68 crewmate, told me that he was puzzled on his first space walk to find that all the handrails in *Atlantis*'s payload bay were installed 90 degrees from the orientation he had expected. Several minutes passed before he realized that the unexpected appearance of Earth beyond the payload bay had misaligned his brain's sense of where "down" was by 90 degrees. Most of the anecdotes I had heard were just amusing, but a few astronauts had been assaulted by a sudden falling sensation when getting outside for the first time. On Jerry Linenger's 1997 foray outside *Mir*, he fought a powerful impression that letting go of the station's handrails would result in a plunge to the dark ocean 200 miles below. As I climbed, I kept my eyes zeroed in on my gloves, the handrails, and the intricate texture of PMA-3's exterior.

In Earth-centered terms, I was hand-walking away from the planet, heading straight toward the top of the ISS. But my brain told me I was gliding horizontally, left to right, along PMA-3 toward the Node, just as if I were back in the pool at the NBL. The ISS mock-ups there were laid out horizontally across the floor of the tank, forty feet down, and I had always moved sideways in my traverse from *Atlantis* to the Node. And so it was in space, where free fall enabled habit to overrule orbital reality. Halfway up the starboard side of the Node, I caught my first glimpse of Marsha's robot arm, holding position just past my extended legs and boots. With PMA-2's grapple fixture clamped firmly in the grasp of the arm's end effector, Marsha had positioned the tunnel within a few feet of its parking spot on Z1. The tunnel's gaping mouth loomed like an oversized megaphone held steady against the black sky.

The whir of the suit fan was now just a whisper tugging at a corner of my awareness. The cool oxygen flowing past my face brought with it a faintly medicinal smell from the plastic and metal suit lining. I reached out across a gap between the Node and Z1 that in training had always snagged my tools or left me hanging on by just a finger or two, and I was pleased to find that in free fall my arms easily spanned the distance. Keeping my boots canted slightly away from the Station, I checked that my tools were still clear of the delicate wiring and plumbing running across the truss surface. Turning toward the arm and PMA-2, I moved into position to guide Marsha's approach.

The robot arm vanished. Like a heavy curtain dropping across the stage, night rushed headlong over the Space Station. In less than ten seconds, the sun's arc-light brilliance faded into yellow, orange, and red, then disappeared completely as Earth's shadow enveloped us. Roman had given Beamer and me a heads-up a few moments before, but the rapid onset of nightfall still surprised me. The dazzling silver and white insulation of the surrounding Station had vanished into an abyss. Faint illumination filtered up from the payload bay floodlights below, but everything nearby was in shadow. Reaching up on each side of my face-plate, I snapped on a pair of helmet lamps. Aimed by hand, the small halogen lights restored a comforting, yard-wide circle of visibility.

In the warm glow thrown by the lights, I checked the settings on a portable foot restraint, a small platform with a set of boot stirrups left in place for me by Joe Tanner and Carlos Noriega during their shuttle EVA two months before. Swinging my boots carefully up and into the toe clips, I wound up facing the docking tunnel, the foot restraint firmly anchoring me to the right side of the Z1 truss. By flexing my knees and arching backward, I could easily see the open mouth of the docking tunnel a few feet away. I gave Marsha the go-ahead for bringing its berthing ring the last four feet into its parking spot.

Inside *Atlantis's* cockpit, Marsha checked her arm displays. The or-biter computer's best estimate of the distance remaining was accurate only to within a few inches, so I would provide the final distance calls as she slowly closed the range. We had practiced this routine about a dozen times back in the Virtual Reality Lab, with her operating the arm's joy-sticks and me using a bug-eyed pair of goggles to simulate my view of the Station and PMA-2. In the VR Lab, I had been able to peer into nooks and crannies and around corners to guide Marsha's approach, rehearsing the close eye-and-radio coordination needed to execute the docking. In training, it all seemed too easy. Would the actual berthing be as smooth?

As Marsha inched the tunnel toward me, I leaned left and right in the stiff suit, watching for any misalignment between the tunnel mouth and the berthing ring. She was aiming for a stopping point just at my shoulder level, close enough so that the berthing ring latches on Z1 could reach out and grab the tunnel's docking ring. Marsha "flew" the arm closer while I watched carefully, ready to call a halt should any chance of a collision develop. *Three feet to go.* The blunt, shark-toothed alignment vanes on the tunnel slid smoothly toward corresponding gaps in the berthing ring. "Two feet," I called. Marsha was relaxed, even laughing over the radio: "How does he *do* that?" I was just grateful that my estimate confirmed what she was seeing on her computer readout. *Steady . . . steady, Marsha. . . .* A foot to go, and my eyes scanned quickly across the PMA for signs we were headed off track. No . . . everything was clear. She was headed straight down the pipe about six inches away.

"Approaching ready-to-latch position, stand by for a stop," I called. I gave her a verbal count as Marsha's left hand erased the last few inches of the gap: "three . . . two . . . one . . . ," *Wait for it . . .* "stop!"

The PMA bobbed up and down an inch or so, then eased to a halt in the grip of the remote manipulator system. Marsha was spot on. I popped my boots out of the foot restraint and eased around the rim of the two berthing rings now hanging about four inches apart. The align-ment was good; there was no debris to interfere with the latches. We were ready to lock the PMA to its Z1 berth.

A foot in front of my chest, the mirror-like radiator on one of Z1's electronics boxes scattered pinpoint reflections from my helmet lights across the Station's skin. I swung the big pistol-grip tool (PGT, a battery-powered socket wrench) up into play from its bracket near my right hip, slipped it onto the shaft near the berthing ring, and drove the four latches home. The tunnel was now parked snugly against the Z1 ring. Our first task was complete: PMA-2 would stay there until the second space walk, when we planned to relocate it to the front of the Lab.

We were about an hour into the EVA, and Beamer was finishing up his removal of the seal covers over *Destiny*'s aft hatch. Marsha swung the RMS over to the payload bay to begin lifting *Destiny*. I dropped back down the truss to inspect the Node hatch just cleared by Marsha's removal of the PMA. In the stark shadows cast upward from *Atlantis*'s floodlights, I peered in at the hatch surface, checking to see that the berthing mechanism was clear of obstructions. From inside the circular hatch window, Marsha's illuminated docking target projected an eerie red glow, anticipating the Lab's arrival. I imagined Shep, Sergei, and Yuri hard at work behind the dull gray of the forward Node berthing port and hatch.

Finished with the Lab seal contamination covers, Beamer stationed himself at the aft end of the bay and cleared Marsha to grapple and lift *Destiny*. I headed skyward again, up to the P6 truss, the structural bridge supporting the twin solar arrays five stories overhead. Folded flat against the right side of the P6, its silvery panels stacked like an accordion, was the starboard radiator. Those panels housed hundreds of feet of ammonia tubing to carry waste heat away from the Lab's systems. We planned to deploy the radiator four days later on EVA 3, but first I had to release six fittings, called cinches, that had held the radiator in its folded position since its December launch. My job was to loosen a bolt atop each clamp fitting and remove a retention cable, freeing the radiator. Heading up the Z1 toward the work site, I was startled to see my life-sized reflection appear in the polished surface of a small avionics radiator. Its mirror-like tiles broke my reflection into a glittering mosaic, my gold visor, the flag's stars and stripes, and the suit's dazzling fabric caught against Earth's lovely blue and white horizon. The beauty of the scene took my breath away. *Can I really be working out here?*

We were right on schedule, ninety minutes into the EVA: Marsha driving the arm, Beamer heading for work on the P6 truss, and Roman seemingly perched on our shoulders via radio, riding shotgun on our checklist. I realized I was enjoying this. I felt challenged by the delicate nature of the work but reassured by our success so far. Still tense inside the suit, I made a deliberate attempt to relax.

Setting up for the cinches, I puzzled over a minor problem with the right arm of my suit. At each sunrise, when I needed to lower my gold visor, I found I could barely reach the knob on the right side of my helmet. Ordinarily I had plenty of reach up there, but I discovered that moving my right arm in close to my chest or helmet brought my lower forearm against an unfamiliar pressure point just inside and below the elbow joint on the suit's right arm. It was little more than an annoying distraction so far, but the problem had never occurred in training, and I didn't need surprises out here.

While removing the first cinch bolt, I heard Marsha's call that she was lifting *Destiny* out of the payload bay. Glancing downward, I could barely detect any arm motion—so careful was her progress. Popping the first bolt free and tucking it into its nearby clip, I next inched my way skyward along the P6 handrails. My body was suspended horizontally, as if from a flagpole, the soles of my boots heading into the coming sunrise at five miles per second. What a sensation! Hanging confidently by my fingertips, I looked down to see the thin steel safety tether snaking along behind me in the darkness. I had a heightened sense of isolation up there on the truss, so in the darkness I stayed zeroed in on my hands and the next few inches of handrail. The underwater training back in the NBL kicked in again; my internal roadmap of the Station guided me confidently into the blackness beyond my headlamps. Dodging a big steel trunnion projecting from the truss, I floated effortlessly within the surrounding void, my body invisible. For the moment, all that mattered were brain and fingertips.

The third bolt had been torqued down so tightly before launch that my PGT could not budge it. Even with the help of my body restraint tether, a sort of stiff "third arm" that I could lock to a nearby handrail, the frozen bolt twisted the power tool right out of my hands. Muttering to myself about Murphy and his all-pervasive law, I yanked the drifting PGT back in on its tether and tried again. Joe Tanner and Carlos Noriega had encountered this problem in December on the forward radiator, and I followed their lead on my second attempt. With the PGT in "ratchet" mode, I used the entire power tool as a hand wrench, but the job was quite a balancing act. I held the PGT with two hands to keep it on the bolt, my body restraint tether providing a wobbly connection to the Station handrail. Balanced on these three contact points, I tried to rotate the PGT while pushing in on the spring-loaded bolt, attempting to break it free. It was like balancing on a unicycle while using both hands to spin a lug nut off a flat tire. *Can I even do this?* After a few false starts, I got the bolt loosened enough to finish up with the power tool. The technique was slow, but it worked.

I lost about a half hour finishing up the rest of the cinches, and Beamer went on ahead of me to the $K_u$ antenna boom, the big strut supporting the main ISS communications dish on the port side of the P6 truss. I'd meet him there shortly to complete the activation of the antenna, but as I maneuvered around the starboard radiator I was rewarded with a spectacular view of the payload bay below. Marsha had by then hoisted the Lab high, the bus-sized canister poised on the slender robot arm ready for what Marsha called the "big flip." Below *Destiny*'s gleaming hull, *Atlantis*'s wings sliced backward across the gorgeous blue of the Earth, 240 miles below. I took a quick snapshot with the Nikon and turned to join Beamer.

Up on P6, Beamer had already shinnied out on the golden antenna boom to the underside of the parabolic $K_u$ dish. Ever since Bill McArthur and Leroy Chiao had swung it into place four months earlier on STS-92, the dish had been staring uselessly skyward, frozen in place by its two metal launch locks. Beamer and I were to remove the locks and free the antenna, giving ISS the ability to send video and science data transmissions back to Earth. Grasping the handrails just beneath the dish, Beamer removed first one bracket, then the other, handing the hardware to me. Both of us dangled far out over the Station's port side, suspended from the boom like mountain climbers, Earth sweeping by a couple of hundred miles beneath our boots. As our bodies drifted together and then apart, I found it a relief to be working as a team again, face-to-face after an hour and a half of solo space walking.

We inched our way back down the boom and were brought up short by Marsha's progress. *Destiny* pointed straight up out of the cargo bay, halfway through its grand rotation toward the Node's forward port. Turning almost imperceptibly, the Lab's meteoroid shields caught and reflected the dazzling sunlight.

We hurtled into darkness again as I slipped around the P6 truss to pick up an antenna blanket Beamer had left for me earlier. Negotiating this girder-and-fabric surface took some real attention because there were no handrails on this section of the truss. The only way to reach the antenna cover on this orbital climbing wall was to pick my way by fingertip over insulation blankets and ammonia lines, finding purchase on a stray bolt head or two. Negotiating this tenuous path, I remembered Story's advice from my Ascan years: a pinch from two fingers was all I needed to drift along the ISS, but misjudging my momentum could easily cost me my grip. I dreaded winding up adrift, hauling myself in darkness back to the Station from the end of my safety tether. My floodlights shone spookily on the white insulation and exposed aluminum beams as I located the blanket and clipped it to my tool rack.

Back on the handrails, I was soon sixty feet above the payload bay, heading for the "P6 trash bag" on the forward face of the truss up near the solar arrays. I peeled back the Velcro flap covering the hamper-like compartment, the designated home for the blanket until needed by another crew. Shoving the thermal blanket deep into the cavity, I pulled my head and shoulders out of the bin and turned to head back to the cargo bay. I had just started back down the truss when . . . *twang!* . . . my body jerked to a halt.

Looking around, I discovered I had run to the end of my fifty-foot tether, the taut steel wire snaking off to my left into the darkness. *Isn't this the right way down?* In the blackness surrounding my helmet lights, I couldn't see the rest of the truss or even the orbiter, but the strand of undulating wire gave me a clue. "Down" must be along the tether, since it was secured to the port side of the orbiter's payload bay. *That makes sense.* I paused for a minute, making sure, before swinging my body back into the familiar NBL orientation. I must have twisted around while working in the trash bag, gotten disoriented, and then just taken off in the wrong direction. Soon the radiator appeared in my helmet lamps, and I was back in business.

As I descended the truss I looked past my left shoulder and saw Beamer working atop the Node, getting ready for the EVA's top priority: plugging *Destiny*'s power and cooling connections into the Station. We heard Marsha report the completion of the Lab's rotation, and within a few minutes she had slipped it accurately into its mated position. Pressing ahead, she and Taco sent commands to latch *Destiny* in place and drive home the Node's sixteen motorized bolts into their corresponding nuts on the new module. The ISS command and control software ran smoothly through this sequence, with none of the lock-ups or laptop crashes we had feared. With the Lab now locked rigidly to the Node and the bolts providing a permanent airtight seal between the hatches, Beamer and I clambered aboard *Destiny*'s exterior to begin our crucial utility work.

While I was occupied up on the P6 truss, Beamer had finished much of the prep work for our task with the electrical and ammonia lines. We rendezvoused at the Z1 utility tray, a fold-down shelf laced with wiring and coolant lines that extended toward the newly berthed Lab. The tray would be the focus of our work for the next two hours as we pulled these utility lines over to the Lab's power and fluid receptacles. We were about halfway through the six-and-a-half-hour EVA.

I gulped some lukewarm water from my drinking tube and took a minute to assess our progress. I was relieved to find us just a few minutes behind schedule, with plenty of reserves in our suit backpacks to accomplish all of our first day's work. Station flight controllers had verified that the power and fluid lines were in a safe configuration for mating, and Roman and Beamer were ready to proceed. This was work only astronauts could do: no robot had the dexterity and reach to tackle this job.

Beamer took his place atop the Lab's port side to begin his work on the ammonia cooling connections. He was soon deep in conversation with Roman, reviewing the steps needed to mate the fluid connectors between the Station's radiators and the Lab's thermal control system. Meanwhile, I had to get *Destiny* back on life support. Low on the starboard side near the junction with the Node, I located the small storage box that housed the Lab's launch-to-activation heater cable. This circuit fed Station power to a series of strip heaters lining the Lab's hull, keeping critical systems from freezing in the cold of orbital night. The heater circuit couldn't function, though, until I ran the Lab cable along the exterior to its receptacle on the Node.

My right forearm was beginning to ache, and each contact with the pressure point inside the suit arm sent an unpleasant, tingling jolt along my wrist and arm. I tried to compensate by doing most of my work out in front of my chest pack, which eased the pressure against my forearm. Using the body restraint tether to stabilize myself on a handrail, I popped open the panel covering the box, peeled back an insulation blanket, and removed the spool of heater wire. Holding the spool in my right hand, I pulled off about ten feet of cable and watched as the spool floated easily on its wire tether. Just lifting that heavy reel of wire in our Lab tests back at the Cape was a two-handed job.

Pulling the cable over to the Node, I got Roman's go to attach the heater cable, and I recited back to him the litany we had learned through constant repetition in the pool: "Connectors clear, no bent pins, good bend radius on the cables." The engineers wanted to make absolutely sure these electrical hookups were good. I plugged the heater cable into the Node receptacle and snapped the locking bail over the plug to guarantee a good mate. "Do you have a good view, Roman?" Inside *Atlantis*'s cabin, he checked the monitor showing the scene from my helmet camera and quickly responded: "Looks great, Tom. You're cleared to wire-tie the cable down to handrail 108."

One of the Russians' *Mir* innovations adapted for ISS was the copper wire tie used to anchor equipment firmly to the Station's exterior. Each flexible tie, about eighteen inches long and made of heavy-gauge copper wire, had thumb-sized tether loops at each end that made it easy to twist them around a handrail to anchor cables or spare equipment. I had four of them coiled around the trunk of my body restraint tether, and I pulled the first one off and wrapped it around both the Node handrail and the heater cable. Then, using a gloved pinkie to keep the tether holding the wire tie to my suit out of the way (there was *always* a tether in the way out there), I put a couple of twists into the copper wire and tucked the thumb loops under the handrail. Satisfied with the first anchor job, I began to work my way back along the cable to secure the slack in the heater wire.

"Uh-oh." It was Beamer, his voice flat and matter-of-fact. He was just getting started on the fluid connections. With my head deep within the three-foot gap between the outer hulls of the Lab and Node, I was tightly focused on the placement of the wire ties. If he had a problem, it had yet to register.

"Um, I think I've got a leak here." Beamer was still unexcited, so I continued with my next wire tie. I imagined Beamer holding an ammonia line that was slowly bleeding some vapor . . . perhaps like the first wisp from a simmering teapot. *No big deal: just get it attached to the Lab quickly and we're in business.*

His next transmission to Roman finally grabbed my attention: "Yeah, I've got a pretty good stream coming out here. I'm going to go to the F-1 side and cut off the feed."

Beamer was invisible to me, hidden from view around the cylindrical hull. I decided I'd better get up there and see what was going on. As I moved onto a handrail, I peered up through the top of my faceplate at black space and saw a silvery, fluttering stream of snowflakes jetting out from Beamer's work area. The flakes were being driven outward by wisps of a faintly visible vapor, like a fine sea spray caught in a gust of wind. Sunlight sparkled off the fat flakes as they tumbled outward to form an expanding cloud on the Station's starboard side; they swirled and darted in the gas stream, glowing fireflies in the inky void. *It's beautiful!* But even as my eyes marveled at the sight, my stomach knotted with anxiety. The Station's ammonia supply was limited. If we couldn't choke off this leak, at least one of the Lab's two thermal control circuits was going to lose all of its coolant, imposing an enormous constraint on operation of the Lab. Worst case: we might not have enough cooling capacity for Lab activation. Our crew would have delivered a billion-dollar cripple, limping along at much-reduced capacity until a later mission could attempt repairs.

I reached for the next handrail in line and hauled myself skyward.

# 17
# HIGH STEEL IN ORBIT

*I'm sure it was the most fun I'll ever have in my life.*
SALLY RIDE, AFTER STS-7, 1983

The sky overhead was full of ice crystals, tumbling and glittering like stars against the black nothingness above. *What was going on?* I knew only that Beamer had disconnected from the Z1 tray the first umbilical, a one-inch-thick coolant line circulating anhydrous ammonia from the Lab to the Station's radiators. One of the two connectors must have stuck open.

Pressurized ammonia forced its way through the faulty valve, hit space, and instantly crystallized. Peering through the spray, Beamer knew that if he opened the F-1 connector just upstream, he would cut the flow to the leaking valve. Our EVA team had discussed this emergency action in a meeting a few weeks before launch.

This connector was just as resistant to opening as the first, stiffened by the cold of space. If he didn't get the locking bail open, one of the ammonia coolant loops was sure to be lost. My partner, about 200 pounds and built like an NFL linebacker, mustered every ounce of force and put it into the bail, his body flexing the foot restraint under him to its limits. Nothing. He pulled again, straining against the bail. The F-1 began to move a little.

As he forced the bail to retract, the connector poppet valves snapped shut and the ammonia flow ceased. Beamer remembered after the flight that it seemed he was struggling "for a half hour. It was three minutes. Three minutes to move that valve." But Beamer had stopped the leak.

My helmet peeked over the hull of the Lab just as Beamer choked off the ammonia flow, the last ice crystals bouncing away into the void. We had no idea how much coolant had escaped, but at least we had some time to plot our next move. MCC had us tackle the other pair of connectors while waiting for the ammonia frost on the F-1 and F-3 connectors to dissipate in the brilliant sunshine. I joined Beamer in brushing away the crystals from the second pair, and together we routed them successfully to the Lab.

I took a look at my partner. A thin ammonia frost dusted the outside of his suit. Most of it would soon evaporate, but we both knew we couldn't risk bringing any of the toxic chemical inside *Atlantis*. While MCC worked on a plan to get us cleaned up, I moved off to finish the heater wiring. We still had another two hours of critical work ahead connecting about a dozen data and electrical power umbilicals to the Lab. *At least those can't leak. . . .*

After tying down the heater wire, I rejoined Beamer at the Z1 tray to begin the electrical connections. The frost had vanished, and we made short work of the thinner, more flexible wiring hookups. What had been a terribly aggravating task under water, requiring me to hang upside down from a foot restraint while clumsily hoisting heavy switch boxes into position, proved to be almost a pleasure in free fall. Even though the suit was stiff and its right arm ground into my own forearm whenever I raised it too high, I found I could do this work.

I finished four of five electrical connections on the starboard side as Beamer installed the port ones, but the connector for the last circuit interrupt device, a cubical switch box about the size of a tissue carton, gave me trouble. With my left hand on a Z1 handrail, I tried to guide the CID connector into place with my right. But the bulky CID box on the other end of the connector cable drifted in front of my faceplate and stayed there, totally blocking my view. I needed a third hand to do the job; both of mine were occupied. For a moment I wondered how I was going to finish. Suddenly I realized: "Beamer's here. He can help me. It's no foul to call Beamer over!"

He was beside me a minute later, and with a few easy moves, he reached in and made the connection while I kept the box clear. The job was done, and for the rest of my time outside, I never forgot we were a team. It was my most valuable space-walking lesson of the flight.

After locking down the last connector, the two of us took stock. All the vital fluid and electrical connections to the Lab were in place. We had lost some ammonia from one cooling loop; how much, we didn't know. Our backup heater was functional, and the P6 radiator was nearly set for extension on EVA 3. Five and a half hours into the space walk, we were bruised and a little tired, but the important work had been done. One more task remained: it was time to get Beamer cleaned up. MCC ordered him to tether to the top of the Lab and stay there for an entire daylight pass, baking any remaining ammonia from his space suit.

Meeting him about halfway down the top of *Destiny*, I borrowed a brush Beamer had retrieved earlier from the Node tool bag. Like a pilot fish attending a shark, I scoured the surface of his suit for any signs of stray ammonia ice. It was hard work, free-floating around Beamer while using his suit as a handhold. While I worked, Beamer alternated between worrying about the leak and taking in the amazing view for the first time.

I shared Beamer's worries, and I had a few of my own. I could no longer raise my right hand high enough to operate either my helmet lamp or visor. I wasn't worried about the lamp: its battery was rechargeable, and I had another on the left side. But I couldn't work in the intense sunlight without the gold visor down, and I finally had to ask Beamer to lean in and lower it for me. Putting one's own visor down is the most trivial space-walking skill imaginable, but Beamer didn't question the favor. He was simply my partner out there.

Six and a half hours into the EVA, Beamer's bake-out and brush-off were complete, and we were more than ready to head inside. We stowed equipment and tools and soon eased our way back inside the airlock. Fighting fatigue, I got the thermal cover back into place, again with an assist from Beamer, wedged securely near the airlock ceiling. My faceplate began to fog over with the effort required to wrestle the stiff hatch closed, and it wasn't until we reattached our orbiter cooling umbilicals that I felt comfortable again. But the EVA held one more surprise.

MCC was still worried about residual ammonia on our suits and called for additional decontamination precautions. They radioed up a plan to repressurize the chamber to 5 psi, vent to vacuum again to flush out any ammonia, then pressurize again. Taco and Roman would then open the hatch briefly, toss in some wet towels, and have Beamer and me wipe down our suits and the airlock walls.

Unable to bring my sore right forearm in close to my chest and forced to use my left hand to throw all the switches on my suit, I wasn't enthusiastic about the manual wipe-down. The decontamination procedures would stretch our already lengthy EVA to beyond seven and a half hours, and I found myself running out of patience. Checking to make sure I was transmitting only on intercom, I called our commander: "Taco, I'm beat. My arm's giving me real problems, and I just want to get out of this suit. Can we streamline all of this somehow?" I could hear the sympathy in his voice, but he couldn't afford to take chances with ammonia: "I hear you, TJ. Let's just run through these procedures, and once we're sure everything's clean, Roman and I will get you two out fast."

That was good enough for me, and we began the repress to 5 psi. But as the pressure rose, a new problem appeared. My headphones faded out, and I could no longer hear Roman or Beamer on the intercom. At 5 psi, Roman stopped the repress, we vented once more to vacuum, and he started the pressure back up again. As he called out our next suit switch action, I simply couldn't hear him. Oblivious to the problem, I floated, waiting for the next command.

Facing the floor, it occurred to me that things were mighty quiet. Then Beamer, who noticed I wasn't answering, reached down to my backpack, rolled me over, and flipped my suit's oxygen supply lever to the correct setting. Realizing then that I was missing the intercom calls, I followed Beamer's lead; I knew most of the checklist by heart. Nodding in thanks to Beamer through my faceplate, I felt a tired jubilation at what we had managed to pull off. Reaching through the tangle of floating tools and umbilicals, I gave him the firmest handshake of congratulations and gratitude my spent right arm could muster.

Roman burst into the airlock in his oxygen mask, flipping me a weightless towel. Comfortable again thanks to the orbiter's cooling capacity, I joined Beamer in swabbing down the airlock surfaces and our suits. A few minutes later, Roman was back with Taco to bag up the towels, and God bless 'em, they removed our gloves and helmets. I sucked in a cool breath of cabin air and mopped my face with my hands in relief. Gloves off, the source of my right arm trouble was now easy to see. I had gone outside with my suit's right elbow joint on the *inside* of the arm. In the rush of getting into the suit, I had neglected to flex the arm and check that the joint bent the right way. Working against the metal bellows at the suit's elbow, I had jarred and scraped my forearm every time I flexed it. A Band-Aid took care of the raw skin, but it was weeks before the numbness left my right thumb and forefinger.

Soon, we were out of our suits and quizzing Taco and Marsha on their Lab systems activation work. While we were outside, ISS flight controllers had discovered that the thermostat on the Lab's hull heater circuit had failed. Temperatures inside *Destiny* had risen to more than 100°F. Realizing that the heat could damage the computers, electronics, and life-support systems inside *Destiny*, flight director Andy Algate's team sent word to Taco that activation of the internal cooling system was now urgent. Marsha and Taco got busy, commanding each step from laptops on *Atlantis*'s flight deck. MCC needed the crew to reduce the activation time from a period of several hours to as little as forty-five minutes. In response, the pair worked fast, sending instructions to turn on the Lab's command and control computers, using them in turn to activate the various subsystems.

We soon learned the importance of Beamer's prompt fix of the coolant leak. Telemetry indicated that the Lab had lost only two-and-a-half pounds of ammonia, just 5 percent of the coolant quantity in that system. The loss didn't significantly affect the cooling capacity, and once the system was activated, the Lab's interior temperature dropped back to normal. It had been a tense day for all five of us, but when we reopened the hatches to the ISS that evening, we were able to tell Shep and crew that *Destiny* would be ready for handover in the morning.

After an early breakfast, both crews met in the Node's comfortable foyer adjacent to the new Lab. After a handover ceremony in which Shep signed with mock seriousness the government's Form DD250 accepting Lab delivery, the two commanders opened the hatch into *Destiny*. Sailing in together wearing stars-and-stripes socks that Marsha had packed for the occasion, we reveled in the US Lab's roomy white expanse, making ample use of the long central aisle for some space aerobatics. The eight of us spent the entire day outfitting the Lab interior with air filters, communications cables, laptop computers, and emergency gear. Turning to system checkouts, we tackled the biggest job: installation of the air revitalization system rack. Imagine moving a 500-pound refrigerator across *Destiny*'s eight-foot aisle and sliding it delicately into its permanent rack location using just a few fingers and toes for guidance and stability. Our crew was the first to demonstrate the way future expeditions would install science racks in *Destiny*'s research spaces.

When we wrapped up flight day 5, *Destiny* was open for business. The Lab added 3,800 cubic feet of volume to the station and increased the Expedition crew's living space by 41 percent. With the new module, the interior volume totaled more than 13,000 cubic feet—bigger than any space station in history, surpassing *Skylab* and *Mir*. The ISS had grown to a mass of 112 tons.

I discovered the exhilaration of scooting down the Lab's center aisle, through the hatch, and into the adjoining Node. I bounced off walls, somersaulted down the aisle, and flew like Superman for more than forty feet down *Alpha*'s spacious corridors. Ten minutes of "arrested carrier landings"—zooming down the Lab corridor and stopping by hooking my outstretched toes on a bungee cord—matched the zero-g fun of my three previous missions in the smaller orbiter cabin. But there was no doubt who the real aces were aboard the ISS. Truly adapted to free fall, Sergei, Shep, and Yuri could glide from the Service Module to *Destiny* with only an occasional brush with the walls. One push would take them 150 feet.

The Expedition crew put the new Lab to work immediately, Shep commanding its systems from laptops linked to *Destiny*'s computers. The Lab's life-support system pumped fresh air throughout the US and Russian segments. The crew now had a direct link to MCC-Houston via the Lab's high-rate S-band telecommunications system. *Destiny*'s storage racks and empty spaces furnished the room to stow bulky items like spare space suits and tools. For the ISS, the Lab's most important asset was the software—just activated—that would enable *Destiny* to assume control of the Station from the Service Module. This important transition, eagerly anticipated in Houston, was viewed by Moscow with apprehension and mistrust.

# 18
# SKY WALKING

*Make a memory somewhere during the mission. Plant it in your brain and don't ever let it go, and you'll have it with you always.*

BOB CABANA, STS-88, 1998

With *Destiny* in place, flight controllers in Houston and Moscow had a greatly expanded capability to control the International Space Station. Moscow had been the primary ISS control center since the launch of the FGB in November 1998. Now MCC-Houston could assume some of the monitoring and control functions for the ISS. But the new capabilities also reignited a dispute between the two partners about who was really in charge of the Station.

Before the Russians joined the partnership, MCC-Houston had been the planned control center. The Europeans and Japanese would have smaller teams helping manage their science modules, but flight control and crew activities would have been managed from Houston. That changed when Moscow joined the ISS consortium: with the first two modules built in Russia, it made sense to have the TsUP, located in the northern Moscow suburb of Korolev, monitor the Station and manage the first expedition. The experienced *Mir* flight control team had only to shift gears slightly to control the early ISS. Most of the major systems were in the Russian modules, and Shep and his crew lived and worked in the Service Module, using the US-built Node mainly as a storage closet.

The two major partners agreed that Moscow would continue as lead center, supported by Houston, until *Destiny* arrived. The Lab's capabilities would enable a shift in Station attitude control from the Service Module's guidance system and thrusters to the Z1's control moment gyros, four high-speed flywheels that used their considerable angular momentum to keep the ISS oriented properly. This control mode used much less fuel than the Russian thrusters operating alone, greatly reducing the need for propellant sent up on the Progress cargo ships. Moscow had agreed to turn over "lead center" responsibilities to Houston when the Lab assumed attitude control, a milestone that would occur during our 5A mission.

Shortly after we had activated *Destiny*, MCC-Houston brought all four gyros up to their operating speed of 6,600 rpm to check out their control software. *Atlantis* took over attitude control from the SM while MCC-Moscow transmitted new software to its computers, enabling them to work in synchrony with the US gyros. Just a day after we had berthed and activated the Lab, MCC-Houston was ready for *Destiny* to assume attitude control, and with it, primacy in Space Station operations.

But the Russians quietly resisted the notion of formally handing over lead center responsibilities to Houston. In fact, they had written a letter to Andy Algate, the US flight director for 5A, proposing to delay the control transfer pending further procedural and flight control review—meaning indefinitely. They suggested more testing was needed on the Lab software, but the real reason for their hesitancy was their understandable reluctance, with *Mir* mothballed and unmanned, to yield their position as the world leader in space station operations.

I personally looked forward to the shift, as it reflected not only the Americans' hefty funding of the Station but also NASA's leadership role in the partnership. It would be a shift not just in attitude control but in *attitude*; I hoped the control center transfer to Houston would mark a more assertive NASA management stance.

But events on Earth didn't match the shift in orbital capability. Houston quietly assumed the lead on flight control activities while MCC-Moscow continued to act as if nothing had changed. There was no symbolic handover, no ceremony in orbit or at MCC-Houston to mark the occasion. Although Moscow tacitly acknowledged Houston's larger role, it gave no sign that the TsUP had drifted into Houston's shadow. As far as the Russians were concerned, the two centers were equals. Eager to avoid a public disagreement during our mission, NASA dropped any further mention of a shift in ISS control. It would be months before the problem was officially (and quietly) resolved.

Our crew was unaware of the terrestrial tug-of-war as we prepared for our second space walk. Our first EVA day had put *Destiny* in place. Now we had to outfit its exterior for the Station's future expansion. EVA 2 was set for February 12, 2001.

LETTER FROM HOME
SUNDAY AFTERNOON [FEBRUARY 11, 2001]
*Dear Tom,*

*I'm glad the hardest EVA is behind you and the other two will be easier. That's got to be a good feeling. . . .*

*We saw some good shots of you this morning and we saw the opening of the hatch to Destiny live. For the most part, I haven't seen too much of you or heard you this afternoon. Nonetheless, I have maintained a near constant vigil at the TV. I can move the phone into the family room and pull out the chair by the fireplace and watch TV and talk on the phone at the same time. Seems like I've been on the phone a lot this afternoon to one person and another.*

*I missed you in church today. Everyone prayed for the success of the mission and the safety of the crew. I said some of my own prayers of thanksgiving for the success so far. . . .*

*I'll go into Mission Control tomorrow and watch the EVAs. . . . You've been in my thoughts from the moment I wake up until the moment I go to sleep. . . . I hope your sleeping returns to your usual pattern of "close your eyes and go to sleep."*

*All my love,*

*Liz*

Lingering anxiety had kept me awake late into the night after the first space walk. The four to five hours of sleep I did get made flight day 5 in the Lab a long one, but by the next morning I was more than ready for another trip outside. Although the first EVA had proved an ordeal,

Beamer and I had been up to every challenge, thanks to our training, our physical conditioning, and the superb support we had received from our EVA team. I never felt alone out there, not with Kerri and her team helping from the ground and Roman, Marsha, and Taco working with us inside the cabin.

Our major EVA 2 tasks centered on outfitting the Lab to support a growing Station. We would first help Marsha relocate the PMA-2 docking tunnel to the Lab's forward hatch, where it would host future shuttle dockings. Then Beamer and I would install equipment needed for future EVA and robotics work on the Lab's hull. Roman suited us up, loaded us with our tools, and buttoned up the middeck hatch. We were ready to go, and this time both my suit elbows were right-side out.

Everything seemed to click on this second space walk. From the time Beamer opened the hatch until the moment six hours and fifty minutes later when we came back inside, nearly everything went according to plan. Working closely with Marsha, I unlatched PMA-2 from its Z1 truss location, enabling her to move the tunnel directly to its berthing location at the forward end of the Lab. As she eased the PMA away, I spotted Beamer up forward on the Lab, removing a tablecloth-sized insulation blanket from *Destiny*'s forward hatch. The stiff foil and fabric blanket refused to fold into the neat bundle we had seen in the NBL, and as I watched, Beamer waded into the middle of it to lash it down with a couple of tethers. For a moment the disk-shaped cover surrounded Beamer's waist, shimmering in the reflected silver-gray of the Lab's debris shields. As my partner wrestled the cover into submission, I called, "Looks like you're wearing a tutu, Beam."

Marsha next brought the robot arm to the aft end of the Node, where we would join forces to prepare the arm for my ride to our tasks on the Lab. As I attached a foot restraint to the end of the arm, my view took in the delicate sweep of the US solar arrays, an orange-gold bridge painted across the empty black of the cosmos. Although I faced the sky as I worked, my brain was convinced that I was standing upright, putting the rest of the station on its side, the way the mock-ups lay in the NBL. It proved very hard to defeat that underwater indoctrination.

Whenever I asked Marsha to shift the robot arm slightly, I noticed another startling illusion. With each movement I requested, the arm appeared to remain stationary, while *I* drifted in the opposite direction. Once I recognized the optical illusion, I was able to enjoy it, smiling with delight at how free fall and that infinite backdrop above me could so completely fool my perceptions.

Climbing aboard the foot restraint, I was rewarded by Marsha with an awe-inspiring ride across the Space Station's structure. In the cargo bay, hanging from my stirrups on the robot arm, I unbolted a grapple fixture from the sidewall and met Beamer at the Lab to complete its installation. Two months later, this power-and-data grapple fixture would serve as the foundation for the Canadarm II manipulator arm, a billion-dollar construction crane essential for Station assembly and expansion.

Marsha's skill and the shuttle arm's mobility made this part of the EVA a joy so different from some of the grueling physical work of the first space walk. All I had to do was ask to move three inches or three feet, and Marsha smoothly complied. But what seemed effortless to me was a demanding challenge for her. It wasn't a matter of simply looking out the window and nudging a hand controller: "The EVA tasks on the arm were harder than the Lab berthing tasks," she said later. "It took me a year and a half to learn how to do that. . . . Since everything we did with one of you guys on the end of the arm was blind to me for the most part, and since it was going to be more useful to you to tell me 'move me up, or to the left, or to the right,' and since I don't know my left from my right unless I mark my hand, I got Buzz Lightyear."

Buzz was a six-inch-tall flexible action figure from the movie *Toy Story*. Marsha velcroed him on the remote manipulator control panel between her two hand controllers. "I painted his right hand red with nail polish," she said, and then "if I brought up the [laptop] display showing the little guy on the end of it and if I put Buzz in the [same] position," she would be able to visualize how to move the arm. "I always had Buzz with me . . . I used him all the time."

We had some surprises on EVA 2: a tool I expected to find near the FGB was misplaced two dozen feet away on the opposite side of the Node; our socket wrenches didn't fit the bolts on a Lab vent (Beamer had to track down a substitute in the ISS toolbag); and a foot restraint I carefully returned to the airlock for modifications proved to have been one fixed two months earlier by the 4A crew (its serial number had been confused with another restraint attached to the Lab). These minor glitches illustrated the difficulty of keeping track of the thousands of components both inside and outside the expanding outpost—a challenge that would grow right along with the Station.

Despite these problems, our work went so well that we moved ahead of schedule and finished several jobs that had been scheduled for EVA 3. Beamer mated the eight power-and-data connections between the PMA-2 docking tunnel and the Lab, and we then met to unveil *Destiny's* new window set into the bottom of the cylindrical hull. Over the window, Beamer bolted a skillet-shaped protective metal cover opened and closed by a hand crank inside the Lab. Peeling back an insulation blanket, we peered through the window and laughed at the scene inside. There was Sergei in his shirt sleeves, floating in *Destiny's* brightly lit corridor, waving nonchalantly at two spacemen outside his home.

The unveiling of the new window was a high point of the EVA, but as we smiled in at Shep and company, I was completely unaware of a serious problem confronting my partner. Inside Beamer's right boot, a fold in the neoprene bladder material had been forced into the top of his foot by the inflated suit. The innermost of the fourteen layers comprising the space suit's tough fabric, the neoprene liner was stiff and unyielding, and the relentless pressure on his instep soon became excruciating. Beamer later told me:

> *It started as soon as we depressed the airlock. . . . That was kind of nasty. It was pretty excruciating pain for the whole spacewalk. . . . I think that was one of the reasons we got so far ahead, because I was just thinking "I've gotta get out of this suit, and soon."*
>
> *. . . at about the three hour mark, I was ready to cry "uncle!" it was so painful. But I thought, "I can't do that," and then when we got ahead and I got in the APFR [foot restraint], it kinda' pulled back, and I said, "Oh, that feels much better." That actually saved me, because that relieved the pressure on the top of that foot. . . .*
>
> *. . . When I was out of the foot restraint, it was excruciating. I actually lost feeling in my toes for several weeks after that EVA.*

Beamer told no one that he was wrestling with such pain. It was only when he got back inside, after nearly seven hours, that Roman and I learned of his situation. Next time out, Roman got his hands inside each boot, and was able to pull the offending bladder out of the way as Beamer donned them, preventing a recurrence. We were always learning.

On STS-98's space walks, NASA managed to get three brains inside two space suits, and it was a tremendously effective combination. Inside *Atlantis,* Roman mounted the checklists on the wall near the floor on the aft flight deck and just floated inverted while he talked us through the EVA. He had trained under water on our tasks, and using the "helmet cams," the miniature video cameras perched atop each of our suits, he tracked our progress and helped us avoid pitfalls.

Near the end of the space walk, as I hand-walked along the bottom of the Node carrying a portable foot restraint, I heard Marsha in my headphones: "Look down!" I pivoted on a handrail and peered toward Earth. Crowded in a window about ten feet away was my crew! I was looking straight down into the flight deck at Marsha, Taco, and Roman, all of them beaming like kids on Christmas morning. Friendship, I learned, can indeed be transmitted through a vacuum.

Our third EVA came on Valentine's Day, February 14, 2001. This was one last trip out into the bright sunlight, far above Earth, and in some ways it was the best of all. No bruised forearms, no crushed insteps, just challenging work and tremendous, inspiring beauty. With Beamer riding Marsha's arm high over the Lab and up to the Z1 truss, it took only five-and-a-half hours to install a spare radio antenna and transceiver on the Station; route the cables connecting PMA-2 to *Destiny;* unlock the big starboard thermal radiator on P6 (and watch its accordion-like panels extend a full fifty feet); and climb high atop the truss ninety feet above the payload bay to inspect the latches on the solar array support struts.

Beamer and I had reached the very top of the International Space Station, floating between the wings of the extended solar arrays. The view through my faceplate was astounding. Nine stories beneath our boots, the gleaming white fuselage of the space shuttle *Atlantis* sailed tail-first with the Station toward the distant horizon. Stretching up from the orbiter to meet us was the P6 truss: a massive tower of aluminum girders, storage batteries, and delicate radiators. As we worked atop the Station, I stole brief glances at the black velvet sky above, and the cirrus clouds brushed across the cerulean ocean below. Save for the whisper of the suit fan and the occasional crackle of the radio, the only sound was the quiet rhythm of my own breathing.

Following our climb to the top of the Station, I moved on to the cables lacing the black exterior of the PMA-2 docking tunnel, making sure their connections to *Destiny* were secure. That job done, I knew my work outside the Station was just about over. Before beginning my traverse back toward the airlock, I looked up from the handrails to take in the view, and I realized what I had been missing. I called Roman for a favor: "Give me a minute?" I asked. Mario Runco, listening in from Houston, answered: "Go ahead, Tom. You've earned it." I wanted to experience this moment not as a technician but as a human being. My space suit and I were weightless, my movements effortless. Silence prevailed.

Pivoting around my grip on *Destiny*'s forward handrail, I drank in the panorama unfolding around me. Directly in front of me, twenty feet away, the tail of *Atlantis* split the Earth's horizon. Straight up, the glittering solar panels of the Space Station spread like golden wings across the black nothingness of space. To either side of the now-empty payload bay, the royal blue of the ocean and its swirling white clouds rolled past. Behind me, the bulk of the Station plowed forward like a vast, unwavering star cruiser, slicing through the heavens toward a horizon a thousand miles distant.

Never have I felt so insignificant, part of a scene so obviously set by God. Emotions welled up inside: gratitude for how well the mission had gone, humility at my tiny importance in this limitless cosmos, and wonder at His glories revealed. I remembered the voice of Buzz Aldrin, who, returning from the Moon on *Apollo 11*, radioed back the words of David: "When I consider your heavens, the work of your fingers, the moon and the stars, which you have set in place, what is man that you are mindful of him, the son of man that you care for him?"—Psalm 8:3–4.

My five minutes were up. Soon Beamer and I would be heading back inside *Atlantis*. We had another hour or so of test objectives to meet and equipment to stow in the payload bay, but we both felt an immense satisfaction at how well things had gone.

Fate threw us one last curveball outside. Public Affairs had informed us weeks earlier that if one counted up all the EVAs, beginning with Ed White's first trip outside in June 1965, EVA 3 would be the 100th American space walk. Beamer and I planned to say a few words to honor those space-walking pioneers before us. I was thinking about choosing a good time for our little speech when I heard Mario's voice over the radio: "*Atlantis*, Houston, Message 43 is onboard, please let us know when you've read it." A few minutes later when Beamer was near the aft cockpit windows, I heard Roman call: "Beamer, could you come here for a second?" I glanced up and saw Beamer peering intently into the cabin. "I got it," he said over the radio. *Wonder what that's about?*

The answer came quickly. "Hey, guys, this is Roman. I've got the audio on 'private,' so we're only going out to you two. Mario's note says this isn't really the hundredth EVA." Somebody had done a recount and discovered that the *real* hundredth EVA had been two days ago on EVA 2. "Houston just wanted to get us the word so we didn't have egg on our face if somebody checks the numbers. You can read the message up here in the window if you'd like."

This was rich. Here were Beamer and I out in the void, "inches from instant death," as Jay Apt used to say, and Public Affairs informs us they got the count wrong? Sure, it was nice that they caught their mistake, but why cut it so close? Oh, well, we all make mistakes. I could think of a few of my own without much trouble. Beamer and I agreed we could say that the 100th EVA occurred during our mission. The broadcast went fine, we thought, and we used it to sincerely thank our entire EVA team: flight controllers, trainers, NBL divers, suit technicians, the whole gang. They had made America's three latest EVAs possible.

Closing the hatch for the last time was a bittersweet moment. I wasn't ready to say good-bye to EVA; I thought back to those five minutes of reflection up at the front of the Station and thanked God again for having been given the privilege of working outside. How could I close the door on that? Floating into the airlock, I considered these nineteen hours of space-walking an unparalleled gift, and my satisfaction with our success soon overwhelmed any feeling of sadness that our time outside was ending.

Flight day 9 was our final workday aboard the ISS, and our two crews scurried about, setting up equipment in the Lab, filming scenes for the IMAX film *Space Station 3D* in the Service Module, and transferring 3,000 pounds of equipment to the Station. Before the flight, we imagined we would have time to share dinner and perhaps a movie with Expedition One in their living quarters, but the demanding flight plan made that impossible. It was long after bedtime that night when the eight of us finally gathered around Shep's kitchen table for a few minutes of tired but easy conversation. Yuri shared some Russian chocolates he had saved for the occasion, and Marsha presented the trio with one of her trademark chocolate cakes baked at the Cape and launched in our pantry. Surrounded by friends in the cozy warmth of Expedition One's living room, I didn't want to leave, but I was bushed. I cruised down the passageway to *Atlantis* to grab my sleeping bag.

I set up my bunk on *Destiny*'s starboard wall, clipping in parallel to the deck. Roman and Beamer soon settled in nearby. A few minutes later, Taco said good-night and switched off the Lab lights, leaving us in the dim glow from the FGB a couple of dozen feet aft. A few hours later I woke, chilled by the Lab's efficient air conditioner, too much even for my sweater and sleeping bag. Still zipped in the bag, I unclipped the top, bent over to free the bottom fasteners, and drifted free in the aisle. It was the middle of the night on the ISS, and all was quiet. Wrapped in the dark green fleece bag, I tugged myself past my sleeping friends, through the Node's nadir hatch and the PMA-3 tunnel, around a 90-degree turn in the airlock, and into the darkened shuttle middeck. It was warmer there, and I was soon asleep. Tomorrow we would cast off for home.

# 19
## REENTRY

*I'd like to get away from earth awhile*
*And then come back to it and begin over.*

ROBERT FROST, *BIRCHES*, 1916

The STS-98 crew reluctantly said good-bye to our friends, leaving Expedition One to enjoy five more weeks of free fall. Roman was in his element as a pilot, flying *Atlantis* manually as he backed down the undocking corridor below the Station. It was, as he said, "flying the old-fashioned way," with eyes out the window and hands on the thruster controls, without the customary laser-derived trajectory information. The usually reliable TCS system didn't lock on until we had descended about 300 feet. With the rest of us crowded into the windows and taking photographs, he felt a little lonely as he eased *Atlantis* down and away. "Can I get a little help here?" he asked. We realized with a start that we had left our pilot a little short on guidance information, and Marsha began feeding him laser ranges while I punched the data into Roman's laptop display of the undocking corridor.

ISS was still in Earth's shadow, lit only by the glow of our payload bay lights. Roman glanced down for a second at the autopilot panel. When he looked back up at the Station, it had vanished. "Boy, did that get my attention," he told me later. "The Station was just . . . gone!" His mind raced, trying to digest what had happened. "Could I have gotten that far out of position?!" Searching frantically for some way to help Roman, we realized that going through his docking system checklist, Beamer had just turned off the payload bay floodlights. A quick switch throw restored the Station to view, and Roman's heart slowed to a more normal pace. Staring up at the silver-hulled *Destiny*, we could just make out the tiny faces of Shep and company peering down at us through the big window.

The two spacecraft wheeled into sunlight, *Atlantis* climbing above the Station in a half loop that carried us forward, above, and then behind the ISS. The Pacific coast of Chile slid beneath the outpost, the Andes stretching hundreds of miles past the solar arrays on either flank. As Roman took us nearly overhead at orbital noon, the orbiter's shadow swept across the solar arrays like that of a hawk drifting across the prairie. From *Atlantis*, we had a breathtaking view of our colleagues' home, sparkling like a jeweled scepter against the even lovelier backdrop of Earth.

By the afternoon of our undocking day, well clear of ISS, all of us were starting to unwind. Laughter came easily. Taco, Roman, and Beamer got busy setting up the exercise bike on the middeck. Except for our EVA work outside, we had all been too busy to get in a serious workout, and the pilots especially wanted to tune up their muscles for landing. They were about to latch the bike to the floor when a missing part brought them up short.

You lose things in zero-g. This time it was a shock mount, a rubber foot that prevents vibration when the bike is in use. I had noticed one of these shock mounts floating in the air next to Taco as they worked, but I lost track of it as I continued with my own tasks.

Five minutes later, all three of them were looking for that fourth shock mount. The bike wouldn't operate without it, so I was soon treated to the spectacle of three highly skilled astronauts searching desperately for a lost piece of rubber the size of a playing card. After fifteen minutes they finally gave up, hoping the missing component would slowly migrate to the cabin air intake. But after another quarter hour with no sign of it anywhere, I jokingly suggested we put something of similar size just where the shock mount was lost and wait to see where it ended up. Soon the four of us were quietly floating in the middeck, all eyes focused on the selected stand-in, a sealed foil package of beefsteak.

Marsha stuck her head down from the flight deck about then, and heaven only knows what went through her mind when she saw her four crewmates staring intently at a package of beefsteak. At first we joked that it was a "guy thing." "You wouldn't understand," we told her. As we filled her in, we speculated that perhaps the beefsteak wasn't mimicking the shock mount's behavior because we were *watching* it (as in quantum physics, where the very act of observing an experiment can alter its outcome). By now Marsha was laughing, too, and suggested we all go upstairs and watch the steak on a TV monitor, thereby ensuring that it would be totally unaware of our surveillance. With the group looking self-consciously at anything *but* the steak package, I took out a flashlight and squirmed behind an elastic net holding our suit storage bags and helmets. Deep in the pile, up against the cabin wall behind the space suits, I finally found the shock mount. My triumphant discovery earned me the inaugural ride on the bike that afternoon.

A more successful experiment was Marsha's unveiling of a new "soft phone," a software-based telephone for future ISS crews that enabled a laptop to dial terrestrial phone numbers. Using the shuttle's radio link, the software bypassed the normal shuttle-to-MCC radio loop and enabled us to dial home right from the keyboard with no one listening in at the control center.

LETTER FROM HOME
SATURDAY MORNING, FLIGHT DAY 11:
*Dear Tom,*

*I just have time for a quick note. It's 10 a.m., the children are packing, I still need to pack. . . .*

*Amazing—the phone just rang and it was you. The children are all excited now. I'm going to have to calm them down again to get them to do their work. It was wonderful talking to you and hearing your voice. . . .*

*I love you!*
Liz

Spouses and kids would be on their way to the Cape later on Saturday in anticipation of our landing the next day. Our space plane would head for KSC on Sunday morning, but aboard *Atlantis*, whose windows had faced away from Earth for the last week, there was some pent-up demand for viewing the planet.

All right, here I am, it's MET 8 days, 22 hours, 22 minutes. We're coming in over the northwest coast of Italy, at night. I can see from the big bay that comes out of the northwest limb of Italy; [it] cuts up to the north where Monaco is, around to the major cities—Milan—there's ... Torino! And then down, you can see the dark spine of Italy where the mountains are; there [are] no lights. And then you can see all the way out to Venice, on the eastern coast, the Adriatic.... You can see Pisa, and Rome and Naples ... the island of Capri, and where Vesuvius is there's a dark spot right above Naples. We're heading down the boot of Italy, on a descending pass now. This is marvelous. Come on over here—can you guys see this?

And there's all of Sicily with Palermo, and the city on the southeastern coast, Syracuse, is that right?

> BEAMER: No, no, um, ... Sigonella.
> TOM: And then here's North Africa, over here right underneath our shoulders.... Here's Greece, and if you come over here by me you can see Athens, all the way over there.
> BEAMER: You know what's cool, you can look through the atmosphere and see some of the little stars.
> TOM: Yep, through the air glow, yeah. That faint horizon there.... I can't pick out any of these constellations, you know. We're looking north, you think we should be able to see some of them....
> BEAMER: Too many stars! It confuses the situation.
> TOM: It clutters things up! I think there's Cassiopeia over there, in the Milky Way. Or Perseus.
> BEAMER: What are all these lights in the middle of the Libyan desert?
> TOM: Oil wells, the orange flares, those are ... flare stacks.... And then nothing, all the way through Egypt there, 'til you get to the Nile. It's just a string of pearls, like a necklace going down there.
> BEAMER: Wow. That was cool.

To Marsha, these Earth observation passes late in the mission, with all of our noses pressed up against the windows, were like "one giant continuous bedtime story."

We spent the balance of the day stowing equipment, strapping down returning Station cargo in the middeck, and getting our suits ready for entry the next morning.

> **MET 10/23:38.** It's flight day 12, and we waved off today . . . after high crosswinds at Cape Canaveral. Just outside the shuttle's cross-wind landing limits. Think the winds were at 030° at 13 peak 19 [knots], which is right at the 15-knot crosswind limit. Anyway, we attempted one deorbit, and they waved us off from that burn, and before we could get around to try the second burn they waved us off from that and said it just wasn't going to improve. So they just had us knock off for the day, and try again tomorrow. . . .

Suiting up for entry a second day, our landing opportunities were again thwarted by the Cape's high crosswinds and rainy weather at Edwards. Marsha was getting desperate. It wasn't due to our bare pantry (we still had plenty of rice pilaf, Bran Chex, and Italian vegetables) or a shortage of clean clothes (we could always rummage through the dirty laundry bags stowed in the airlock). She dreaded instead the ordeal of getting into and out of her launch and entry suit. Her snug suit fit was exacerbated by her increased height in free fall, and the prospect of squeezing through the neck's tight rubber seal was daunting. "Getting into and out of that thing had me in tears every day," she said. After that second wave-off, Taco sent this note to flight directors Bob Castle and LeRoy Cain.

> 19 FEBRUARY 2001
> MET 11/23:56:58
> *Day 3 of the STS-98 Crew hostage crisis. We're out of film, food is running low, and Tom is starting to look good. The alleged deorbit opportunities today did not come to fruition, and the crew is convinced that there is some sort of diaper DSO [medical experiment] in the works. If we put Marsha in the suit again, no one is willing to risk taking her out.*
> *The boys have stopped quoting lines from bad movies and have begun to quote each other. . . . Tom has begun sketching Earth obs views with crayon and pages out of FDF, while babbling bits of geographical trivia. No one else wants to be on*

*the flight deck with him. Last night, Tom was overheard muttering softly in his sleep, "Balboa, Magellan, where's the beef?" No one is quite sure about this reference.*

*We're down to our last three feet of gray tape and Marsha is working her way loose.*

*We're afraid that we'll have forgotten most of our debrief items unless we get to the ground soon.*

*I fear I will not be able to hold the crew together if we wave off another day. The mere mention of the phrase "working the weather" might put them over the edge. I feel the end is near. Tell our families that we loved them.*

*Taco*

*PS: First Officer (aka PLT) addendum. The CDR can't find his entry pocket-checklist and has been sulking about the cabin all day. I'm not sure of his ability to fly a HAC. I, on the other hand, am quite capable should you decide that this course of action is advisable.*

On flight day 14, after getting into the suit for the third straight day, Marsha joined the pilots on the flight deck. "Taco, we're going to land *somewhere* today 'cause I'm not getting out of this damned suit again!" She meant it. But that morning, we still didn't have a place to land.

Early on the morning of Tuesday, February 20, Liz turned on the TV to find a new landing site being discussed: White Sands Space Harbor in New Mexico, an alternate runway used only once in space shuttle history. She resigned herself to once again watching on TV at KSC while her husband's crew landed somewhere else. MCC debated weather forecasts and options. With rain and clouds in Florida, the obvious choice was Edwards Air Force Base in California, but midlevel clouds and gusting winds there made conditions less than desirable. For a few hours New Mexico was a real possibility, until Brian Duffy, flying a T-38 over Edwards, found the weather there improving. We at last got the go for the deorbit burn; California it would be. We fired *Atlantis*'s orbital maneuvering system engines one last time, beginning our long fall toward the atmosphere.

Sitting to Marsha's right on the aft flight deck, I watched with detached interest as her checklists slowly descended to the ends of their tethers, the subtle but clear indication that our two-week stay in free fall was ending. Since our entry occurred in daylight over the Pacific, we could see very little plasma out the overhead windows, just a trace of flame against the lightening sky. As *Atlantis* rolled left, still well off the California coast, we got a dramatic demonstration of our speed. Roman looked past Taco to "see us coming up on an airliner with a tremendous overtake." We were still at Mach 16 at about 200,000 feet. "Wow! Did I feel low!" he remembers.

Twenty-two minutes after entry interface, our first sensible encounter with the atmosphere, we crossed the coastline north of Los Angeles. Our nose-high attitude blocked most of the view from the rear seats, but Marsha and I stole glances at the monitor showing the scene straight ahead through the video camera mounted on Roman's dashboard. We looked past the Sierras to where the desert lakebeds and the Edwards runway should have been . . . and they were socked in! The clear weather that Duffy had scouted from his T-38 had given way to a fast-moving overcast, and we plunged toward our invisible landing site on computer guidance alone. Feeding in crosswind corrections as we spiraled down over the field, Taco punched through the clouds at about 8,000 feet. My eyes got wider still as we encountered more ragged clouds on the final approach, but Roman imperturbably passed on a steady stream of altitude and airspeed information to Taco.

Compensating for the strong headwinds, Taco carried a bit of extra airspeed as he flared low over Runway 22. "It was a fast touchdown, a little short of the aim point," Taco recalled, but it was insurance against a sudden gust catching us low on energy just above the concrete. Rollout complete, he called, "Wheels stop, Houston."

Scooter Altman answered from his console in Mission Control: "*Atlantis. . . .* Welcome back to Earth after placing our *Destiny* in space." It was a tired but jubilant crew who inspected *Atlantis* shortly after touchdown. While we relaxed and replayed the landing video for the medical staff two hours later, President Bush phoned in with his congratulations on the successful mission. I was impressed by the gesture. His brief call did much to salve the disappointment of landing for a third time away from my family.

It was too late to get back to Houston that day, so I got my first hot shower in two weeks at the Edwards Air Force Base VIP quarters. After a quick nap, I felt rested enough to join the rest of the crew and our astronaut and VITT colleagues for a good dinner. Still feeling a bit heavy in my chair, I nevertheless appreciated a great Mexican meal at Domingo's a few minutes down the road in Boron, California. We swapped stories for a couple of hours until, tired but elated, we headed back to our quarters and some much-needed rest.

The next morning the five of us were very proud to be able to walk somewhat skillfully up the road to the cafe at the base golf course. Over my bacon and eggs, I looked around the sunny room: a retired military man was sipping coffee over his paper, a few golfers drifted in for a snack before teeing off, and a crew of shuttle astronauts sat relaxed, eating their breakfast with a gusto not commonly observed among the cafe's patrons. I was back on Earth after a voyage of 5.3 million miles, but until I covered the final 1,300 miles to the arms of my family, the journey was incomplete.

## 20

# SHOCK WAVES

*The conquest of space is worth the risk of life.*

GUS GRISSOM, COMMANDER OF *APOLLO 1*, 1967

On a lazy Saturday morning two years after I returned from the Space Station, I sipped coffee and worked on a few chores, paying half attention to NASA Television's coverage of another shuttle's return. On February 1, 2003, the orbiter *Columbia* (once *my* ship) was heading home with seven colleagues after sixteen days in space. The familiar view of the MCC flight control room on the screen showed a routine entry, and I remembered the sights and sensations of my own exuberant voyage home aboard *Columbia* six years earlier. As usual, there was little chatter between the crew and the control center; all seemed normal as *Columbia* streaked across north Texas skies on a path to the Kennedy Space Center.

At approximately Mach 18, capcom Charlie Hobaugh relayed to *Columbia* the news that MCC had seen indications of low tire pressures on the left main landing gear. "And *Columbia*, Houston, we see your tire pressure messages and we did not copy your last [transmission]." The same data had showed up on *Columbia*'s computer displays. A return transmission from STS-107 commander Rick Husband was cut off in midsentence: "Roger, buh—." A short minute later, Charlie called the crew again: "*Columbia*, Houston, comm check." No reply. I began to listen more closely. The continued silence was unexpected. Although interference from the orbiter's tail can sometimes break the communications lock between shuttle and relay satellite, a long lapse in radio contact was unusual. Hobaugh began calling more insistently over the backup ultrahigh frequency radio as *Columbia* approached the Florida landing site. "*Columbia*, Houston, UHF comm check." Still nothing. What was just an annoying lapse in communication now became worrisome. I put down my coffee, listening for the reassuring voice of the crew; the TV showed only the expected track of the orbiter toward Florida.

Charlie continued to call *Columbia*. The orbiter should have been well within range of the KSC receivers and tracking radars by now, but over the radio loops there was only silence. Leaning against the couch, I dropped to one knee in front of the TV and said a prayer. Something was terribly wrong.

*Columbia*'s astronauts—seven friends—were already gone.

The catastrophe slammed into the astronauts, their families, and their NASA colleagues like a body blow. Nearly a generation had passed since the loss of *Challenger* seventeen years before, and we had forgotten the terrible anguish of losing good friends. The rest of the weekend was a blur of phone calls to friends, exchanging news and sympathy for the crew's spouses and speculating on how the disaster could have occurred. We shared memories of the seven friends lost in *Columbia*'s breakup, the heartbreaking scene shown repeatedly through several days of nonstop media coverage. Astronaut colleagues were in the field, helping in the search for our friends' remains. Others were caring for their families, helping shield them from the press and well-meaning public, and conveying the fragmentary information streaming in from the accident investigation.

I knew every member of the STS-107 crew. They had waited through years of schedule delays to undertake their science mission aboard *Columbia*. As I watched them work in orbit with such obvious pleasure over the previous two weeks, laughing and enjoying their spacious Spacehab module, I couldn't help feeling a twinge of envy: my friends were in space, and I wasn't.

Rick Husband was a standout test pilot and family man, whose wry guitar ballads of astronaut pleasures and anxieties spiced up our Astronaut Office parties and reunions. His patient example had helped his

STS-107 crewmates weather with good grace the many delays preceding their mission.

I had worked closely with mission specialist Laurel Clark, physician and naval officer. As a member of my Space Station Operations branch for three years, her medical background made her a natural to usher ISS science experiments down the long road from conception to flight. I was impressed not only by her skill but by her persistent good humor as she prepared for her long-delayed first flight. Her *Columbia* mission came more than six years after her arrival in Houston.

Dave Brown was a man I envied—he owned airplanes. I talked with him every week as he worked for me on Space Station experiment development with Laurel. As a "Cape Crusader" astronaut detailed to KSC, Dave flew frequently between Houston and the Cape, and our best moments together were in the T-38. My trips over the Gulf with Brown, a Navy pilot and flight surgeon, showcased both his superb flying skills and his outgoing personality.

Kalpana "K.C." Chawla had shared an office with Mark Lee and me during her 1995 Ascan year. Our teasing always brought forth her dazzling smile, one that cleverly masked her counterpunch until it landed—right on target. Even as she focused on developing new crew equipment for the shuttle and the ISS, she talked about her memories of seeing Earth from above on her first flight and told us how eager she was to experience that view again.

Mike Anderson, skilled pilot and physicist, was *Columbia's* payload commander. Our common flying experiences in the Strategic Air Command—his in tankers, mine in bombers—got our friendship off to a good start. Mike was our astronaut support person on STS-80, and that November afternoon in 1996, he was the last colleague to shake my hand before the pad team closed the hatch on *Columbia*. Before he left the cabin, Taco impulsively pulled off Mike's flight suit name tag, and it stayed right on the instrument panel with us throughout that record-setting journey.

Willie McCool was once a tough cross-country competitor at the US Naval Academy, and during my STS-80 quarantine he caught up with me on a run around the JSC perimeter. Having arrived at JSC just three months earlier, he slowed to jog along with me, and we talked about what quarantine and spaceflight were like. I told him that what I loved most about a mission was the satisfaction of doing your best in the company of good friends. When his own flight came six years later, I knew his crewmates were lucky to have him as their pilot.

Israeli Air Force colonel Ilan Ramon had come to JSC in 1998 for payload specialist training. I grew to know this combat-hardened fighter pilot, who had flown the strike against Saddam Hussein's nuclear reactor in 1981, as he trained for the shuttle and helped his American colleagues with Space Station experiment testing. Quiet and unassuming, his experience made him as much at home aboard *Columbia* as any of his career astronaut colleagues.

Five days after the accident, Liz and I joined the NASA family at the crew's memorial service at the National Cathedral in Washington. Gathered to remember our fallen friends and support their families, we offered what comfort we could to the spouses and their children, only imagining the pain they were going through. Fate had spared us the emotions we saw playing on their stricken faces. Over the following weeks, we joined Willie's family for his memorial at the Naval Academy chapel in Annapolis and said good-bye to Laurel, Dave, and Mike as they were interred at Arlington National Cemetery. On the snow-blanketed ground near the white headstones, we stood silently and honored these friends as rifle salutes cracked in the damp winter air.

Although the nation and NASA mourned *Columbia*'s astronauts, the shock of the loss struck the small astronaut community in Houston hardest of all. The corps had grown from about 90 when I arrived to nearly 160 in early 2003, but shared experience and dreams kept the group tight-knit.

My colleagues in Houston reacted to the accident by working with terrible intensity to find its cause and to care for the families. Dozens joined the search in northeast Texas until the remains of the crew and the fragments of the cabin were recovered. The casualty assistance control officers chosen by each *Columbia* crew member before the flight led small groups that kept watch with the families and helped with funeral arrangements. In Houston and at the Cape, astronauts joined the flight control and engineering investigations to pursue the suspected cause, a debris strike at launch that may have fatally damaged the orbiter's left wing.

The astronauts' coworkers couldn't escape the anguish of the loss, either. My secretary on STS-98, Roz Hobgood, was the STS-107 crew's secretary, and she had stood next to the runway at KSC with their landing guests waiting for the orbiter as the long minutes of silence blanketed them in a slow, horrifying realization. After chief astronaut Kent Rominger arrived to confirm the worst, she stayed to console the brothers, sisters, parents, and friends of the crew. For weeks afterward the sight of the crew's empty office in Building 4 affected Roz deeply.

And then there were the spouses of the other astronauts, those wives and husbands who had always dreaded the possibility of such a tragedy and prayed it would not strike them. As they gathered around the *Columbia* families to comfort and help in any way possible, they remembered their own fears on the roof of the launch control center or at the landing site. Some breathed a secret sigh of relief that their astronaut spouse would likely not fly again. The majority faced the reality that when their spouse headed for space again, it would be with the keen realization that a figurative roll of the dice was all that separated them from the pain so recently witnessed.

Within a few weeks of the accident, evidence from *Columbia's* scorched wreckage pointed toward the probable cause of the accident: a left wing damaged during liftoff by the impact of foam insulation torn from the external tank. The *Columbia* Accident Investigation Board's report issued in August 2003 laid out the fatal sequence of events. A briefcase-sized chunk of external tank foam had torn loose during launch and struck the leading edge of the orbiter's left wing, punching a hole in the heat-resistant carbon-carbon panels. From the time it first encountered the atmosphere at Mach 25 over the Pacific, *Columbia's* left wing had been under assault by the searing entry plasma surrounding the orbiter. The superheated atmosphere cut like a 2,800°F blow torch through the wing's aluminum structure into the left landing gear well and out through the wing's trailing edge. The extreme heat melted the wing's structural spars and cut the wiring to several sensors in the wing and wheel well—damage only hinted at in Mission Control. As the orbiter slowed through Mach 18, about 12,000 mph, the weakened left wing gave way under aerodynamic forces and sheared away. As talented as Rick and Willie were as aviators, they could do nothing to save themselves at these speeds. Once the wing gave way, neither their piloting skills nor the orbiter's computerized autopilot could bring the crippled *Columbia* home again. Even as *Columbia's* computers and thrusters fought to restore the proper entry attitude, the ship tumbled and came apart. The crew cabin broke up under the same hypersonic loads, making survival impossible.

Admiral Hal Gehman's accident investigation panel clearly showed that NASA's managers had failed to recognize the signs of incipient disaster; didn't communicate effectively in the face of a serious threat to the shuttle; and failed to act decisively to prevent the tragedy. Shuttle managers, astronauts, engineers, and flight controllers had ignored crucial evidence that the orbiter was vulnerable.

For example, each time I walked around my orbiter after landing, I could see evidence that foam and ice had struck the orbiter during launch. Even small pieces of the external tank insulation would cause minor but widespread damage to the thermal protection tiles. Yet none of us saw this as a potential catastrophe, only a long-term maintenance problem with annoying repair costs. The tank's foam insulation was not performing as designed, yet NASA reacted too slowly to the possibility that a larger strike in a critical area could cause a disaster. Both those circumstances materialized on *Columbia*.

Even after engineers discovered a few days after launch that a foam strike had occurred, shuttle managers had failed to listen to worries expressed by working-level engineers and flight controllers, chillingly reminiscent of the organizational communications failures that preceded *Challenger*. I know the people—experienced and competent—who worked on *Columbia*'s flight and managed it from the ground. It was not so much a refusal to listen, but a failure to *hear* unexpected warnings. NASA had forgotten how to think critically about spaceflight's hazards. Shuttle managers, who had risen through the ranks of flight controllers, engineers, and flight directors, had mentally classified foam loss from the external tank as outside the realm of those things that could destroy a shuttle. This mindset fatally crippled the agency's decision making during STS-107. The agency ignored the design vulnerabilities of the shuttle's fragile thermal protection tiles and neglected to order satellite images of the foam damage that might have warned of danger. The bottom line was that good people, talented and dedicated, did not listen, evaluate, or act. Seven astronauts died.

Although NASA actively promotes safety awareness in the workplace, in aviation, and in space, the successful years since *Challenger* had dulled the agency's ability to think critically. It will be difficult, as the Gehman board pointed out, for NASA to operate the space shuttle safely in the next decade without a complete reorganization of its human spaceflight management and the promotion of a vigorous independent safety organization. Both new leaders and a fresh, self-critical approach to operating the shuttle are essential to the prevention of a third catastrophe. The agency's new administrator, Michael Griffin, seems intent on implementing this approach.

Lost with orbiter and crew was my own naive sense that we had anticipated the most serious of the shuttle's vulnerabilities, that we understood the technology and science of conquering space. A jammed hatch? Sure, it happened, but the worst consequence had been a canceled space walk. The reentry that I once anticipated as a marvelous sleigh ride through the plasma was in reality a knife-edge balancing act between our knowledge and the uncaring laws of physics. Liz's worries before and during each flight had been closer to the mark than my own sunny optimism in the face of an astounding technical challenge. She knew that a 1995 study for NASA by Science Applications International Corporation had concluded that each mission faced a 1-in-76 chance of a catastrophic failure—a statistic that her training as an accountant made impossible to ignore.

Future shuttle crews and their families will approach their next mission with the clear-eyed realization, born of painful experience, that physical danger is an irreducible element of their profession. *Columbia* showed us that the limitations of human judgment will prevent us from ever making spaceflight completely safe. With full awareness of the risks, we can return to space determined to choose goals commensurate with the seriousness of the undertaking.


COLUMBIA'S ORDEAL WAS STILL two years in the future when I returned from the International Space Station on *Atlantis* in February 2001. With the addition of *Destiny*, the ISS seemed well on its way to completion. The arrival of the Laboratory inaugurated the start of serious scientific research aboard aimed at discoveries in both fundamental science and applied technology. My friends on the first few Expedition crews finally stepped off the seemingly endless training treadmill to undertake the challenge of long-duration spaceflight. My classmates Susan Helms, Dan Bursch, and Carl Walz were hopeful that the scientific and operational advances made on the ISS would lead directly, if incrementally, to more ambitious human voyages beyond low Earth orbit, the limited realm of the shuttle and Station just a few hundred miles above our planet.

There were some hopeful signs pointing in that direction. In March 2001 the White House followed up on the landing-day phone call President Bush had made to our crew at Edwards Air Force Base, inviting us to visit him during our postflight visit to the capital. I had learned from Story and others that Presidents Ronald Reagan and George H. W. Bush regularly invited crews to the White House, but that had been eight years ago. Despite its backing of the US–Russian ISS partnership, the Clinton White House was clearly only peripherally interested in space exploration; only a handful of astronauts met the president between 1993 and 2000. The invitation from the new President Bush was an honor, and we eagerly looked forward to the meeting.

The invitation had arrived on the heels of some momentous shocks to the ISS program. As Shep, Yuri, and Sergei prepared for their return to Earth, an internal review found that the cost of the next four years of assembly and science operations would exceed the planned budget by $5 billion. News of the looming overrun prompted a move astronauts considered as unlikely as the Station rocketing off on a voyage to the Moon: NASA administrator Dan Goldin removed George Abbey from his post as center director.

The ballooning Space Station costs had led Abbey to mothball the prototype X-38 Crew Return Vehicle, which was the US replacement for the Soyuz lifeboat, and halt plans for an ISS habitation module. He also pulled the plug on technology development for Moon and Mars exploration. But the belated cost-control measures were not enough to save Abbey's job. Goldin, angling to keep his job with the new Bush administration, had to move quickly to demonstrate his control over ISS costs. Six years after returning Abbey to power in Houston, Goldin exiled him to a desk at NASA headquarters.

The news of Abbey's removal broke just after we returned from STS-98. When our crew met with President George W. Bush on the afternoon of March 28, 2001, it was Goldin who escorted us into the Oval Office. Sporting a pair of shiny black cowboy boots, the administrator looked confident and polished in his dark business suit. His staff had insisted that we astronauts wear our NASA flight suits for the visit—attire we thought was inappropriate for both the setting and our host's stature. Taco had protested that bright blue aviation coveralls were ill-suited for the historic Oval Office, but Goldin's chief of staff sent word back through our chain of command: the STS-98 crew will wear flight suits, and they certainly won't bother us with any more unsolicited advice.

As we lined up to enter the Oval Office, the president, just inside the door, greeted Goldin, then turned to stare at us in surprise. Following just behind Taco and Marsha, I clearly heard the chief executive's words: "We usually consider a coat and tie the minimum attire for visitors to the Oval Office, but I guess in this case we can make an exception." The five of us privately fumed.

But the president made us feel welcome. He enjoyed our account of the mission and our impressions of spaceflight. For nearly half an hour he listened but shared few of his own thoughts on space exploration. After modestly comparing his own experiences as a fighter pilot with our high-flying exploits, he thanked us for our work on *Atlantis* and the Station. We left with hopes that his interest in the nation's space exploration program would grow.

WHEN LIZ AND I returned from the White House visit, it was time to take stock of the last eleven years and talk about our hopes for the future. I surveyed my future prospects in the Astronaut Office. NASA's funding had declined steadily through the eight years of the Clinton administration, and the agency's budget challenges meant that the pace of shuttle flights and Station assembly would slow. Abbey's practice of hiring large groups of new astronauts to supplement JSC's engineering talent had created a large backlog of rookies waiting for their first flight. In early 2001 the number totaled more than sixty. Although my EVA experience gave me some prospect of flying again in the next three years, the wait for my fifth mission might be a long one.

I wanted a new professional challenge. Ever since my brief years as a research scientist in Arizona and at NASA headquarters, I had wanted to become more involved in exploration planning and policy: Where should we go next in space, and how do we plan successfully to get there? My succession of assignments in Houston had always prevented my getting involved with JSC's exploration group. My next Astronaut Office job was rumored to be chief of the capcom branch—very close to spaceflight but totally focused on current operations. To look to the future, I had to free myself from the astronauts' necessary focus on the present. The price was walking away from the personal rewards of another spaceflight and the opportunity to again visit the Space Station.

We had been in Houston since 1990, and for more than seven of those eleven years I had been training for the next shuttle flight, the most intense and stressful lifestyle I could impose on my family. Training and office technical assignments found me frequently on the road, and my hours were famously irregular. Liz would automatically add one hour to whatever estimate I'd give her for my return, particularly when flying. For years Annie and Bryce had paid the price for my long days, frequent travel, and preoccupation with the critical work at hand. I wanted to regain control over the balance of career and home life. As difficult as it was to contemplate, only by leaving the astronaut corps could I spare Liz and the children another inevitable confrontation with Launch Day, when my family's fears and future were distilled and concentrated into the final anxious seconds of the countdown.

Geography was yet another factor. Liz and I, both Maryland natives, had spent more than eighteen of our twenty years together living west of the Mississippi. Our children barely knew their relatives and had only a vague notion of what life might be like in a place with forested mountains, tumbling streams, and four distinct seasons. Before their high school years arrived, we wanted to offer Annie and Bryce these experiences.

All these considerations combined to help us decide on a new direction. I wanted to set down some of my thoughts about the astronaut experience before the memories became too blurred by time, and I wanted the freedom to indulge in a luxury that astronauts seldom possess: time to think and reflect. The most attractive avenue open to me was to consult, write, and speak about the ideas that might carry Americans, after thirty-five years in Earth orbit, out into the solar system again.

In August 2001 we established ourselves close to family in the suburbs of Washington, DC. Setting my own schedule and working frequently from my home office was a refreshing contrast to the daily routine in Houston. Life there was in the hands of the astronauts' true masters: the crew training manager and the Office schedulers. They controlled where you went, what you did, and when you did it, and though the schedulers were good at their job, there was nothing like the new freedom to set my own pace and direction.

The trying part of leaving Houston was not giving up spaceflight or the T-38—it was leaving behind close friends. We followed their adventures with interest, and I enjoyed a vicarious thrill each time an astronaut colleague took on his latest challenge on the shuttle or Space Station. In space again less than eighteen months after STS-98, Taco joined Linda Godwin to return Dan Bursch, Carl Walz, and Yuri Onufrienko from the Station. Meeting their returning crews at Ellington Field on one of our visits to Houston, I found that their happiness eased my own sense of loss at not being in space with them. My memories and their continuing friendship are enough.

ABOARD *ATLANTIS* IN 2001, I carried with me a memento that gave me a thought-provoking glimpse into my own past and future. I ran across it on my last full day in orbit, when, floating downstairs through the hatchway into *Atlantis*'s middeck, I found Marsha Ivins deep in the pages of a novel. In the chill of a cabin cooled to 60°F to prepare for the warmth of reentry the next day, she was bundled in a sweater, keeping her toes warm by floating with her feet in the oven. She had borrowed my decades-old hardback copy of Arthur C. Clarke's *2001: A Space Odyssey*, reading it for the first time on the first space shuttle mission of that very year. I had first read the novel after seeing Stanley Kubrick's film version as an eighth grader in Baltimore. With the crew of Clarke's fictional *Discovery*, I had traveled to the dawn of the next century and the edge of the solar system, carried along by the author's vision of our future in space. After landing, when the eighty-three-year-old Clarke found out through a mutual friend that I had carried his book to orbit, he e-mailed a brief note of congratulations. I replied that his *Space Odyssey* had sparked my own ambition to be an astronaut. I also admitted that in 1968 I hadn't understood the ending of either the film or his novel. "That's okay," Clarke answered. "I always give people my *2001* prescription: read the novel, see the movie . . . then repeat as often as necessary."

Many of the elements of Clarke's vision have been realized. We do have a reusable space shuttle, though the government, not Pan Am, operates it. There is a permanent outpost circling Earth, though not the great wheel envisioned by Wernher von Braun and Clarke. And the age of tourist travel to Earth orbit seems almost to have arrived. But humanity, far from operating a string of far-flung bases on the Moon, hasn't visited the place in thirty-five years. An astronaut expedition to Jupiter, or the novel's Saturn, is as distant a dream now as it was in 1968, when Kubrick's film was released. Since the last Apollo landing, in 1972, Earth's space-faring nations have been content to circle the Earth occasionally; our ISS is nothing like an orbiting Hilton, and Clarke's cocktail lounge with its breathtaking view of the globe still waits for an entrepreneur with vision and a surplus of investment capital.

Long before I visited the ISS in early 2001, I had been thinking about what would follow. Is the Station a stepping-stone to a vigorous future exploration of the solar system, as in Clarke's novel, or a long-term holding pattern for humanity's dreams? As the space program recovers from the disheartening loss of *Columbia*, the country's policymakers are still debating the proper course for America and its partners in space exploration. How do the aging shuttle and the still uncompleted ISS figure in those plans? Is NASA capable of leading the United States and its partners out of low Earth orbit, as President Bush has directed? And if we leave, where shall we go?

# EPILOGUE

## A New Odyssey

*Don't tell me that man doesn't belong out there. Man be-
longs wherever he wants to go—and he'll do plenty well
when he gets there.*

WERNHER VON BRAUN, 1958

Just after sunset on December 17, 2001, I jogged along the Atlantic
surf in Cocoa Beach, Florida. The lights of the Cape Canaveral
gantries winked behind me as I ran easily southward. Low in the
southwestern sky hung a new Moon, its slender crescent edging a shad-
owed disk just visible in earthshine. Mars glowed above the sands ahead
of me, seemingly close enough to touch. It was my first visit to the Cape
since leaving NASA, and all the old memories came flooding back: the
heightened sense of the salty air, the cool breeze, and the roar of the
surf a few feet away. How many astronauts had run along this beach,
wondering what the heavens held in store for them? By the grace of
God, my dreams of spaceflight had been realized, and I had returned
from orbit to rejoin my family and drink in the sea air along the darkened sands.

When I had jogged on the Cape sands as a new astronaut, the Moon
had seemed again within our grasp. In 1990 I had expected to rocket to
orbit within a few years, and there was the beckoning promise of a trip
to the Moon or beyond. A bit more than a decade later, my fifty-three

days in space had taken me no farther than 250 miles above this beautiful beach. The Moon was as tantalizing as ever, but it still belonged to the men of Apollo, to John Young and his colleagues. By 2001 we had lost our ability to reach the world they once walked.

The *Columbia* tragedy persuaded both the president and Congress to end the decades-long uncertainty over the country's space objectives, and it accelerated a policy review already under way in the White House that resulted in a sharp new focus for NASA. In his January 2004 speech, "A Renewed Spirit of Discovery," President Bush laid out his vision of our space exploration future. Made thirty-one years after Apollo's explorers returned from the Moon, his decision to send Americans beyond low Earth orbit was a long-needed statement of our nation's belief in the benefits of exploration and discovery. But the new vision leaves out some important details and misses other opportunities for sustaining and strengthening our exploration program. A few essential changes will make our new space program more exciting, more popular with the public, and more likely to attract large-scale commercial activity to space.

First, we should move much more quickly toward retiring the space shuttle and replacing it with a new spacecraft. Although the shuttle is due to retire when the International Space Station is finished, about five years after flights resume, the thirty or so missions required to fully assemble the ISS might reveal other flaws in the aging fleet. An expensive grounding and refit, or problems with the ISS itself, might force NASA to keep the shuttle in service until 2012 or later.

The solution is to accelerate the development of the new Crew Exploration Vehicle (CEV), putting the Earth-orbit version into service by 2010, much sooner than NASA's original target of 2014. The agency's administrator, Michael Griffin, moved smartly in 2005 to do just that. A safe, suitable new booster for the capsule-style spacecraft might well be a modified shuttle solid rocket booster. The SRB has flown more than 175 times without a failure since *Challenger,* and adding a second stage would give the CEV plenty of capacity for crew and a modest cargo. With US access to low Earth orbit and the ISS assured, the shuttle's retirement would reduce the risk to astronauts and free up funds for the exploration effort.

Second, NASA should use the shuttle's massive external tank and reliable boosters to develop a heavy-lift cargo ship. A powerful new rocket based on these components could be available quickly enough to replace the orbiter's cargo capability in time to help with the final stages of ISS assembly. We will need such a booster eventually for lunar, asteroid, and Mars expeditions, and it will be cheaper if we don't begin designing it from scratch. NASA will need more funding to bring the new booster along than the president's plan allots, but its readiness beyond 2010 will give us a means to complete the Space Station without using the shuttle and lift the heavier spacecraft required for voyages beyond low Earth orbit. Driving launch costs down may also open up opportunities for establishing large-scale tourism and industrial activity, such as orbital hotels and lunar and asteroid resource refineries.

Third, NASA should not wait until Station assembly is complete to make the ISS a laboratory for exploration technology. Much sooner than 2010, it could host an advanced crew habitat designed for lunar or deep-space voyages, perhaps an inflatable structure like the TransHab developed at JSC in the late 1990s but shelved for lack of funds. The extra living space would increase science productivity aboard the Station while testing the structures and life-support systems that must function reliably far from Earth-based maintenance and support.

These advanced life-support systems, recycling gases, water, and even crew waste, would reduce the demand for supplies from Earth while being tested for, say, a three-year round-trip to Mars. The Station could also serve as a laboratory for studying the interaction of five-, six-, or seven-person crews, the likely complement at a lunar outpost or on voyages to the nearby asteroids and Mars.

Already the Station is teaching us how to run international training programs, to coordinate multinational operations, and however painfully, to manage a complex but resourceful partnership. Not everything we've tried in this global space enterprise has worked, and problems in jointly managing the Station continue to surface. But the success of an ambitious international exploration program will build upon the lessons learned so painstakingly at the ISS.

These initial steps will get us well along the path toward the Moon, the asteroids, or Mars by the middle of the next decade. Should the Moon be our choice for our next destination in space, as called for in NASA's new vision? We should think hard about its benefits and disadvantages before sending astronauts back to its surface. Landing people, equipment, and supplies there, even in the Moon's surface gravity of one-sixth g, is expensive. Much of the value of testing advanced space systems on the Moon can be gained at the ISS or in hostile Earth environments like Antarctica. Rovers and nuclear-powered systems can be checked out robotically under control from the ground. Only if the Moon hosts significant natural resources, such as recoverable water ice at its poles, should we make a major investment in sending people to live and work there for the long term.

In 2004, I participated in a study by the Planetary Society that called for a three-stage approach to human space exploration. After completion of the ISS and the introduction of an operational CEV, our group recommended a second phase, in which astronauts would voyage out of Earth orbit to several nearby destinations before we finally commit to landings and surface exploration. Visits to lunar orbit, the Sun–Earth Lagrange points (gravitationally balanced destinations where space installations can operate with little propellant use), selected near-Earth asteroids, and even the Martian moons will build up our experience and knowledge before we take on the expensive and risky task of setting up operations on the surface of another world.

The near-Earth asteroids (NEA), those fragments of larger bodies in the main asteroid belt between Mars and Jupiter, may serve as an intermediate destination between the Moon and distant Mars. These city- or mountain-sized objects, which approach or cross Earth's orbit, offer real advantages to our exploration and commercial efforts in space. They are likely to be our next in-depth exploration target if the Moon proves to harbor no easily extractable water resources. In the past, some of these ancient objects have collided with our planet, and others will do so in the future. Their Earth-approaching orbits also make them comparatively easy to reach: we already know of round-trip trajectories to some NEAs that require less rocket power than a one-way trip to the lunar surface. Astronauts could make the round-trip voyage to some NEAs in a year or less. A very few Earth-approaching asteroids can be reached with round-trip times of less than six months for the same modest propulsion requirements. We expect that ongoing telescopic searches will identify even more attractive asteroid targets and shorter flight times.

These "weekend getaways" to NEAs would take less time than current astronaut stays aboard the ISS. Quick asteroid visits would be of far shorter duration than a Mars expedition, minimizing exposure to the potentially hazardous physiological effects of free fall and space radiation. And since NEA-bound astronauts would never get farther from home than a few million miles, they could deal with a severe systems failure by executing an abort directly back to Earth.

The water and other mineral resources that we know are present on some NEAs could help reduce the long-term costs of exploring the Moon, Mars, and the rest of the solar system. And in the course of exploring them, we can test the technology needed to divert any asteroid on a collision course with Earth. One group, the B612 Foundation, whose board of directors includes *Apollo 9* astronaut Rusty Schweickart and shuttle and ISS veteran Ed Lu, has already proposed plans to NASA for a robotic demonstration mission that would significantly alter the orbit of a small asteroid by 2015.

From my spacefarer's perspective, the most attractive idea about "astronauts to asteroids" is that such voyages represent a natural progression in difficulty, more challenging than the dash-for-the-Moon Apollo missions but less daunting than the multiyear duration of a Mars-landing expedition. Think of an NEA mission as a shakedown cruise, the twenty-first-century equivalent of *Apollo 8*, the first piloted craft to orbit the Moon, or *Apollo 10*, which descended to within 50,000 feet of the lunar surface on a dress rehearsal for *Apollo 11*'s successful first landing. The upgraded Crew Exploration Vehicle on an asteroid expedition will have to perform nearly every task required for a Mars mission, save for enduring the trip time and executing the landing itself. We will have to scale up the propulsion and life-support capability, but when it's time to go to Mars, the near-Earth asteroids will have already given us invaluable deep-space experience.

The Planetary Society study proposed that after gaining experience with expeditions to lunar orbit, nearby asteroids, and perhaps the Martian moons Phobos and Deimos, the United States would then decide with its international partners which body—the Moon, Mars, or both—merits in-depth human exploration. Only then would we develop the landing vehicle and accept the associated risks of working and living on the surface. This commitment should come with an understanding of the price tag. Travel to the planets won't be cheap, but neither will the cost be astronomical. We estimated that a thirty-year program from 2014 to 2044, culminating in nine Mars expeditions spread over the final twenty years, would cost roughly $125 billion. By comparison, the Apollo lunar landing program cost about $130 billion in 2004 dollars. Going to Mars will cost about $4 billion per year—about 25 percent of NASA's budget, and the amount we spend annually on the shuttle program. Moving beyond Earth orbit will certainly cost more than the initial $11 billion over ten years that President Bush had initially proposed, but much of the needed funding will become available once the shuttle retires from service.

I believe that once under way, human exploration beyond low Earth orbit cannot stop short of Mars. Its atmosphere, permafrost, subsurface water, and possible volcanic heat are resources we can use to explore and colonize. Its ancient surface may preserve the fossils of Martian organisms or even maintain the conditions for hardy life forms to survive today. Any traces of a Martian biology can tell us much about the origins of life on our own world. With a surface area equal to that of Earth's continents, Mars promises decades of exploration and discovery to tempt the restless talents of our youth. The planet beckons our imaginations today the way the New World must have fascinated European society in 1500. We may find only intellectual wealth there, but until our machines and our fellow explorers go there in earnest, we have no real idea what we'll discover.

From my home, it's an easy exercise to search the sky and watch the ISS soar overhead. It's a short leap of the imagination from there to that Cape beach, or to those precious hours spent watching Earth and the stars from the cockpit of a space shuttle. Today's young people deserve to reach beyond the shuttle, the Space Station, even beyond the Moon. We have begun, I think, to create such an opportunity for our children. I believe that if we as a nation put our minds to it, within the next two decades another crew of astronauts will walk the Cape sands near the Beach House. It will be a restless night, their last on Earth for perhaps a few months: at dawn the next morning they will climb aboard a rocket bound for orbit.

But their destination won't be the International Space Station. A few hours after liftoff, they will fire engines that will sever their gravitational bonds to Earth. I can't know which world they will visit, but that detail is unimportant, for that crew will carry with them something more valuable than any planetary samples they return: the dreams of a new generation of explorers embarked on an odyssey to the stars.

# APPENDIX *Mission Statistics*

| Shuttle Mission | STS-59 | STS-68 | STS-80 | STS-98 |
|---|---|---|---|---|
| **PAYLOADS AND EXPERIMENTS** | Space Radar Lab 1 | Space Radar Lab 2 | Orbiting and Retrievable Far and Extreme Ultraviolet Spectrometer–Shuttle Pallet Satellite II (ORFEUS-SPAS II); Wake Shield Facility 3; EVA Developmental Test Flight-05 | International Space Station Assembly Flight 5A, US *Destiny* Lab |
| **ORBITER** | *Endeavour* | *Endeavour* | *Columbia* | *Atlantis* |
| **LAUNCH** | April 9, 1994;. 7:05 a.m EDT | September 30, 1994; 7:16 a.m. EDT | November 19, 1996; 2:56 p.m. EST | February 7, 2001; 6:13 p.m. EST |
| **LANDING** | April 20, 1994; 12:54 p.m. EDT Edwards Air Force Base | October 11, 1994; 1:02 p.m. EDT Edwards Air Force Base | December 7, 1996; 6:49 a.m. EST Kennedy Space Center | February 20, 2001; 3:33 p.m. EST Edwards Air Force Base |
| **MISSION DURATION** | 11 days, 5 hours, 49 minutes | 11 days, 5 hours, 46 minutes | 17 days, 15 hours, 53 minutes | 12 days, 21 hours, 21 minutes |
| **COMMANDER** | Sidney M. Gutierrez | Michael A. Baker | Kenneth D. Cockrell | Kenneth D. Cockrell |
| **PILOT** | Kevin P. Chilton | Terrence W. Wilcutt | Kent V. Rominger | Mark L. Polansky |
| **MISSION SPECIALIST** | Jerome "Jay" Apt | Steven L. Smith | F. Story Musgrave | Robert L. Curbeam |

| Shuttle Mission | STS-59 | STS-68 | STS-80 | STS-98 |
|---|---|---|---|---|
| **Mission Specialist** | Michael R. "Rich" Clifford | Daniel W. Bursch | Thomas D. Jones | Marsha S. Ivins |
| **Mission Specialist** | Linda M. Godwin | Peter J. K. "Jeff" Wisoff | Tamara E. Jernigan | Thomas D. Jones |
| **Mission Specialist** | Thomas D. Jones | Thomas D. Jones | | |
| **Orbits of Earth** | 183 | 183 | 279 | 202 |
| **Orbital Altitude** | 121 nautical miles | 120 nautical miles | 189 nautical miles | 173 nautical miles |
| **Orbit Inclination** | 57 degrees | 57 degrees | 28.5 degrees | 51.6 degrees |
| **Distance Traveled** | 4.7 million miles | 4.7 million miles | 7.0 million miles | 5.3 million miles |
| **Middeck Payloads** | Space Tissue Loss/ National Institute of Health—Cells; Visual Function Tester; Shuttle Amateur Radio Experiment; Get Away Special | Commercial Protein Crystal Growth; Biological Research in Canisters; Cosmic Radiation Effects and Activation Monitor; Chromosomes and Plant Cell Division in Space Experiment | Physiological and Anatomical Rodent Experiment; CCM-A (bone cell experiments); Biological Research in Canisters; Visualization Experimental Water Capillary Pumped Loop | Biological Protein Crystal Growth—Enhanced Gaseous Nitrogen Dewar |

# Glossary

**ACRV**—Assured Crew Return Vehicle. Space station lifeboat designed to return the crew to Earth if the station becomes unsafe; never flown.

**APFR**—Articulating portable foot restraint. An adjustable work platform with stirrups for an astronaut's boots that can be relocated to work sites on the shuttle or station. It enables an astronaut to work with his or her hands free.

**BFS**—Backup flight system. An independently developed software routine that runs on the orbiter's fifth general-purpose computer during ascent and entry. If the four primary computers or the primary flight software fails, the BFS can assume control to achieve a safe orbit or landing.

**Capcom**—Capsule communicator. The astronaut in Mission Control who makes all voice transmissions to the crew during a mission.

**CDR**—Commander.

**CEV**—Crew Exploration Vehicle. NASA's proposed replacement for the space shuttle, a piloted spacecraft that can carry a crew of three or more to low Earth orbit. Later versions of the CEV are planned to be capable of flights to lunar orbit or beyond.

**CIA**—Central Intelligence Agency.

**CID**—Circuit interrupt device. Electrical switch box installed outside the ISS.

**Contingency**—Unexpected or serious condition.

**DCM**—Display and control module. The chest-mounted control panel on a shuttle astronaut's space suit.

**EMU**—Extravehicular Mobility Unit. The shuttle space suit, also used on the Space Station.

**ET**—External tank. The shuttle's huge orange-brown propellant tank. Mounted below the orbiter, it carries liquid hydrogen and liquid oxygen propellants; it is discarded just before the shuttle achieves orbit.

**EVA**—Extravehicular activity. Work conducted outside a spacecraft by a space-suited astronaut. A space walk.

**FCOD**—Flight Crew Operations Directorate. The Johnson Space Center organization that manages the Astronaut Office, spaceflight crew assignments, and Ellington Field aircraft operations.

**FCR**—Flight control room. The familiar Mission Control facility directly involved in monitoring and controlling a shuttle or Station mission. There are separate FCRs for the shuttle and Station at the Johnson Space Center.

**FGB**—Russian ISS module, the "Functional Cargo Block." Also named *Zarya*.

**Free fall**—The ideal falling motion of a body that is subject only to Earth's gravitational field, as in a spacecraft in orbit about Earth. Falling around Earth under only the force of gravity, astronauts perceive that they and all objects around them are floating. Less precise synonyms are zero gravity (physically incorrect) and weightlessness.

**g**—g-force, or acceleration. 1 g equals the normal gravitational acceleration we experience on Earth's surface.

**GLS**—Ground launch sequencer. The Kennedy Space Center computer system that controls the final minutes of the space shuttle countdown. Orbiter computers take over at the T-31 second mark.

**GPC**—General-purpose computer. One of five identical computers that control the orbiter during launch, orbit, and entry. One of the five runs the backup flight system software during ascent and entry as a hedge against a failure of the other four GPCs.

**HAC**—Heading alignment cone. The descending spiral flown by the orbiter as it lines up with the landing runway.

**HUD**—Heads-up display. Aircraft cockpit display that positions vital flight information directly in the pilot's line of sight through the windscreen. The HUD enables the pilot to observe the runway, for example, without shifting his or her gaze down to the instrument panel.

**HUT**—Hard upper torso. The shuttle space suit's torso section built from fiberglass and steel. The HUT mates to fabric arms and trousers and supports the helmet assembly.

**ICM**—Interim control module. A backup ISS module developed by NASA to provide station control in case the Russian Service Module failed. The ICM was never launched.

**IFR**—Instrument flight rules; typically, flying in conditions that force reliance on instruments.

**IMAPS**—Interstellar Medium Absorption Profile Spectrograph. Instrument flown on STS-80's ORFEUS-SPAS satellite to analyze the composition of gas clouds between the stars.

**ISS**—International Space Station.

**IV**—Intravehicular. Inside the shuttle or station.

**JPL**—Jet Propulsion Laboratory, Pasadena, California.

**JSC**—The Lyndon B. Johnson Space Center. NASA's center in Houston that oversees human spaceflight operations.

**KSC**—The John F. Kennedy Space Center. NASA's Florida launch site.

**L–**—Time until launch date. For example, "L–3" means "launch minus three," or three days before launch.

**LCC**—Launch control center. Kennedy Space Center facility that controls shuttle countdowns.

**LCVG**—Liquid Cooling and Ventilation Garment. The water-cooled long underwear or body suit worn by astronauts inside the space suit's pressure bladder.

**LES**—Launch and entry suit. The orange protective pressure suit designed to keep astronauts alive in case of a shuttle cabin leak. The

LES also serves as a survival suit for bail-out and ocean recovery. It is equipped with a water-cooling system, survival gear, emergency oxygen supply, and parachute harness. An updated version called the Advance Crew Escape Suite (ACES) began flying in 1995.

**LOC**—Loss of control.

**MAPS**—Measurement of air pollution from satellites. Atmospheric carbon monoxide mapping experiment flown on SRL-1 and SRL-2.

**Max Q**—Maximum dynamic pressure. The point during launch of greatest stress on the shuttle, where the combination of increasing speed and decreasing air density is at maximum.

**MCC**—Mission Control Center. The facility in Building 30 at the Johnson Space Center, Houston, where mission operations are conducted. MCC is responsible for mission planning and controlling the shuttle and Station in orbit. Another control center in Moscow assists with control of the ISS.

**MDM**—Multiplexer-demultiplexer. Orbiter component that transmits data from sensors to computers and commands from computers to servomotors or other mechanisms.

**MECO**—Main engine cutoff. The point at the end of ascent when the shuttle's main engines shut down. A normal MECO velocity is just over 25,000 feet per second, about 17,000 miles per hour.

**MET**—Mission Elapsed Time. The time since liftoff.

**Middeck**—The lower compartment in the shuttle crew cabin. Contains sleeping, eating, storage, experiment, and bathroom facilities.

**MLP**—Mobile launch platform. The steel structure that supports the shuttle at the launch pad. It is carried with the shuttle to the pad by the crawler-transporter.

**MMT**—Mission management team. Any of several NASA panels that make decisions on launch or flight operations.

**MRE**—Meals Ready to Eat. Precooked, nonrefrigerated meals heated in a foil pouch. Borrowed or adapted by NASA from the Army field rations of the same name.

**MS**—Mission specialist. NASA astronaut primarily responsible for payload and experiment operations aboard the shuttle or Station. Additional duties include robot arm and space-walk operations.

**NASA**—National Aeronautics and Space Administration.

**NBL**—Neutral Buoyancy Laboratory. The 6-million-gallon water tank used for space-walk training at the Johnson Space Center.

**Nominal**—Operating normally.

**NTD**—NASA test director. NASA official responsible for coordinating shuttle launch operations at the launch control center.

**O&C**—Operations and Checkout Building. Kennedy Space Center facility used for spacecraft testing, also housing astronaut crew quarters.

**OMS**—Orbital maneuvering system, the pair of small engines at the rear of the orbiter that enables the shuttle to achieve initial orbit, maneuver to a new orbit, and slow down for reentry. Each OMS engine produces 6,000 pounds of thrust.

**ORFEUS**—Orbiting and Retrievable Far and Extreme Ultraviolet Spectrometer. The 1-meter telescope and spectrometer package flown on STS-80's 1996 mission, among others. The instrument was carried on the reusable shuttle pallet satellite, SPAS.

**OSHA**—Occupational Safety and Health Administration, a workplace safety bureau of the US Department of Labor.

**OTC**—Orbiter test conductor. The senior contractor representative on the shuttle test team for launch countdown, the OTC conducts and integrates all orbiter testing activities required in preparation for flight.

**O$_2$**—Molecular oxygen.

**PAO**—Public Affairs Office.

**PGT**—Pistol grip tool. Battery-powered socket wrench used in Space Station assembly. The PGT resembles a bulky cordless drill.

**PHRR**—Payload high-rate recorder. Digital tape recorder used on the Space Radar Lab to capture radar data.

**PLT**—The shuttle's pilot, who with the commander is responsible for operating the space shuttle orbiter during launch, orbit operations, and entry.

**PMA**—Pressurized mating adapter. The docking tunnels used to join the space shuttle to the International Space Station. PMA-1, 2, and 3 are currently at the ISS. *Atlantis* docked at PMA-3 on STS-98. PMA-2 is the current (2006) docking site for shuttles, at the forward end of the Station.

**POCC**—Payload Operations Control Center. The work area adjacent to Mission Control's flight control room, where science team members manage the operation of a shuttle science payload. During Space Radar Labs 1 and 2, the POCC was operated on the second floor of Building 30 by scientists and engineers from the Jet Propulsion Lab and the German and Italian space agencies.

**psi**—Pounds per square inch. The standard English unit of pressure measurement.

**RCS**—Reaction control system. The shuttle's rocket thrusters used for attitude control and small orbit changes.

**R-dot**—Closure rate during rendezvous.

**RHC**—Rotational hand controller. Shuttle attitude control joystick.

**RMS**—Remote manipulator system. The shuttle's robot arm system, also known as the Canadarm.

**RSLS**—Redundant set launch sequencer. Automatic software routine controlling final countdown. It permits booster ignition and liftoff only if all critical shuttle systems are functional.

**RTLS**—Return to launch site. Shuttle abort mode where the orbiter returns to the Kennedy Space Center immediately after liftoff.

**SAC**—Strategic Air Command. US Air Force major command responsible for nuclear deterrence and strike missions. It existed from 1946 to 1992.

**SAF**—Spacecraft assembly facility. Jet Propulsion Laboratory clean room for spacecraft assembly and testing.

**SAIC**—Science Applications International Corporation.

**SAR**—Synthetic aperture radar.

**SAREX**—Shuttle amateur radio experiment. Small ham radio set sometimes carried aboard shuttle.

**SAS**—Space adaptation syndrome. The condition afflicting astronauts just after arrival in free fall, characterized by headaches, sinus congestion, fatigue, and nausea.

**Scrub**—Cancel or postpone a launch.

**SES**—Shuttle engineering simulator. An astronaut training and engineering development simulator located at the Johnson Space

Center, noted for its high-fidelity visual presentation of the space shuttle, Space Station, robotic arm, and orbiter landing scenes. Shuttle crews use the SES regularly for advanced training in rendezvous, robotic arm operations, and abort situations.

**Sim**—Simulator or simulation. Astronauts train for spaceflight in high-fidelity, computer-operated simulators that replicate shuttle or Station controls. The Shuttle Mission Simulator (SMS) is the prime example.

**Sim Sup**—Simulation supervisor.

**SIR-C**—Spaceborne Imaging Radar-C. Third in a series of synthetic aperture radars for observing Earth's surface, flown twice on the space shuttle in 1994.

**SLF**—Shuttle landing facility. The Kennedy Space Center landing runway for the space shuttle orbiter.

**Slip**—Launch postponement.

**SMS**—Shuttle Mission Simulator. Its two versions, the fixed base and motion base, train astronauts in orbit and launch and entry operations.

**SPAS**—Shuttle pallet satellite, a carrier spacecraft deployed from the shuttle orbiter. *See also* ORFEUS.

**SRB**—Solid rocket booster, the shuttle's white, solid-fueled strap-on boosters positioned on each side of the external propellant tank.

**SRL**—Space Radar Laboratory.

**SRO**—Supervisor, Range Operations.

**SSME**—Space shuttle main engine. Liquid oxygen- and hydrogen-burning rocket engine, three of which are installed at the base of the orbiter's tail. They power the shuttle for eight and a half minutes, drawing propellant from the external tank.

**SSPF**—Space Station Processing Facility. The test and assembly building for Station hardware at Kennedy Space Center. The SSPF is located on Merritt Island just east of the Operations and Checkout Building.

**SST**—Single Systems Trainer, a low-fidelity shuttle cockpit trainer that replicates the basic functions of individual shuttle systems. This is the first crude simulator in which astronaut candidates train.

**STA**—Shuttle Training Aircraft. NASA's Grumman Gulfstream II executive jet modified to handle like a gliding shuttle orbiter.

**STS**—Space Transportation System. Official name of the space shuttle.

**SVS**—Space Vision System. Video-computer system designed to aid astronauts in precise positioning of the shuttle or Station robot arms.

**Swing arm**—The narrow, retractable catwalk that bridges the gap between the launch service structure, or gantry, and the space shuttle orbiter. The arm swings back against the service structure a few minutes before liftoff.

**T−**—Time until liftoff. "T–0" is literally "T minus zero," zero seconds until launch or liftoff time.

**TAEM**—Terminal Area Energy Management. Orbiter guidance software routine used to bring the shuttle to final approach and landing.

**TCDT**—Terminal Countdown Demonstration Test. Final countdown rehearsal for space shuttle crews at the Kennedy Space Center.

**TCS**—Trajectory control system. Laser tracking system that provides range and closure data to shuttle crews during rendezvous.

**TDRS**—Tracking and Data Relay Satellite. One of several NASA satellites in geostationary orbit that relay communications and science data from the shuttle, Station, and satellites to the ground.

**THC**—Translational hand controller. Orbiter joystick that operates thrusters to control the lateral, vertical, and fore-and-aft position of the shuttle.

**$T_i$**—Terminal initiation burn. Critical rocket firing one orbit before rendezvous.

**TsUP**—Moscow Mission Control Center. Russian acronym for "Center for Spaceflight Control."

**VAB**—Vehicle Assembly Building. The Kennedy Space Center structure where Saturn V Moon rockets were assembled, now used to stack the orbiter, boosters, and external tank. The shuttle is hauled over a three-mile-long roadway from the VAB to the launch pad on the back of a diesel-electric crawler-transporter.

**VITT**—Vehicle integration and test team. The astronauts' representatives at the Cape, indispensable engineers who participate in orbiter and payload tests and coordinate the crew's prelaunch and postlanding schedules at Kennedy Space Center.

**WCS**—Waste collection system. The orbiter's toilet housed in a tiny compartment on the middeck's left side just aft of the side hatch. A swing-out door and folding screen provide privacy during orbit operations.

**Weightlessness**—The floating sensation experienced by an astronaut in free fall as all objects accelerate toward the center of the Earth. Although objects in free fall have no weight, they retain their mass, or inertia.

**WETF**—Weightless Environment Training Facility. A water tank in Building 29 at the Johnson Space Center, twenty-five feet in depth, used until 1996 to train shuttle astronauts on space-walking tools and tasks.

**WSF**—Wake Shield Facility. NASA and University of Houston satellite flown for the third time on STS-80. Wake Shield's mission was to test the production of high-quality semiconductor wafers in the vacuum conditions of orbit.

**X-SAR**—X-band Synthetic Aperture Radar. Built by German and Italian space agencies for flight on SRL-1 and SRL-2.

# BIBLIOGRAPHY

Air and Space Smithsonian Magazine. *Space Shuttle: The First 20 Years*. Edited by Tony Reichhardt. New York: DK Publishing, 2002.

Apt, Jay, Michael Helfert, and Justin Wilkinson. *Orbit*. Washington, DC: National Geographic Society, 1996.

Burrough, Bryan. *Dragonfly*. New York: Harper Collins, 1998.

Burrows, William E. *Deep Black*. New York: Random House, 1986.

Burrows, William E. *This New Ocean*. New York: Random House, 1998.

Chaikin, Andrew. *A Man on the Moon*. New York: Viking, 1994.

Evans, Diane, et al. *Seeing Earth in a New Way: SIR-C/X-SAR*. NASA-JPL 400-823. Pasadena: NASA, Jet Propulsion Laboratory, 1999.

Ezell, E. C., and L. N. Ezell. *The Partnership: A History of the Apollo–Soyuz Test Project*. NASA Special Publication 4209 in the NASA History Series. Washington, DC: NASA, 1978. Available at http://www.hq.nasa.gov/office/pao/History/SP-4209/toc.htm.

Linenger, Jerry M. *Off the Planet*. New York: McGraw-Hill, 2000.

McConnell, Malcom. *Challenger: A Major Malfunction*. Garden City, NY: Doubleday, 1987.

Morgan, Clay. *Shuttle-Mir: The U.S. and Russia Share History's Highest Stage*. NASA SP-2001-4225. Houston: NASA, 2001.

Murray, Charles, and Catherine Bly Cox. *Apollo: The Race to the Moon.* New York: Simon and Schuster, 1989.

NASA Human Spaceflight Home Page. NASA. Available at http://www.spaceflight.nasa.gov/.

SIR-C/X-SAR: Space Radar Images of the Earth. Pasadena: NASA, Jet Propulsion Laboratory. Available at http://www.jpl.nasa.gov/radar/sircxsar/.

Smith, Marsh S. 2001. *NASA's Space Station Program: Evolution and Current Status.* Testimony before the House Science Committee. April 4, 2001. Available at http://www.hq.nasa.gov/office/pao/History/smith.htm.

Stafford, Thomas P., and Michael Cassutt. *We Have Capture.* Washington, DC: Smithsonian Institution Press, 2002.

Zimmerman, Robert. *Leaving Earth.* Washington, DC: Joseph Henry Press, 2003.

The author's website is www.astronauttomjones.com.

# INDEX